AF768258

TROPOSPHERIC OZONE

Human Health and Agricultural Impacts

Edited by

David J. McKee

LEWIS PUBLISHERS
Boca Raton Ann Arbor London Tokyo

Library of Congress Cataloging-in-Publication Data

Tropospheric ozone: human health and agricultural impacts / edited by
David J. McKee
p. cm.
Includes bibliographical references and index.
ISBN 0-87371-475-X
1. Ozone—Health aspects. 2. Plants, Effect of ozone on. 3. Air—Pollution.
I. McKee, David J.
RA577.097T76 1993
615.9′02—dc20 93-13474
CIP

This book contains information obtained from authentic and highly regarded sources. Reprinted material is quoted with permission, and sources are indicated. A wide variety of references are listed. Reasonable efforts have been made to publish reliable data and information, but the author and the publisher cannot assume responsibility for the validity of all materials or for the consequences of their use.

Neither this book nor any part may be reproduced or transmitted in any form or by any means, electronic or mechanical, including photocopying, microfilming, and recording, or by any information storage or retrieval system, without prior permission in writing from the publisher.

All rights reserved. Authorization to photocopy items for internal or personal use, or the personal or internal use of specific clients, may be granted by CRC Press, Inc., provided that $.50 per page photocopied is paid directly to Copyright Clearance Center, 27 Congress Street, Salem, MA 01970 USA. The fee code for users of the Transactional Reporting Service is ISBN 0-87371-475-X/94/ $0.00+$.50. The fee is subject to change without notice. For organizations that have been granted a photocopy license by the CCC, a separate system of payment has been arranged.

CRC Press, Inc.'s consent does not extend to copying for general distribution, for promotion, for creating new works, or for resale. Specific permission must be obtained in writing from CRC Press for such copying.

Direct all inquiries to CRC Press, Inc., 2000 Corporate Blvd., N.W., Boca Raton, Florida 33431.

© 1994 by CRC Press, Inc.
Lewis Publishers is an imprint of CRC Press

No claim to original U.S. Government works
International Standard Book Number 0-87371-475-X
Library of Congress Card Number 93-13474
Printed in the United States of America 1 2 3 4 5 6 7 8 9 0
Printed on acid-free paper

DEDICATION

This book is dedicated to the memory of Dr. Ronald L. Bardow (1935–1993), pioneer and leader in air pollution control research.

PREFACE

This book was written to provide scientists, regulators, industry and public interest groups, students of the environmental sciences, and the general public with an overview of the vast array of information associated with effects, control, enforcement, and public policy related to tropospheric (ground-level) ozone (O_3). The information contained in the chapters of this book has been compiled by individuals who have become some of the world's leading experts in air pollution and, in particular, have been recognized for their work on O_3. Many of the authors are affiliated with government, industry, universities, or public interest organizations; however, neither the material contained in any chapter individually nor in the book as a whole represent a policy statement by any of the organizations with which the authors are affiliated.

CONTRIBUTORS

Mariska L. G. Absil
Energy Systems Engineer
Analysis Branch
Decision and Information Sciences Division
Argonne National Laboratory
Argonne, Illinois

David V. Bates
Professor Emeritus of Medicine
Department of Health Care and Epidemiology
University of British Columbia
Vancouver, British Columbia
Canada

Willis P. Beal
Office of Air Quality Planning and Standards
U.S. Environmental Protection Agency
Research Triangle Park, North Carolina

Francis L. Bunyard
Environmental Engineer
Ambient Standards Branch
Office of Air Quality Planning and Standards
U.S. Environmental Protection Agency
Research Triangle Park, North Carolina

Howard Feldman
Senior Environmental Scientist
American Petroleum Institute
Washington, D.C.

Stanley R. Hayes
Principal
Environ Corporation
Emeryville, California

G. Thomas Helms
Chief
Ozone/CO Programs Branch
Office of Air Quality Planning and Standards
U.S. Environmental Protection Agency
Research Triangle Park, North Carolina

Andrew A. Herstrom
ManTech Environmental Technology, Inc.
Environmental Research Laboratory
Corvallis, Oregon

Timothy Hogan
Senior Regulatory Analyst
American Petroleum Institute
Washington, D.C.

Steven M. Horvath
Emeritus Professor
Environmental Stress Laboratory
Neuroscience Research Institute
University of California, Santa Barbara
Santa Barbara, California

Daniel S. Jacobs
Trial Attorney
Environmental Enforcement Section
Environmental and Natural Resource Division
U.S. Department of Justice
Washington, D.C.

E. Henry Lee
ManTech Environmental Technology, Inc.
Environmental Research Laboratory
Corvallis, Oregon

Thomas R. McCurdy
Physical Scientist
Office of Air Quality Planning and Standards
U.S. Environmental Protection Agency
Research Triangle Park, North Carolina

David J. McKee
Physical Scientist
Office of Air Quality Planning and Standards
U.S. Environmental Protection Agency
Research Triangle Park, North Carolina

Paula J. Narducci
Software Engineer
Analysis Branch
Decision and Information Sciences Division
Argonne National Laboratory
Argonne, Illinois

Anne Nikbakt
Office of Air Quality Planning and Standards
U.S. Environmental Protection Agency
Research Triangle Park, North Carolina

William Ollison
Senior Scientist
American Petroleum Institute
Washington, D.C.

David M. Olszyk
Environmental Research Laboratory
U.S. Environmental Protection Agency
Corvallis, Oregon

Harvey M. Richmond
Environmental Protection Specialist
U.S. Environmental Protection Agency
Research Triangle Park, North Carolina

Arlene S. Rosenbaum
Senior Scientist
Systems Applications, Inc.
San Rafael, California

Roger P. Stewart
Executive Director
Environmental Protection Commission of Hillsborough County
Tampa, Florida

David T. Tingey
Environmental Research Laboratory
U.S. Environmental Protection Agency
Corvallis, Oregon

Thomas S. Wallsten
Professor and Director
Cognitive Psychology Program
L. L. Thurstone Psychometric Laboratory
University of North Carolina at Chapel Hill
Chapel Hill, North Carolina

Ronald G. Whitfield
Manager
Analysis Branch
Decision and Information Sciences Division
Argonne National Laboratory
Argonne, Illinois

Robert L. Winkler
James B. Duke Professor
Fuqua School of Business
Duke University
Durham, North Carolina

Terry F. Yosie
Environmental Development Group
Washington, D.C.

CONTENTS

SECTION I

Ozone Effects, Exposure, Risk, and Regulation

INTRODUCTION

TROPOSPHERIC OZONE: WHAT IT IS AND HOW IT IS FORMED

Ozone (O_3) is a highly reactive gas composed of three atoms of oxygen. It is most predominant and important in two different areas of the atmosphere — the stratosphere and the troposphere. In the stratosphere, miles above the earth's surface, O_3 provides a protective shield by filtering out dangerous ultraviolet radiation from the sun. Stratospheric O_3, as it is often called, is being depleted by chlorofluorocarbons (CFCs) and has been the focus of much research in recent years. It will clearly be one of the major environmental issues of the 1990s and beyond but should not be confused with the health and environmental problems associated with direct exposure to tropospheric O_3.

Tropospheric O_3 is formed in the lower atmosphere and exists in the ambient (i.e., outdoor) air to which people, agricultural crops, forests, and ecosystems are exposed and must respond every day. Because it is not emitted directly into the atmosphere, tropospheric O_3 is described as a secondary or transformation pollutant rather than a primary pollutant. Ozone is formed in ambient air by various chemical reactions involving nitrogen oxides (NO_x), reactive volatile organic compounds (VOCs), and atmospheric oxygen (O_2) in the presence of sunlight. These factors alone, however, are not sufficient to produce elevated levels of O_3. High ambient O_3 levels generally occur only when ambient air temperatures exceed 80 to 90°F. An exception to this can be found in cities at very high elevation (e.g., 10,000 ft) such as Mexico City where exceedingly high levels of O_3 ($\geq$ 0.3 ppm) have been reported even when air temperatures were in the low 70s (ϒF). This is likely due to the highly intense solar radiation (hη) which pushes the chemical equilibrium toward higher levels of O_3.

In the unpolluted troposphere, naturally produced (i.e., biogenic) VOCs interact with NO_x and atmospheric O_2 to continuously produce O_3, albeit at low levels. This occurs even when conditions of formation are not favorable (e.g., low temperatures, low light intensity, high wind speed), such as those which commonly occur in the winter months. In summer months when conditions of O_3 formation improve (e.g., high temperatures, higher light intensity, wind stagnation), O_3 levels typically rise even in remote rural areas. Simplistically, tropospheric O_3 is produced in the ambient air by the addition of an oxygen atom (O) to a molecule of O_2. This occurs through a process called photolysis (i.e., breakdown by sunlight) of nitrogen dioxide (NO_2) into nitric oxide (NO) and O. The NO formed then reacts with the O_3 produced in the above reaction to produce O_2 and NO_2, resulting in an equilibrium or series of cyclic reactions between O_3, O_2, O, NO_2, and NO. This equilibrium can be illustrated by the following series of chemical equations:

a. Photodissociation of NO_2 by near-ultraviolet solar radiation

$$NO_2 + h\eta(295 \leq \lambda < 430nm) \rightarrow NO + O$$

b. Reaction of O and O_2 in the presence of M, where M is a molecule (e.g. nitrogen) which removes excess energy of reaction

$$O + O_2 + M \rightarrow O_3 + M$$

c. Absent any competing reactions, NO reacts rapidly with O_3

$$NO + O_3 \rightarrow NO_2 + O_2$$

The equilibrium or steady-state condition which results among the above chemical species can be expressed by the following:

$$[O_3] = K[NO_2]/[NO]$$

where the equilibrium constant (K) depends on sunlight intensity.

Because reaction (c) is very rapid, urban atmospheres tend not to see increased O_3 levels until most of the NO reacts to form NO_2. This may partially explain air quality measurements of O_3 which are lower in city centers where there are substantial NO_x emissions from automobiles and industrial or utility sources but which are higher in downwind suburban or rural areas. Downwind transport of the chemical mixture eventually results in photodissociation and O_3 formation at sites sometimes quite distant from the original source of the precursor emissions.

Reactions which produce NO_2 without destroying O_3 will disrupt the above equilibrium and shift it toward a net increase in ambient O_3 levels. Oxidation of photochemically reactive VOCs results in the formation of highly reactive chemical species called free radicals, which are capable of converting NO to NO_2 without destruction of O_3. Disruption of the chemical equilibrium eventually terminates and equilibrium returns unless continued emissions of NO_x and reactive VOCs prevent stabilization. As the vast array of reactions involving a multitude of different chemical species has been studied in recent years, it has become clear that the original simplifying assumptions used to develop relationships between O_3 levels and VOC/NO_x ratios are no longer applicable. Information gained through the use of air quality models such as the Urban-Airshed Model and the Regional Oxidant Model has improved understanding of these relationships and permitted better planning to control ambient O_3 levels. For a more thorough discussion of O_3 formation and related reactions in the ambient air, the reader is referred to numerous sources of information.[1-6]

THE DANGERS OF TROPOSPHERIC OZONE

With the advent of industrialized society and increased use of coal and petroleum products, emission of VOCs resulting from fossil fuel combustion has resulted in increased photochemical smog levels in most urbanized areas of the world. Concomitant with increases in urban air pollution has been a centralization of populations in large cities. These events have produced a situation in which large numbers of people are exposed to high levels of O_3 with a resulting increased risk of both acute (short-term) and chronic (long-term) health effects. As human exposure to O_3 has increased, there also has been an increase in exposure and damage to agricultural crops, forests, and ecosystems, as well as to materials such as rubber, paints, and dyes. Essentially anything that is susceptible to oxidation is a potential target for the adverse effects of O_3.

Early efforts to understand the extent of the air pollution problem in the United States began with passage of legislation in 1955. The Clean Air Act of 1963 encouraged states and local air pollution agencies to develop approaches to curtail pollutant emissions from local sources with limited support from the Federal government. During the late 1960s when it became clear that pollution control efforts had become so highly variable between states, it was decided that a national approach was needed. In 1970, Congress created the U.S. Environmental Protection Agency (U.S. EPA) and passed the Clean Air Act, which established a process for setting national ambient air quality standards (NAAQS) and for attaining the NAAQS. The Clean Air Act Amendments of 1977 subsequently mandated a 5-year review cycle for the NAAQS and deadlines for attaining the NAAQS. Although there was little success during the 1980s in meeting requirements for review of NAAQS, passage of the Clean Air Act Amendments of 1990 focused on implementation and not on updating the NAAQS. This may have been an unfortunate oversight, particularly with regard to the O_3 NAAQS, since thousands of new studies of health and welfare effects of O_3 provide evidence of a potential need to revise both the primary (health-based) and secondary (welfare-based) standards. The U.S. EPA was subsequently sued by the American Lung Association and others to set an O_3 NAAQS and take into account new information. Details of the O_3 NAAQS review requirements and process are provided later in Chapter 1. While Chapter 7 traces the evolution of the Clean Air Act and related efforts to control air pollution at the federal level, Chapter 9 describes a local program's initiatives to reduce O_3 levels. These programs are underway today and exemplify the multibillion dollar effort to help reduce air pollution throughout the U.S.

As discussed in Chapter 2, tropospheric O_3 concentrations fluctuate seasonally, generally reaching peak levels during the summer months. Although these peaks have been reduced in most major metropolitan areas of the U.S., evidence is mounting to suggest that seasonal averages, which may be better indicators of chronic health effects, may not be falling. There are long-term trends which suggest that global troposhperic background O_3 levels are on the rise and indirectly may be producing effects on human health and the environment by contributing to global warming. Although atmospheric scientists are far from developing a full understanding of the role of tropospheric O_3 in global change, excellent discussions of potential causes and implications can be found in *Global Alert: The Ozone Pollution Crisis*[7] and *Tropospheric Ozone and the Environment*.[8]

During the past two decades, a vast literature of research on acute and chronic health effects of O_3 has been published. Literally thousands of studies have described a wide array of effects ranging from transient reductions in lung function and symptoms (e.g., cough, chest pain) to the irreversible changes in lung structure which may result in accelerated decline in lung capacity and premature mortality. Chapter 3 provides an overview of mechanisms of effect, factors of susceptibility, susceptible populations, and the variety of effects which have been documented in clinical, epidemiological, and animal toxicological studies. From the perspective of one of the world's leading health scientists, Chapter 8 offers a viewpoint which implies that the large and growing literature base on O_3 health effects may raise more questions than answers; however, the convergence of data discussed in the chapter provides a foundation for considering the wide range and seriousness of these health effects.

A discussion of O_3 exposure data and state-of-the-art O_3 exposure modeling efforts during the past decade is presented in Chapter 4. These efforts have been undertaken in part to gain a better understanding of the pattern of exposures humans encounter in their daily activities. In addition, however, it has become increasingly apparent that future decisions regarding control of O_3 will depend on not only effects and exposure information but also on the ability to incorporate such information into a framework for assessing health risks associated with both acute and chronic exposure to O_3. Chapter 5 focuses on the U.S. EPA's efforts to assess such risks and to provide decisionmakers with better tools for making very difficult judgments about air standards and air pollution control.

Air pollution accounts for an estimated several billion dollar crop loss every year in the United States alone. Research and analysis suggests that about 90% of this crop loss may be directly or indirectly attributed to O_3. Chapter 6 provides not only an overview of mechanisms and types of vegetation effects of O_3 but also discusses the National Crop Loss Assessment Network, the major research effort conducted to assess crop loss in the United States. Finally, evaluation of various exposure indices is used to assess adequacy of current standards to protect crops, predict crop loss with current O_3 levels, and discuss alternate forms of the O_3 standard.

Costs associated with implementation of O_3 control programs have increased as efforts have intensified in recent years to bring ambient O_3 levels under control. Numerous studies have attempted to assess the magnitude of costs required to bring most of the U.S. metropolitan areas into attainment with the NAAQS for O_3. Chapter 10 summarizes those studies conducted by the U.S. EPA which have analyzed the impact of meeting Title I and Title II provisions of the Clean Air Act Amendments of 1990 relevant to O_3 nonattainment.

Avoiding the costs associated with installation or operation of control equipment has resulted in numerous cases of individuals or organizations taking actions which violate legal requirements, thus subjecting them to civil or criminal penalties. Because the regulatory control scheme of the Clean Air Act can only be effective if its requirements are enforced, the federal role in enforcing Titles I and II provisions has become an integral part of the total effort to control emissions of precursors to O_3 in the ambient air. Chapter 11 provides an overview of enforcement provisions of the Clean Air Act and offers several cases in which enforcement action has been necessary to reduce ambient O_3 concentrations.

The final two chapters of this book provide very diverse perspectives of the O_3 problem and what needs to be done about it. Chapter 12 focuses on major public health problems associated with exposure to ambient O_3 and the past failures of control programs to adequately reduce O_3 levels. The chapter also makes some recommendations for changing the NAAQS, expanding research, improving control strategies, and increasing congressional oversight. Chapter 13 tends to focus more on the many factors which affect ambient O_3 levels such as regional differences, NO_x/VOC ratios, natural emissions, and stratospheric intrusion. A discussion of the health effects debate is provided, and the chapter concludes with an American Petroleum Institute perspective on control programs and future strategies.

REFERENCES

1. U.S. Environmental Protection Agency. "Air Quality Criteria for Ozone and Other Photochemical Oxidants," Draft Final, Environmental Criteria and Assessment Office EPA Report No. EPA-600/8-84/020a to 020e. Available from NTIS, Springfield, VA; PB 87-142949 (1986).
2. Demerjian, K. L., Kerr, J. A., and Calvert, J. G. "The Mechanism of Photochemical Smog Formation," in *Advances in Environmental Science, Vol. 4*, J. N. Pitts, and R. L. Metcalf, Eds. (New York: John Wiley and Sons, Inc., 1974), pp. 1–262.
3. Finlayson, B. J., and Pitts, J. N., Jr. "Photochemistry of the Polluted Troposphere, *Science* 192:111–119 (1976).
4. Logan, J. A., Prather, M. J., Wofsy, S. C., and McElroy, M. B. "Tropospheric Chemistry: A Global Perspective," *J. Geophys. Res.* 86:7210-7254 (1981).
5. Whitten, G. Z. "The Chemistry of Smog Formation: A Review of Current Knowledge," *Environ. Int.* 9:447–463 (1983).
6. Atkinson, R., and Lloyd, A. C. "Evaluation of Kinetic and Mechanistic Data for Modeling of Photochemical Smog," *J. Phys. Chem. Ref. Data* 13:315–444 (1981).
7. Fishman, J., and Kalish, R. *Global Alert: The Ozone Pollution Crisis*, Plenum Press, New York (1990).
8. Berglund, R. L., Lawson, D. R., and McKee D. J. "Tropospheric Ozone and the Environment: Papers From an International Conference," TR-19, Air & Waste Management Association, Pittsburgh, PA (1991).

CHAPTER 1

LEGISLATIVE REQUIREMENTS AND OZONE NAAQS REVIEW

David J. McKee

LEGISLATIVE REQUIREMENTS AFFECTING NAAQS REVIEW*

Two sections of the Clean Air Act govern the establishment and revision of national ambient air quality standards (NAAQS). Section 108 (42 U.S.C. 7408) directs the Administrator to identify pollutants which "may reasonably be anticipated to endanger public health and welfare" and to issue air quality criteria for them. These air quality criteria are to "accurately reflect the latest scientific knowledge useful in indicating the kind and extent of all identifiable effects on public health or welfare which may be expected from the presence of [a] pollutant in the ambient air . . . ."

Section 109 (42 U.S.C. 7409) directs the Administrator to propose and promulgate "primary" and "secondary" NAAQS for pollutants identified under Section 108. Section 109(b)(1) defines a primary standard as one "the attainment and maintenance of which, in the judgment of the Administrator, based on the criteria and allowing an adequate margin of safety, [is] requisite to protect the public health." A secondary standard, as defined in Section 109(b)(2), must "specify a level of air quality the attainment and maintenance of which, in the judgment of the Administrator, based on [the] criteria, is requisite to protect the public welfare

* The following Section is based closely on language used in the final decision on the O_3 NAAQS signed by the EPA Administrator on March 1, 1993.

0-87371-475-X/94/$0.00+$.50
© 1994 by Lewis Publishers

from any known or anticipated adverse effects associated with the presence of [the] pollutant in the ambient air." Welfare effects as defined in Section 302(h) [42 U.S.C. 7602(h)] include, but are not limited to, "effects on soils, water, crops, vegetation, manmade materials, animals, wildlife, weather, visibility and climate, damage to and deterioration of property, and hazards to transportation, as well as effects on economic values and on personal comfort and well-being."

The U.S. Court of Appeals for the District of Columbia Circuit has held that the requirement for an adequate margin of safety for primary standards was intended to address uncertainties associated with inconclusive scientific and technical information available at the time of standard setting. It was also intended to provide a reasonable degree of protection against hazards that research has not yet identified. [*Lead Industries Association v. EPA*, 647 F.2d 1130, 1154 (D.C. Cir. 1980), *cert. denied,* 101 S. Ct. 621 (1980); *American Petroleum Institute v. Costle*, 665 F.2d 1176, 1177 (D.C. Cir. 1981), *cert. denied,* 102 S. Ct. 1737 (1982)]. Both kinds of uncertainties are components of the risk associated with pollution at levels below those at which human health effects can be said to occur with reasonable scientific certainty. Thus, by selecting primary standards that provide an adequate margin of safety, the Administrator seeks not only to prevent pollution levels that have been demonstrated to be harmful but also to prevent lower pollutant levels that may pose an unacceptable risk of harm, even if the risk is not precisely identified as to nature or degree.

In selecting a margin of safety, the EPA considers such factors as the nature and severity of the health effects involved, the size of the sensitive population(s) at risk, and the kind and degree of the uncertainties that must be addressed. Given that the "margin of safety" requirement by definition only comes into play where no conclusive showing of adverse effects exists, such factors, which involve unknown or only partially quantified risks, have their inherent limits as guides to action. The selection of any particular approach to provide an adequate margin of safety is a policy choice left specifically to the Administrator's judgment. [*Lead Industries Association v. EPA*, supra, 647 F.2d at 1161-62.]

Section 109(d)(1) of the Act requires that "not later than December 31, 1980, and at 5-year intervals thereafter, the Administrator shall complete a thorough review of the criteria published under Section 108 and the national ambient air quality standards . . . and shall make such revisions in such criteria and standards . . . as may be appropriate . . . ." Sections 109(d)(2)(A) and (B) require that a scientific review committee be appointed and provide that the committee "shall complete a review of the criteria . . . and the national primary and secondary ambient air quality standards . . . and shall recommend to the Administrator any . . . revisions of existing criteria and standards as may be appropriate . . . ." Figure 1 shows the many steps involved in the NAAQS review process, and Table 1 is a list of current NAAQS including levels and averaging times.

States are primarily responsible for ensuring attainment and maintenance of ambient air quality standards once the EPA has established them. Under Title I of the Act (42 U.S.C. 7410), states are to submit, for EPA approval, state implementation plans (SIPs) that provide for the attainment and maintenance of such

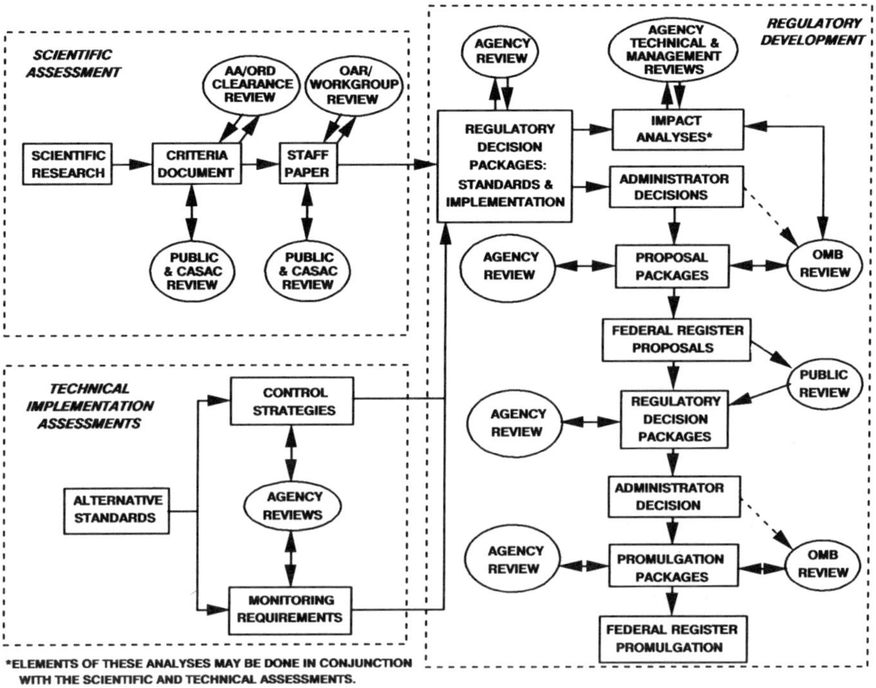

Figure 1. NAAQS process flow chart.

standards through control programs directed to sources of the pollutants involved. The states, in conjunction with the EPA, also administer the prevention of significant deterioration program (42 U.S.C. 7470–7479) and the visibility protection program (42 U.S.C. 7491–7492) for these and other air pollutants. In addition, federal programs provide for nationwide reductions in emissions of air pollutants through the federal motor vehicle control program under Title II of the Act (42 U.S.C. 7521–7574), which involves controls for automobile, truck, bus, motorcycle, and aircraft emissions; the new source performance standards under Section 111 (42 U.S.C. 7411); and the national emission standards for hazardous air pollutants under Section 112 (42 U.S.C. 7412).

HISTORY OF THE OZONE NAAQS REVIEW

On April 30, 1971 the EPA promulgated primary and secondary NAAQS for photochemical oxidants under Section 109 of the Act (36 FR 8186). These were set at an hourly average of 0.08 ppm total photochemical oxidants not to be exceeded more than 1 h per year. On April 20, 1977 the EPA announced (42 FR 20493) the first review and updating of the 1970 Air Quality Criteria Document

Table 1. National Ambient Air Quality Standards

Pollutant	Primary standards	Averaging time (h)	Secondary standards
Carbon monoxide	9 ppm (10 mg/m^3)	8-h[a]	None
	35 ppm (40 mg/m^3)	1-h[a]	
Lead	1.5 μg/m^3	Quarterly average	Same as primary
Nitrogen dioxide	0.053 ppm (100 μg/m^3)	Annual (arithmetic mean)	Same as primary
Particulate matter (PM-10	50 μg/m^3	Annual (arithmetic mean[b]	Same as primary
	150 μg/m^3	24-h[c]	
Ozone	0.12 ppm (235 μg/m^3)	1-h[d]	Same as primary
Sulfur oxides (SO_2)	0.03 ppm (80 μg/m^3)	Annual (arithmetic mean)	— —
	0.14 ppm (365 μg/m^3)	24-h[a] (primary) 3-h[a] (secondary)	0.5 ppm (1300 μg/m^3)

[a] Not to be exceeded more than once per year.
[b] The standard is attained when the expected annual arithmetic mean concentration is less than or equal to 50 mg/m^3, as determined in accordance with Appendix K.
[c] The standard is attained when the expected number of days per calendar year with a 24-h average concentration above 150 mg/m^3 is equal to or less than as determined in accordance with Appendix K.
[d] The standard is attained when the expected number of days per calendar year with maximum hourly average concentrations above 0.12 ppm is equal to or less than 1, as determined in accordance with Appendix H.

for Photochemical Oxidants in accordance with Section 109(d)(1) of the Act. In preparing the Air Quality Criteria Document, the EPA provided a number of opportunities for external review and comment. The EPA made two drafts of the document available for public comment, and these drafts were peer reviewed by the Subcommittee on Scientific Criteria for Photochemical Oxidants of the EPA Science Advisory Board. The EPA published the final revised Air Quality Criteria for Ozone and Other Photochemical Oxidants on June 22, 1978.

Based on the 1978 revised Air Quality Criteria Document (U.S. EPA, 1978) and taking into account the advice and recommendations of the Subcommittee, on June 22, 1978, the EPA proposed (43 FR 16962) revisions to the then-current primary and secondary NAAQS for photochemical oxidants. The proposed changes included raising the primary standard to 0.10 ppm, retaining the 0.08 ppm secondary standard, changing the chemical designation of the standards from photochemical oxidants to O_3, and changing to standards with a statistical form (i.e., expected exceedances) rather than a deterministic form (i.e., not to be exceeded more than x number of times per year).

After taking into account public comments and following a careful review by the Office of Management and Budget (OMB), the EPA announced its final decision on the proposed revisions to the 1971 standards (44 FR 8282, February 8, 1979). The final rulemaking revised the level of the primary standard from 0.08 to 0.12 ppm, set the secondary standard identical to the primary standard, changed the chemical designation of the standards from photochemical oxidants to ozone (O_3), and revised the definition of the point at which the standard is attained to when the expected number of days per calendar year with maximum hourly average concentrations above 0.12 ppm is equal to or less than one as determined by Appendix H.

On March 17, 1982 (47 FR 11561) the EPA announced that it was planning to revise the 1978 Air Quality Criteria Document for Ozone and Other Photochemical Oxidants on August 22, 1983 published in the *Federal Register* a notice (48 FR 38009) indicating that review of primary and secondary standards for O_3 had been initiated. The EPA subsequently provided opportunities for public review and comment on the draft Criteria Document, announcing availability of two drafts of the revised criteria document (49 FR 29845; 51 FR 11339), prepared by the EPA's Environmental Criteria and Assessment Office (ECAO), and holding two public meetings of the Clean Air Scientific Advisory Committee (CASAC) (March 4–6, 1985 and April 21–22, 1986). The EPA considered the numerous and often extensive comments received from the public and the CASAC members in preparing the final Criteria Document. On October 22, 1986 the CASAC sent the Administrator a "closure letter," outlining key issues and recommendations and indicating that it was satisfied with the final draft of the 1986 Criteria Document as a basis for making a decision on the standards.

Following closure, a number of scientific articles were published or accepted for publication and appeared to be of sufficient importance concerning potential health and welfare effects of O_3 to warrant preparation of a supplement to the criteria document. By early 1988, ECAO had begun preparation of a draft supplement to the 1986 Criteria Document and made draft copies available to the CASAC and the public in November 1988.

After the CASAC meeting on March 4–6, 1985 the EPA's Office of Air Quality Planning and Standards (OAQPS) began work on the first draft of the Staff Paper (Review of the National Ambient Air Quality Standards for Ozone: Assessment of Scientific and Technical Information — OAQPS Staff Paper). The CASAC reviewed first and second drafts of the Staff Paper at public meetings of the CASAC held on April 21–22, 1986 and December 14–15, 1987. Numerous written and oral comments were received on the drafts from the CASAC, state agencies, representatives of industry and environmental organizations, individual scientists, and other interested members of the public. However, CASAC concluded that sufficient new information existed to recommend incorporation of relevant new information into a third draft of the Staff Paper to be developed concurrently with the draft Supplement. In early 1988, the EPA began working on a third draft of the Staff Paper and provided copies to the CASAC and public in November 1988.

The CASAC held a public meeting on December 14–15, 1988 to review the draft Supplement and draft Staff Paper. Major issues included: the definition of adverse health effects of O_3, the significance of health studies suggesting that exercising individuals exposed for 6 to 8 h to O_3 levels at or below 0.12 ppm may experience inflammation and transient decreases in pulmonary function, the possibility that chronic irreversible effects may result from long-term exposures to elevated levels of O_3, and the importance of analyses which indicate agricultural crop damage may be better defined by a cumulative seasonal average than by a 1-h peak level of O_3. In its "closure letter" of May 1, 1989 the CASAC indicated that the draft Supplement and draft Staff Paper "provide an adequate scientific basis for the EPA to retain or revise primary and secondary standards for ozone." The CASAC further suggested that the EPA should proceed as rapidly as possible to complete its review of the NAAQS. Following numerous discussions and briefings within EPA on the need for modified and possibly more stringent standards, it was determined that any such changes might be disruptive of the ongoing Clean Air Act negotiations and that such action should be delayed.

On October 22, 1991 the American Lung Association and other plaintiffs filed suit under Section 304 of the Act to compel the EPA to complete its review of the criteria and standards for O_3 under Section 109(d)(1) of the Act [*American Lung Association v. Reilly*, No. 91-cv-4114 (JRB) (E.D.N.Y.)]. The U.S. District Court for the Eastern District of New York subsequently issued an order requiring the EPA to sign a *Federal Register* notice announcing its proposed decision on whether to revise the standards for O_3 by August 1, 1992 and to sign a *Federal Register* notice announcing its final decision by March 1, 1993.

On August 10, 1992 (57 FR 35542) then-Administrator William Reilly announced his proposed decision under Section 109(d)(1) that revisions to existing 1-h primary and secondary standards were not appropriate at that time. Consistent with the order in the American Lung Association case, Administrator Reilly signed the proposed decision on August 1, 1992. The notice explained in some detail (see 57 FR 35546) that the proposed decision would complete the EPA's review of information on health and welfare effects of O_3 assembled over a 7-year period and contained in the 1986 Criteria Document and its Supplement. The review included an evaluation of key studies published through early 1989, the 1989 Staff Paper assessment of the most relevant information in these documents, and the advice and recommendations of the CASAC as presented both in the discussion of these documents at public meetings and in the CASAC's 1986 and 1989 "closure letters."

Under Section 109(b) of the Act, primary and secondary NAAQS are to be based on the air quality criteria issued under Section 108. Additionally, under Section 109(d), the EPA must periodically conduct a "thorough review" of the criteria taking into account the advice and recommendations of the CASAC, as the basis for periodic decisions on whether revisions of NAAQS are appropriate. When Congress enacted the latter requirement in 1977, it was well aware that implementation of the NAAQS can have profound economic and social, as well as environmental, consequences. Understandably, it required that the Administrator's periodic decisions on whether to revise the NAAQS be based on scientific studies that had been rigorously assessed and incorporated into air quality criteria and whose implications for public health and welfare had been carefully considered by both the EPA and CASAC. In view of this, the August 10, 1992 notice made clear that Administrator Reilly did not take into account more recent studies on the health and welfare effects of O_3 because these studies had not been assessed in the 1986 Criteria Document nor its Supplement, nor had they undergone the rigorous review process (including CASAC review) required to incorporate them into a new Criteria Document.

Because of the scientific and technical complexity of such assessments, the EPA estimated that 2 to 3 years would be necessary to rigorously assess the over 1000 new studies and incorporate key information into a revised Criteria Document, to evaluate the significance of the key information for decision-making purposes, to develop staff recommendations for the Administrator, and to provide appropriate opportunities for the CASAC review and public comment. Given the importance of these new studies and concern about the health and welfare effects of O_3, the August 10, 1992 notice also outlined the EPA's plans to proceed as rapidly as possible with the next review of the criteria and standards for O_3.

The Primary Standard

In reaching his August 10, 1992 proposed decision that revisions of the existing O_3 primary standard were not appropriate, Administrator Reilly considered the health effects information assessed in the 1986 Air Quality Criteria Document, the Supplement that updated that information, the 1989 Staff Paper, and the advice and recommendations of the CASAC in its 1989 "closure letter." Based on his review of this information, Administrator Reilly concurred with the staff and the CASAC conclusions that the preliminary information on effects of prolonged exposures to O_3 was not sufficient to support the establishment of a new 6- to 8-h standard to protect against prolonged exposures, or a seasonal or other long-term standard to protect against chronic effects. In reaching this proposed determination, Administrator Reilly recognized that a number of new studies, particularly on 6- to 8-h exposures to O_3, had been published in the scientific literature since early 1989. Mindful of this, as well as research in progress on the chronic effects of O_3, the EPA made clear in the proposal that it intended to proceed with the next periodic review of the air quality criteria as rapidly as possible. The notice added that when this new information had been incorporated into the air quality criteria, a more informed decision could be made as to whether adding a new 6- to 8-h standard and/or a seasonal or other long-term standard would be appropriate.

Administrator Reilly also carefully considered the health effects information on 1- to 3-h exposures to O_3 contained in the air quality criteria. Based on these assessments and taking into account the advice and recommendations of the CASAC in its 1989 "closure letter," Administrator Reilly in August 1992 reached the proposed determination that revisions of the existing 1-h primary standard were not appropriate to provide increased protection against 1- to 3-h exposures to O_3. The standard level is below the levels where controlled human 1- to 3-h exposure studies found substantial changes in pulmonary function and symptoms. In reaching this conclusion, Administrator Reilly decided that the mean group response observed in the controlled human studies up to 0.15 ppm O_3 should be characterized as mild, and that most of the responders within this population of normal healthy individuals reportedly experienced only mild to moderate responses under very heavy exercise (U.S. EPA, 1989, pp. VII-53 to VII-56).

Although there was a difference of opinion among the EPA's scientific advisors as to the significance of decrements in lung function in the range of 10 to 20% when accompanied by symptoms (CASAC, 1989), it was Administrator Reilly's judgment that the lesser effects associated with exposure to O_3 in the range of 0.12 to 0.15 ppm observed in the controlled human studies did not constitute adverse effects for purposes of Section 109 of the Act.

Administrator Reilly also considered other sensitive population groups whose response to O_3 had not been fully characterized. Although some epidemiology studies considered in the 1986 Criteria Document and its Supplement suggested that exposure to O_3 at ambient concentrations may result in the aggravation of asthma and preexisting respiratory disease, Administrator Reilly in August 1992 concurred with the staff's view that the direct use of these studies was limited by uncertainties about individual exposure levels and the role of other pollutants. In addition, although individuals with preexisting lung disease are not more responsive to O_3 than healthy persons, the same small change in pulmonary function may have more impact on people whose lung function is already compromised. While these epidemiology studies suggested that these sensitive groups may be at somewhat greater risk at levels of 0.12 ppm O_3 and higher, compared to normal healthy individuals in controlled human exposure studies, in Administrator Reilly's judgment these studies did not provide a sufficient basis for lowering the existing standard. The Administrator also considered and concurred with the staff recommendations that O_3 should remain as the surrogate for controlling ambient concentrations of photochemical oxidants and that the existing form of the standard should not be revised.

The Secondary Standard

In reaching the proposed decision that revision of the existing 1-h O_3 secondary NAAQS was not appropriate, Administrator Reilly considered the welfare effects information assessed in the 1986 Air Quality Criteria Document and its Supplement, the 1989 Staff Paper assessment, and the advice and recommendations of the CASAC in its 1989 "closure letter." A principal reason for the proposed decision was Administrator Reilly's judgment that there was insufficient information in the air quality criteria to specify a new form, averaging period, and level of a secondary standard that would be more protective of forest tree species as well as agricultural crops. The notice added that when information had become available from research currently under way on key aspects of O_3 exposure dynamics that are important for assessing the effects of O_3 on forest tree species and had been incorporated into the air quality criteria during the next review, a more informed judgment could be made as to whether revision of the secondary standard is appropriate.

Administrator Reilly also considered the available information on the effects of O_3 on agricultural crops alone. Although the National Crop Loss Assessment Network (NCLAN) studies have provided extensive data on the effects of O_3 on crops, the appropriateness of the seasonal mean exposure indicator used in these studies had been subject to much criticism during the development of revised air quality criteria. Because of this and other shortcomings of this exposure index that are discussed more fully in the August 10, 1992 proposal notice, the EPA concluded that the use of the 7-h seasonal mean derived from NCLAN data for standard-setting purposes would be inappropriate. The CASAC also recognized this and recommended that retrospective analyses be undertaken in order to identify a more appropriate exposure index that would offer protection from both repeated O_3 peaks of concern and long-term O_3 exposures. While these analyses had identified several indicators that showed promise, Administrator Reilly concurred with the view that it would be premature to base a change in the form and averaging time of the secondary standard on the preliminary results presented in the Supplement to the 1986 Criteria Document and the Staff Paper.

Administrator Reilly also considered tightening the current secondary standard as an interim measure. He noted that, throughout the review of the air quality criteria and staff assessment, no consensus had been reached on an appropriate range of alternative 1-h standards. The staff had great difficulty throughout the review in developing and justifying alternative levels below that of the current standard due to the lack of data (U.S. EPA, 1989, p. XI-13). In the end, while the staff relied on the preliminary results of the Lee et al.[4] study to conclude that the upper-end of the proposed range (0.12 ppm) offers little protection for vegetation (U.S. EPA, 1989, pp. XI-16 to XI-18), the staff also determined that the study was too preliminary to serve as a basis for recommending changes in the form and averaging time of the standard. Even if the results of the Lee et al. (1988) study provide a sufficient basis for revising the standard downward from 0.12 to 0.10 ppm, as has been suggested, it was Administrator Reilly's judgment that such a change would provide only marginal improvement because a 1-h averaging period is not the most appropriate exposure indicator for the full range of exposures.

Given the above information, it was Administrator Reilly's judgment that the most prudent course of action was to retain the current secondary standard until a more informed decision could be made during the next standard review.

SUBSEQUENT DEVELOPMENTS IN THE OZONE NAAQS REVIEW

After the defeat of George Bush in the election of November 1992, President William Clinton appointed Carol Browner to be the new Administrator of EPA. Given that the court-ordered deadline for signing the final decision was March 1, 1993, only a few weeks after confirmation of Administrator Browner, it was clear that time constraints prevented a careful review by the new Administrator of the vast amount of new information before the deadline. The decision not to revise the O_3 standards was signed on March 1 and published in the *Federal Register* on March 9 (58 FR 13008). The American Lung Association, on May 10, 1993, petitioned for review of the decision. Development of a new Criteria Document was initiated in 1992 and should be available for public review by 1994. Completion of the next review of O_3 NAAQS is not expected for several more years.

REFERENCES

1. Clean Air Scientific Advisory Committee. Closure Letter to Lee Thomas, October 22, 1986.
2. Clean Air Scientific Advisory Committee. Transcript of meeting held in Washington, D.C., December 14–15, 1988.
3. Clean Air Scientific Advisory Committee. Closure Letter to William K. Reilly, May 1, 1989.
4. Lee, E. H., W. E. Hogsett, and D. T. Tingey, "Evaluation of Exposure Indices in Exposure-Response Modeling," *Environ. Pollut.,* 53:43–62 (1988).
5. U.S. Environmental Protection Agency. "Air Quality Criteria for Ozone and Other Photochemical Oxidants," Office of Research and Development, Washington, D.C., EPA Report No. EPA-600/8-78-004 (1978).
6. U.S. Environmental Protection Agency. "Air Quality Criteria for Ozone and Other Photochemical Oxidants," Draft Final, Environmental Criteria and Assessment Office, EPA Report No. EPA-6008-84/020a to 020e. Available from NTIS, Springfield, VA, PB 87-142949 (1986).
7. U.S. Environmental Protection Agency. "Review of the National Ambient Air Quality Standards for Ozone Assessment of Scientific and Technical Information," OAQPS Staff Paper, Office of Air Quality Planning and Standards, Research Triangle Park, NC, EPA Report No. 450/2-92-001 (1989).
8. U.S. Environmental Protection Agency. "Summary of Selected New Information on Effects of Ozone on Health and Vegetation: Supplement to Air Quality Criteria for Ozone and Other Photochemical Oxidants," Office of Health and Environmental Assessment, Environmental Criteria and Assessment Office, Research Triangle Park, NC, EPA/600/8-88/105A. Available from NTIS, Springfield, VA (1992).

CHAPTER 2

CONCENTRATIONS OF OZONE IN THE LOWER TROPOSPHERE (AMBIENT AIR)

Thomas R. McCurdy

INTRODUCTION

Data presented in this chapter are intended to provide a context for the health and ecological effects information provided in subsequent chapters. Thus, data on monitored concentrations of O_3 are provided for a variety of geographic areas and averaging times. While most of the data come from the U.S., data from other countries are given for comparative purposes. Recent O_3 data for other countries, frankly, are difficult to obtain in the U.S. except in summary format or for limited special-study areas. As will be seen, O_3 levels in most other countries are similar to U.S. levels except for the highly polluted areas of southern California. Even then, there are a few urban areas outside of the U.S. that record O_3 concentrations higher than southern California. Mexico City, for instance, experiences O_3 levels higher than Los Angeles.

The literature on O_3 monitoring often divides space, both horizontally and vertically, into three general categories. In the vertical dimension, space is divided into the lower troposphere, the mid-troposphere, and the stratosphere. The lower troposphere is also known as the boundary layer; it is that portion of ambient air closest to the ground. It sometimes is itself divided vertically into that part of the troposphere above the nocturnal inversion layer and that part below the inversion layer. As the label implies, the inversion layer occurs at night and it can be quite

0-87371-475-X/94/$0.00+$.50
© 1994 by Lewis Publishers

close to the ground (within meters). The inversion usually "breaks up" in the morning due to surface heating.

In the horizontal dimension, space often is divided into socioeconomic regions, such as urban, rural, and remote areas. Diurnal O_3 patterns are quite different in these three regions. In addition, these regions are the loci for different "sensitive populations" of concern from an O_3 exposure viewpoint. These populations, for the most part, are people, agricultural crops, and forest stands.

O_3 observed near the ground is the net result of numerous interacting atmospheric processes.[1] These processes are:

1. Local photochemical production in the lower troposphere from the interaction of nitrogen oxides and reactive volatile organic compounds (VOCs) emitted by natural and anthropogenic sources. This O_3 may be transported tens of kilometers from source areas into the surrounding countryside. This type of transport is often known as an "urban O_3 plume".
2. Mixing in the lower boundary layer of O_3-rich air above the nocturnal inversion layer with ambient air such that ground-level O_3 concentrations increase throughout the morning. This increase in O_3 is independent of local photochemical production.
3. Long-range transport of non-locally produced O_3. This transport can occur over very long distances (many hundreds of kilometers). This type of transport is sometimes called "mesoscale transport".
4. Intrusion of O_3 in the stratosphere into the troposphere — even down to the lower boundary layer and, on special occasions, down to the ground. This phenomenon is associated with "tropopause folding events" that are associated with large-scale amplifying air waves and upper-level low pressure troughs that occur on the "back side" of a frontal system. These mechanisms result in a hole being formed in the tropopause that separates the troposphere and the stratosphere. Stratospheric O_3 is transported downward through this hole into the troposphere.
   A recent analysis, however, indicates that stratospheric O_3 represents only a small portion of summertime O_3 — at any altitude. Net stratosphere-troposphere O_3 exchange is minimal during the July-to-October time period, when lower boundary layer O_3 concentrations are the highest.[2] In the words of the scientist who chaired the 1989 international conference on O_3, "transfer of ozone from the stratosphere makes only a minor contribution to tropospheric ozone."[3]
5. Formation of O_3 in the mid-troposphere via photochemical processes and transported, or mixed, downward into the lower boundary layer.
6. Removal of O_3 by chemical scavenging in the troposphere. One of the most potent scavengers is nitric oxide (NO), which is emitted by many of the same sources that produce O_3 precursors. Often, this NO occurs in a plume from power plant stacks. O_3 scavenging in this situation is quite localized.
7. Removal of O_3 by physical scavenging in the troposphere. Rain washout and the temperature-dependent decomposition of certain chemicals that perpetuate the photochemical process are examples of this atmospheric mechanism that reduces ambient O_3 levels.
8. Removal of O_3 by chemical and physical scavenging on surfaces that are in contact with ambient O_3. This type of scavenging usually occurs within a few

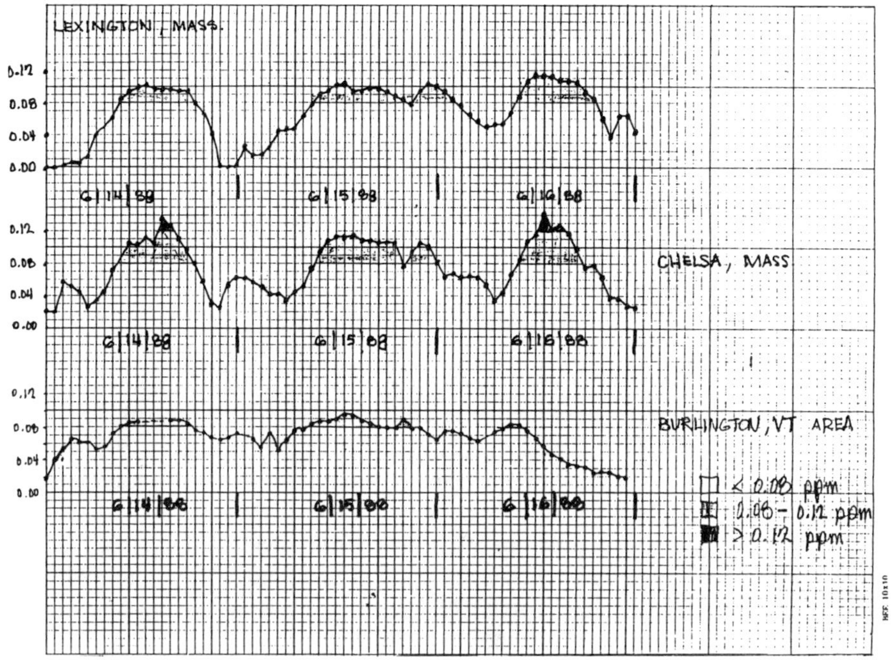

Figure 1. Example of O_3 patterns seen in suburban and rural areas.

meters of the ground, and involves biological and nonorganic surfaces. Vegetation, soil, and man-made polymers are examples of O_3 scavengers.

See Appendix A for a detailed discussion of the formation of ambient O_3. In addition, a concise explanation of O_3 formation process appears in Reference 4. An interesting attempt to allocate O_3 patterns to transport, scavenging, and local photochemical production processes is discussed in Reference 5.

In urban areas, high O_3 levels are usually associated with meteorological conditions involving high solar radiation and relatively low wind speeds. These conditions occur most often during warm-weather months of the year, and are enhanced by the presence of a high pressure system. All of these factors result in a diurnal pattern for ambient O_3 in urban areas in which peak concentrations are seen in the mid-afternoon. Such a pattern is shown in Figure 1A — especially on 6/14/88 and 6/16/88 for the Chelsa, MA, site which is downwind of Boston. The O_3 pattern is quite typical of a near-suburb site downwind of a major urban area. However, due to the various O_3-forming processes, high O_3 concentrations have been observed at almost any time of day — even at nighttime. Rural areas downwind of a major urban area often experience a secondary O_3 peak due to long-range transport. This is shown in Figure 1B. The site is in Lexington, MA, and the O_3 pattern is for the same 3 days as used for Chelsa. Note the secondary peak experienced on all 3 days, but especially on 6/15/88. It should be noted that the Chelsa site also had nighttime peaks on the first 2 days, indicating that a regionwide O_3 transport situation occurred on those nights. It should be noted that high O_3 levels can persist for periods longer than 3 days due to stagnating high pressure systems.

Non-urban areas only marginally affected by transported O_3 usually have a flatter diurnal profile, with a lower peak-to-mean daily ratio. Such a pattern is depicted in Figure 1C. The site shown is near Burlington, VT. The same 3 days are used as before. The site has lower daytime peaks of O_3, but its nighttime levels do not decrease as much as the suburban or rural sites depicted. This probably is due to diminished chemical scavenging by NO in the low-NO_x rural/remote areas.

The remainder of this chapter is devoted to a presentation of ambient O_3 data so that the reader can compare "real-world" O_3 concentrations to the adverse levels discussed in the effects chapters. Very little discussion of the data is undertaken, and even fewer words are spent on explaining the causal mechanisms of O_3 formation underlying the data depicted. Most of the data come from the 1980s because data obtained before 1979 suffer from a calibration error. This is especially true for U.S. data, but also affects other countries' data. In addition, there is little reliable O_3 data before 1979, even in the U.S. where the O_3 problem has been a topic of interest since the early 1950s.

OZONE CONCENTRATIONS IN URBAN AREAS

U.S. Urban Areas

This part of the chapter is devoted to characterizing O_3 ambient air quality in the U.S. Reference to the U.S. National Ambient Air Quality Standard (NAAQS) for O_3 is made frequently. The O_3 NAAQS is also referred to as the O_3 standard. The standard is fully defined in Reference 6 and includes the following attributes:

1. A 1-h averaging time.
2. A temporal aggregation period that focuses on the maximum hourly concentration for each day. This usually is referred to as the 1-h daily maximum (or, max.).
3. An "ozone season" that is based upon a calendar monthly period which varies by state. Currently, it varies between 4 months in two northern states to 12 months in the south. Most states have a 7-month O_3 season, and usually those months are April through October.
4. A concentration evaluation level of 0.12 ppm (120 ppb).

Exceedances of the O_3 NAAQS occur whenever the recorded 1-h daily maximum is greater than 0.12 ppm. Since missing data can only reduce the number of days in a year that could possibly exceed the O_3 NAAQS, a correction factor is used to "normalize" the exceedance statistic to a complete O_3 season. This normalization is accomplished by the following (simplified) formula:

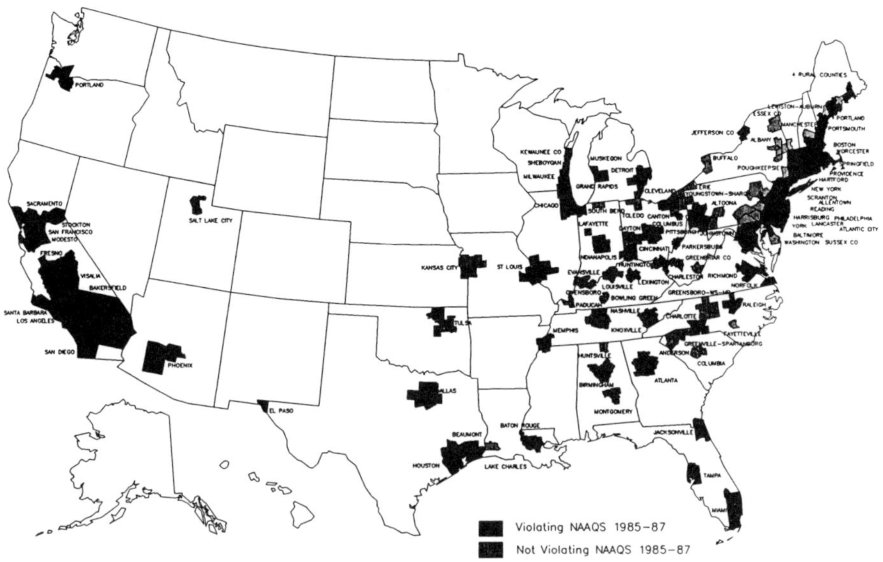

Figure 2. O_3 areas violating standards, 1987–1989.

$$O_3ExEx = \frac{v * (N - z)}{n}$$

where $ExEx$ = expected exceedances of the O_3 NAAQS
v = number of measured 1-h daily maximum values >0.12 ppm
N = number of days in the season
z = number of missing or invalid daily maxima assumed to be $\leq$012 ppm
n = number of days having a valid 1-h daily maximum O_3 value

This formula, and its application, is described in an EPA guidance document.[7] It has received criticism from statisticians outside of EPA,[8] and does not account for annual variation in meteorological conditions. Note that "position-dependent" air quality indicators (e.g., the 2nd highest 1-h daily maximum) are not adjusted for missing values—only the expected number of exceedances are adjusted.

The O_3 NAAQS is violated when the 3-year mean of the annual ExEx is greater than 1.05. Areas violating the O_3 standard in 1987–1989 are shown in Figure 2.

The U.S. data presented below focus on the following O_3 air quality indicators:

- The expected number of exceedances of the current 0.12-ppm O_3 NAAQS (hereafter often cited as the 1-h ExEx index)
- The second-highest 1-h daily maximum O_3 concentration (hereafter often cited as the 2nd-high daily max index)
- The number of days in an O_3 season with an 8-h daily maximum average greater than 0.08 ppm
- The maximum 3-month mean of 8-h daily maximum averages (hereafter often cited as the 3-month mean indicator)

The first two indicators are those normally used by the USEPA to ascertain if an area exceeds the current O_3 NAAQS. The remaining two indicators are potentially useful formulations for considering possible alternative longer-term O_3 NAAQS. The 8-h index is useful for describing how many days have an 8-h exposure greater than an O_3 concentration level thought by some researchers to cause adverse agricultural effects[8] (for a slightly different averaging time)[9] or adverse human health effects.[10] The last indicator is a simplified formulation of the long-term exposure index used by many U.S. research scientists interested in understanding O_3 impacts on agriculturally important crops. It could also be a surrogate indicator of adverse long-term or chronic human health effects.[11]

There are 331 metropolitan statistical areas (MSAs) in the 50 states. Ozone air quality data for every MSA were reviewed against the 75% completeness criteria noted below.* Sites meeting these criteria were determined to have "valid data" and were used to determine which site in an MSA had the highest 2nd-highest 1-h daily maximum value. This site was used to represent O_3 air quality in the MSA for the time period 1980–1987. (More recent air quality data are available, but were not analyzed in the manner discussed below.) Most of the data came from 1985–1987, although older data had to be used in 37 MSAs. Including these areas, valid O_3 data are available for 224 MSAs — about 68% of all U.S. metropolitan areas. The data are summarized in Table 1 in a cumulative frequency format.

The data indicate that 101 MSAs (45.1% of those with data) had more than 1.05 expected exceedance per year of the NAAQS. Thus, a significant proportion of U.S. urban areas exceeded the existing O_3 standard concentration level during the time period analyzed.

Approximately 100 million people live in these high-ozone MSAs.** That is slightly below one half of the total U.S. population (using a 1988 estimated population of 260 million people). Over 16 million people live in MSAs that had 20 or more estimated expected exceedances in at least 1 year between 1985 and 1987. Over 18 million people live in MSAs that had between 10 and 20 expected exceedances.

The median number of 1-h expected exceedances is 1.1; however, there is a lot of variability in the data. Although not directly observable in Table 1, 37% of the MSAs investigated have 0 exceedances. The mean of the sample, almost 5 exceedances, is much greater than the median due to the very high number of expected exceedances in the worst areas. The worst MSA is Los Angeles, with about 125 exceedances.

---

* A valid year of data is one that contains a valid daily maximum 1-h average for 75% of all the days in the state-defined O_3 season. A valid day has data for 75% of the hours between 9 a.m. and 9 p.m. (9 h).

** This does not mean that all persons living in these MSAs experience an O_3 concentration >0.12 ppm as they go about their daily lives. People living in these MSAs have the *potential* for doing so, but many people living in even moderately polluted areas never experience O_3 levels greater than 0.12 ppm. They are, for example, indoors in an air-conditioned space during the peak O_3 period, or in a low-O_3 area of the MSA at the time of peak O_3. For a discussion of this, see Reference 13. Human exposure to O_3 is covered in Chapter 4.

Table 1. Cumulative Frequency Descriptive Statistics Associated with Peak and Multiple-Hour O_3 Air Quality Indicators in Urban Areas

Statistic	Expected exceedances of 0.12 ppm 1-h daily maximum	2nd-high 1-h daily maximum (ppm)	Observed exceedances of 0.08 ppm 8-h daily maximum	Maximum 3-month mean of the 8-h daily maximums (ppm)
Mean	4.6	0.126	14.9	0.056
Std. dev.	13.1	0.036	20.9	0.012
Minimum	0.0	0.039	0	0.016
Median	1.1	0.119	9	0.055
75th per.	4.0	0.140	18	0.061
90th per.	10.0	0.167	29	0.067
95th per.	14.8	0.180	38	0.072
Maximum	125.3	0.340	150	0.132
Sample size	224	224	224	222[a]

Note: Std. dev. = standard deviation; per. = percentile.

[a] This indicator cannot be computed for two MSAs due to lack of a valid 3-month mean for the monitor used for the areas.

From Ambient Standards Branch. "Review of the National Ambient Air Quality Standard for Ozone: Assessment of Scientific and Technical Information," U.S.EPA Report (1989). Hereafter cited as the "Staff Paper". With permission.

There have been numerous analyses to ascertain if there is a trend over time in the ExEx indicator. The USEPA's trends analysis of O_3 indicates that between 1979 and 1988, the composite average of ExEx decreased about 10% for 388 sites that have a long record.[14] Other analysts looking at trends for individual sites *and* correcting for annual fluctuations in meteorology also find a decreasing trend in expected exceedances of the O_3 standard.[15,16]

Table 1 indicates that the median 2nd-high 1-h design value for the sample of 224 MSAs is slightly lower than the current O_3 NAAQS level. The mean, 0.126 ppm, is higher than the current standard because of very high values seen in a few MSAs. For instance, 10% of the MSAs have a 2nd-high daily maximum >0.167 ppm. The worst 5% have a value >0.180 ppm. The highest sites are located in MSAs near or adjacent to the New York, Los Angeles, and Chicago major metropolitan areas.

As seen in Table 1, there are many expected exceedances of an 8-h daily maximum >0.08 ppm. There is also wide variability in their number — from 0 for 63 MSAs (28% of the sample) to 150 for the worst MSA. The mean is about 15, and the standard deviation of the sample is larger than the mean.

Turning to the long-term O_3 indicator presented in Table 1, the large 3-month means of the top 5% MSAs in the sample should be noted. The median value is 0.055 ppm, close to the mean value of 0.056 ppm. The worst area has a maximum 3-month mean of 8-h daily maximum averages of 0.132 ppm — higher than the

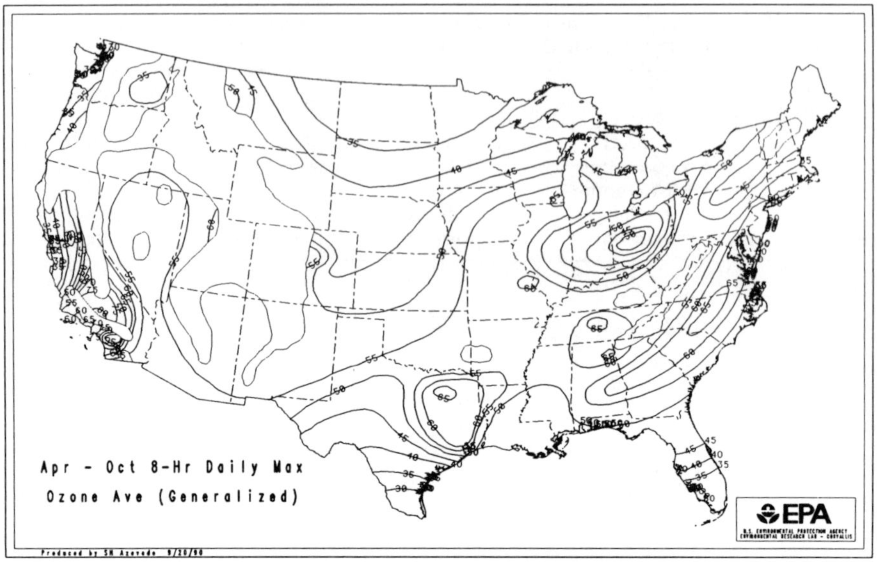

Figure 3. O_3 averages, April–October (8-h daily maximum).

current 1-h daily maximum NAAQS! (The area is Los Angeles.) A map generalizing interpolated long-term O_3 data in the U.S. appears as Figure 3.

Non-U.S. Urban Areas

Summary O_3 data for non-U.S. areas are difficult to obtain, especially for short-term averaging periods. There is more "foreign" data available on annual means or monthly means, for instance, than on short-term averaging periods. What short-term data could be obtained are depicted in Tables 2 and 3.

Table 2 shows exceedances data for two O_3 air quality "cutpoints", 0.12 ppm and 0.08 ppm. The Mexico City data show (1) great annual variability and (2) quite high exceedances. In fact, the 1988 stations 2 and 3 data indicate that Mexico City has a worse O_3 problem than Los Angeles, the MSA in the U.S. with the most severe O_3 problem. Exceedances in Taipei in 1986 are similar to those seen in Los Angeles. Most of the other urban areas shown in Table 2, however, have better O_3 air quality than most U.S. areas.

Table 3 presents maximum O_3 concentrations, comparable to the data for the U.S. shown in the second column of Table 1. Again, Mexico City has very high levels — higher than any urban area in the U.S. The remaining cities shown in the table have similar or lower maximum O_3 concentrations than most urban areas in the U.S. The Canadian cities, by the way, have quite variable yearly maximums over the 1983–1989 time period. Their yearly pattern essentially is identical to that seen in U.S. urban areas.[30]

For the record, reported annual O_3 means monitored in urban areas generally are in the 0.020–0.040 ppm range for lower-altitude regions in the mid-latitudes

Table 2. Short-Term O_3 Exceedances Information for Non-U.S. Urban Areas

Urban area, country	Year	Observed exceedances of 0.12 ppm	Observed exceedances of 0.08 ppm	Ref.
Daily maximum 1-h averages				
Delft, Neth.	1976	—	29	17
	1977	—	6	
Vlaardingen, Neth.	1976	—	48	17
	1977	—	5	
Calgary, Canada	1986,87	—	0	18
60 Canadian sites	1987	—	55% > 1	19
956 Japanese sites	1986	50% > 1	—	20
Mexico City, Mexico				
Station 1	1987	5	—	21
	1988	52	—	
Station 2	1987	94	—	21
	1988	178	—	
Station 3	1987	118	—	21
	1988	264	—	
NW Station[a]	1987	59	—	22
	1988	138	—	
	1989	98	—	
	1990	135	—	
CEN Station[a]	1987	132	—	22
	1988	178	—	
	1989	90	—	
	1990	201	—	
SW Station[a]	1987	158	—	22
	1988	262	—	
	1989	205	—	
	1990	265	—	
1-h average				
Canadian cities[b]				23
8 Brit. Col. sites	1983	—	36	
	1984	—	12	
	1985	—	36	
	1986	—	16	
	1987	—	5	
	1988	—	39	
	1989	—	4	
10 Prarie sites[b]	1983	—	9	
	1984	—	38	
	1985	—	1	
	1986	—	4	
	1987	—	8	
	1988	—	9	
	1989	—	15	
24 Ontario sites[b]	1983	—	119	
	1984	—	139	
	1985	—	77	
	1986	—	75	
	1987	—	75	
	1988	—	223	
11 Quebec sites[b]	1983	—	68	
	1984	—	20	
	1985	—	29	

Table 2. (Continued)

Urban area, country	Year	Observed exceedances of 0.12 ppm	Observed exceedances of 0.08 ppm	Ref.
	1986	—	27	
	1987	—	20	
	1988	—	157	
4 Atlantic sites[b]	1983	—	8	
	1984	—	1	
	1985	—	5	
	1986	—	3	
	1987	—	32	
	1988	—	41	
Delft, Neth.	1976	—	145	17
	1977	—	21	
Vlaardingen, Neth.	1976	—	48	17
	1977	—	5	
Calgary, Canada[b]	1986,87	—	0	18
Taipei, Taiwan	1986	127	—	25
	1987	60	—	
Auckland, New Zealand				26
Station 1	1976	0	0	
Station 2	1976	0	0	
Steronage, England	1987	0	1	27
Central London, England	1987	0	0	27

[a] These exceedance estimates are corrected for missing data using USEPA's procedure described earlier in the text. See Reference 7.

[b] The exceedances are for Canada's 1-h air quality objective of 0.082 ppm. Thus, there would be slightly more exceedances of a 0.08-ppm concentration value.

of the Northern Hemisphere.[18,19,31,32] These levels drop north of the 50°N latitude in North America, but not in Europe. The mid-50 percentile annual O_3 means are between 0.013 and 0.020 ppm in Canada, for instance.[8]

OZONE CONCENTRATIONS IN RURAL AND REMOTE AREAS

As mentioned in the introduction, non-urban areas* usually have less diurnal variability in their O_3 concentrations than urban or suburban areas. They also have less annual variability, although mesoscale transport of aged O_3 or its precursors into a non-urban area greatly affects its diurnal, monthly, and annual patterns. Elevation of the monitoring site also affects O_3 concentrations experienced at a location.[33] These types of issues are well described by Logan,[34,35] and an interesting topology of diurnal O_3 patterns observed at Western U.S. non-urban monitoring sites is presented in Bohm et al.[36]

In the U.S., the non-urban area database includes O_3 monitoring information for 86 "rural" or "remote" sites located in 35 states. The data shown in Table 4 come from the 1982–1985 time period, with almost 80% coming from 1984 (35%)

* This collective term is used to delineate the entire class of areas which are usually labeled as: agricultural, forest, alpine valley or meadow, mountain, rural, or remote.

Table 3. Hourly and Daily Maximum O_3 Concentrations for Non-U.S. Urban Areas

Urban area, country	Year	Maximum hourly O_3 observation (ppm)	2nd-high 1-h daily (ppm)	Ref.
60 Canadian cities	1987	0.150		19
Canadian cities				
8 Brit. Col. sites	1983	0.169	—	
	1984	0.109	—	
	1985	0.124	—	
	1986	0.133	—	
	1987	0.105	—	
	1988	0.213	—	
	1989	0.101	—	
10 Prairie sites	1983	0.100	—	
	1984	0.100	—	
	1985	0.080	—	
	1986	0.090	—	
	1987	0.090	—	
	1988	0.100	—	
	1989	0.100	—	
24 Ontario sites	1983	0.159	—	
	1984	0.161	—	
	1985	0.118	—	
	1986	0.114	—	
	1987	0.170	—	
	1988	0.178	—	
11 Quebec sites	1983	0.170	—	
	1984	0.100	—	
	1985	0.160	—	
	1986	0.130	—	
	1987	0.120	—	
	1988	0.180	—	
4 Atlantic sites	1983	0.110	—	
	1984	0.160	—	
	1985	0.160	—	
	1986	0.080	—	
	1987	0.100	—	
	1988	0.100	—	
French cities				28
3 Mer sites	1989	0.117	—	
6 Paris sites	1989	0.137	—	
Marseille	1989	0.080	—	
Le Havre	1989	0.126	—	
Grenoble	1989	0.146		
Besancon	1989	0.188	—	
8 Urban sites	1989	0.093	—	
Sapporo, Japan				29
Station 1	1987	—	0.095	
Station 2	1987	—	0.100	
Mexico City, Mexico				21
Station 1	1987	0.492	—	
	1988	0.402	—	
Station 2	1987	0.139	—	
	988	0.350	—	
Station 3	1987	0.495	—	
	1988	0.405	—	

Table 4. Cumulative Frequency Descriptive Statistics Associated with Peak and Multiple-Hour O_3 Air Quality Indicators in Non-Urban Areas

Statistic	Expected exceedances of 0.12 ppm 1-h daily maximum	2nd-high 1-h daily maximum (ppm)	Observed exceedances of 0.08 ppm 8-h daily maximum	Maximum 3-month mean of the 8-h daily maximums (ppm)
Mean	1.3	0.11	21.1	0.054
Std. dev.	3.4	0.02	18.1	0.009
Minimum	0.0	0.06	0	0.029
Median	0.0	0.11	17	0.054
75th per.	1.4	0.12	29	0.059
90th per.	3.5	0.14	48	0.063
95th per.	6.8	0.15	58	0.067
Maximum	27.3	0.16	132	0.086
Sample size	86	86	85[a]	85[a]

Note: Std. dev. = standard deviation; per. = percentile.

[a] This indicator cannot be computed for one non-urban area due to lack of a valid valve for the monitoring site.

From U.S. Environmental Protection Agency, Ambient Standards Branch. "Review of the National Ambient Air Quality Standard for Ozone: Assessment of Scientific and Technical Information," U.S. EPA Report (1989). With permission.

and 1985 (43%). More recent O_3 data are available for some of these sites, but are not shown in the table.

The 86 sites are those listed in the USEPA's O_3 database as having one of the following land use codes:

- 32 — Rural or suburban agricultural
- 41 — Strictly remote
- 45 — Other remote

Sites so coded are thought to adequately represent rural (code 32) and forested areas (codes 41 and 45).

It is important to realize that the majority of non-urban sites are located within a U.S. Census bureau-defined MSA. This fact does not contravene the land use coding classification. Rural areas are found within an MSA, as are forested areas; truly remote areas, however, are not, except in unusual circumstances. The sites used for the non-urban analyses reported in this section are described in Reference 37. See that report for additional information on sites included, year of analysis, state, and month of highest 1-h and 8-h averages.

The air quality indicators shown in Table 4 for non-urban areas are those previously used in urban areas. Our discussion begins with short-term peak indicators of O_3.

For the 86 non-urban sites:

- 70% attained the 0.12-ppm O_3 standard in the years analyzed
- 76% had fewer than 1.5 expected exceedances of a 0.12-ppm daily maximum 1-h value
- The largest number of expected exceedances at a site was 27.3 days*

The non-urban distribution is different from the urban distribution: compare Tables 1 and 4. There are many fewer expected exceedances of the O_3 standard in rural/remote areas than in urban areas. This is to be expected, given the urban origin of precursor pollutants that ultimately result in high ambient O_3 concentrations.

Turning to the second-highest daily maximum indicator, it can be seen that there is less variability in the non-urban database than in the urban sample. While non-urban areas have similar, but somewhat lower, mean second-high "design values", there is a large difference in this air quality indicator for the worst areas. The maximum second-high indicator seen in rural/remote areas is 0.16 ppm; it is 0.37 for urban areas.

Some 75% of U.S. non-urban areas have fewer than 30 days with an 8-h daily max greater than 0.08 ppm. Most have about 20 days. Both of these are *higher* than the corresponding percentiles for urban areas. This hints at the differences between urban and non-urban daily O_3 patterns: urban areas experience high daily peaks but *relatively* lower 8-h averages than rural areas. This finding has been identified by other investigators.[39] There are smaller differences in cumulative O_3 indicators between urban and non-urban areas than in peak indicators.

The 3-month mean data indicator shown in Table 4 is marginally different from the one used by researchers to characterize O_3 exposure to agricultural crops. Most of these studies, such as those conducted by the National Crop Loss Assessment Network (NCLAN), use a "growing season" fixed 7-h (9 a.m.–4 p.m.) daily average exposure statistic. (See Chapter 6.)

This is a difficult air quality indicator to analyze because of the variability in growing season across the country. While a 3-month period seems biologically relevant because it approximates the average growing season for most crops in the U.S., there is considerable uncertainty as to whether the fixed daily daylight period captures the time interval of greatest plant sensitivity. It certainly does not capture the time period of maximum O_3 in a day for many rural and remote sites.[37] Since the fixed time period does not capture the maximum 7-h (or 8-h) ozone exposure for many days and probably does not include all periods of plant biological activity, the USEPA states that a more-encompassing, long-term air quality indicator is needed than the one used by NCLAN and others.[12]

---

* This site is Kern County, CA, and it is very different from most of the other rural/remote sites. It is the "outlier", so to speak, for many of the air quality indicators investigated. However, the county definitely is an agricultural area. Johnson et al.[38] indicate that it has the second-highest amount of crop acreage and harvested acreage of the 84 counties in the sample (two counties have two sites). Thus, the site belongs in the sample and it indicates how much variability in air quality is possible at rural sites.

That notwithstanding, a growing-season O_3 indicator is used in the Netherlands. It is the May-September mean of values monitored between 1000 and 1700. Values experienced at Dutch stations generally vary between 0.039 and 0.047 ppm, with a mean for all stations being 0.040 ppm in most years.[40] In years with enhanced photochemical activity, however, these means can get as high as 0.057 ppm.[41] Dutch O_3 seasonal averages are somewhat lower than comparable U.S. station averages.

The growing season length and time-of-day problems associated with the above indicators can be minimized by using a maximum 3-month mean of the 8-h daily maximum averages. However, many other O_3 air quality indicators have been investigated by the USEPA[42] and non-Agency scientists for use in non-urban areas.[27, 43-49]

One type of O_3 air quality indicator that seems to be gaining favor among plant scientists is a cumulative index, where hourly O_3 concentrations are summed over time. The unit of this type of indicator is ppm-h. Concentrations are summed over all hours, or hours with an O_3 concentration equal to or greater than some specified value. Two such concentration levels frequently used are 0.06 and 0.07 ppm. However, there are almost as many alternative O_3 air quality indications being proposed as there are authors on the subject. (See, for instance, the articles cited above and the work by Tiedemann et al.[50]) Some of these indicators are discussed in more detail in Chapters 6 and 7.

Short-term peaks in low-altitude forested areas in the eastern U.S. can reach 0.187 ppm.[48] The number of hours that reach O_3 levels of 0.08 ppm or higher often exceeds 200 per year at these sites during a "bad" O_3 year — such as 1988. As usual, however, there is considerable annual variability in exceedances of 0.08 ppm at these sites. Canadian forested sites generally have fewer exceedances than American forested sites. For instance, at six British Columbia sites, three averaged no exceedances of Canada's 0.082-ppm guideline, and three had only one exceedance, on average, over a 7-year period.[23]

Most forest sites exhibit no discernible trend in O_3 concentration for the 5- or 10-year time periods ending in 1988.[49] No trend was discovered in O_3 data for the 1971–1988 time period at Hohenpeissenberg, FRG either.[51]

Turning to short-term O_3 peaks in non-U.S., low-altitude remote ("clean") areas, an extensive analysis of data undertaken by Lefohn, et al.[44] indicates that maximum 1-h O_3 concentrations are usually in the 0.045–0.080 ppm range. However, peaks as high as 0.133 ppm have been recorded on occasion at these low-altitude clean sites. The site consistently having the lowest O_3 peaks is in American Samoa; its maximum 1-h O_3 concentrations are between 0.03 and 0.042 ppm.[44] Peak O_3 concentrations in two clean areas in Finland usually average 0.045 to 0.055 ppm, but on occasion can get as high as 0.07 to 0.08 ppm.[52]

With respect to short-term peaks in non-U.S. mountainous areas, 1-h daily peak O_3 concentrations are between 0.04 and 0.10 ppm in the Teine Mountains of Japan.[29] At a 2835-m elevated site at the South Pole, they are 0.038 to 0.047 ppm.[44] These values are much lower than the O_3 daily peaks of 0.12 ppm occasionally seen at Whitetop (VA) and Whiteface (NY) Mountains and at Mt. Mitchell (NC) in the U.S.[51,52] The highest 1-h O_3 value reported for any mountain site (defined to be any location with an elevation greater than 1400 m) apparently is 0.18 ppm. This value was recorded at Mauna Loa, HI, in 1983.[44] This high a concentration probably is associated with stratospheric intrusion. On the other hand, Canadian data in remote mountainous areas rarely show peak daily O_3 concentrations exceeding 0.08 ppm.[18]

Dutch data from the remote coastal area of Vlissingen indicate that there are numerous days having peak hourly O_3 concentrations greater than 0.08 ppm.[17] In 1976, 50 days had one or more hourly concentrations of 0.08 ppm or higher. In 1977, however, there were only 4 days with concentrations that high for an hour or more. Obviously, there is a lot of annual variation in daily O_3 concentrations at that site. This is similar to the pattern seen in remote U.S. areas.[55] In addition, the frequency of 0.08-ppm peaks at Vlissingen is comparable to many non-urban areas in the U.S. that are affected by long-range O_3 transport.

There have been a number of studies that monitored year-round O_3 in non-urban areas in order to determine an "all-hours" annual mean concentration. These long-term means vary with altitude, generally being higher in elevated terrain. O_3 annual means thought to be representative of low-altitude "background" sites are:

- 0.026–0.030 ppm (Canada)[56]
- 0.023–0.035 ppm (upper U.S. Midwest)[57]
- 0.025 ppm (Tasmania)[58]
- 0.025 ppm (Washington, U.S.)[59]
- 0.018–0.023 ppm (Preila, Lithuania)[60]
- 0.019–0.030 ppm (northern England)[61]

Annual O_3 means for higher-altitude background sites (>1400 m elevation) that are presented in the literature include:

- 0.028–0.030 ppm (South Pole)[44]
- 0.031–0.038 ppm (White River, UT)[44]
- 0.041–0.048 ppm (Fortress Mt., Alberta)[44]
- 0.038–0.047 ppm (Apache Nat. Forest, AZ)[44]
- 0.036–0.042 ppm (Mauna Loa, HI)[44]
- 0.033–0.042 ppm (Whiteface Mt., NY)[44]
- 0.045–0.053 ppm (eastern Alps)[62]
- 0.024–0.038 ppm (Switzerland)[33,63]

It should be noted that few of the background sites listed above are entirely free of anthropogenic impacts. Thus, their concentration estimates include both "natural" and man-made O_3. Altshuller, a noted U.S. expert in O_3 air quality, states that natural background levels of O_3 include contributions only from biogenic sources and from stratospheric intrusion into the troposphere. Both processes, after netting out O_3 scavenging near the ground, result in an annual O_3 level of approximately

0.02 ppm.[62] The portion of the annual mean due solely to stratospheric intrusion, averaged over the entire northern hemisphere, is estimated to be 0.002 ppm at low altitudes.[2] This contribution could be as high as 0.01 ppm in some localized areas.

O_3 resulting from man's activities is added to the natural O_3 background to comprise the total O_3 burden in our troposphere. Describing the human, vegetation, and ecosystem impacts associated with this burden is the subject of the next five chapters of this book.

REFERENCES

1. Environmental Criteria and Assessment Office. "Air Quality Criteria for Ozone and Other Photochemical Oxidants," U.S. EPA Report 600/8-84/020bF (1986).
2. Austin, J. F., and M. J. Follows. "The Ozone Record at Payerne: An Assessment of the Cross-Tropopause Flux," *Atmos. Environ.* 25A:1873–1880 (1991).
3. Penkett, S. A. "International Conference on the Generation of Oxidants on Regional and Global Scales, Norwich, 3–7 July, 1989: An Overview," *Atmos. Environ.* 25A:1735–1737 (1991).
4. Comrie, A. C. "The Climatology of Surface Ozone in Rural Areas," *Prog. in Phys. Geo.* 14(3):295–316 (1990).
5. Butkovic, F., T. Cvitas, and L. Klasinc. "Photochemical Ozone in the Mediterranean," *Sci. of Total Environ.* 99:145–151 (1990).
6. 40 U.S. Code of Federal Regulations 50.
7. Curran, T. C. "Guideline for the Interpretation of Ozone Air Quality Standards," U.S. EPA Report 450/4-79-003 (1979).
8. Chock, D. "Issues regarding the Ozone Air QualityStandards," *J. Air Waste Manage. Assoc.* 41:148–152 (1991).
9. Lee, E. H., D. T. Tingey, and W. E. Hogsett. "Selection of the Best Exposure-Response Model Using Various 7-H Ozone Exposure Statistics," unpublished U.S. EPA Report (1987).
10. Lioy, P. J., T. A. Vollmuth, and M. Lippmann. "Persistence of Peak Flow Decrement in Children Following Ozone Exposures Exceeding the National Ambient Air Quality Standards," *J. Air Pol. Control Assoc.* 35:1068–1071 (1985).
11. Wallsten, T., et al. "Chronic Lung Injury Risk Assessment for Ozone," paper presented at the *83rd Annual Meeting of the Air and Waste Management Association,* Pittsburgh, PA, June 1990.
12. U.S. Environmental Protection Agency, Ambient Standards Branch. "Review of the National Ambient Air Quality Standard for Ozone: Assessment of Scientific and Technical Information," U.S. EPA Report (1989).
13. Paul, R., et al. "National Estimates of Exposure to Ozone Under Alternative National Standards," U.S. EPA Report (1986).
14. U.S. Environmental Protection Agency, Technical Support Division. "National Air Quality and Emissions Trends Report, 1989" U.S. EPA Report 450/4-91-003 (1991).
15. Wakim, P. G. "1981 to 1988 Ozone Trends Adjusted to Meteorological Conditions for 13 Metropolitan Areas," paper presented at the *83rd Annual Meeting of the Air and Waste Management Association,* Pittsburgh, PA, June 1990.
16. Hemphill, M. W., et al. "Statewide Ozone Trends," paper presented at the *83rd Annual Meeting of the Air and Waste Management Association,* Pittsburgh, PA, June 1990.

17. Guicherit, R. "Concentration and Pattern of Ozone in Western Europe," in *Atmospheric Ozone Research and Its Policy Implications,* T. Schneider, et al., Eds. (Amsterdam: Elsevier, 1989), pp. 167–176.
18. Peake, E., and B. D. Fong. "Ozone Concentrations at a Remote Mountain Site and at Two Regional Locations in Southwestern Alberta," *Atmos. Environ.* 24A(3):475–480 (1990).
19. Environment Canada, Environmental Protection. "National Urban Air Quality Trends, 1978–1987," Environment Canada Report EPS 7/UP/3 (1990).
20. Environment Agency. "Quality of the Environment in Japan 1988," Government of Japan Report, 1989.
21. Bravo, H., et al. "Weekdays and Weekend Days Variation in Ozone and Nitrogen Oxide Concentrations in the Atmosphere of Mexico City," paper presented at the *83rd Annual Meeting of the Air and Waste Management Association,* Pittsburgh, PA, June 1990.
22. U.S. Environmental Protection Agency, Environmental Criteria and Assessment Office. "Mexico City Air Pollutants," U.S. EPA Report ECAO-R-504 (1991).
23. Environment Canada, Canadian Council of Ministers of the Environment. *Management Plan for Nitrogen Oxides (NO_x) and Volatile Organic Compounds (VOCs)* (Ottawa: Environment Canada, 1990).
24. G. V. R. D. (Greater Vancouver Regional District) Pollution Control. "Ambient Air Quality, 1985 Annual Report," G. V. R. D. Report (1987). Also used were the 1986, 1987, 1988, and 1989 Annual Reports.
25. Liu, C.-M., S. C. Liu, and S.-H. Shen. "A Study of Taipei Ozone Problem," *Atmos. Environ.* 24A(6):1461–1472 (1990).
26. Farkas, E. "Surface Ozone Variations in the Auckland Region," *N.Z. J. Sci.* 22:63–76 (1979).
27. Bower, J. S., et al. "Surface Ozone Concentrations in the U. K. in 1987–1988," *Atmos. Environ.* 23:2003–2016 (1989).
28. Ministere Charge de l'Environnement. *Pollution Atmospherique en France: Resultats de Mesure, Annee Civile 1989* (Paris: Service de l'Environnement Industriel, 1991).
29. Murao, N. "The Causes of Elevated Concentrations of Ozone in Sapporo," *Atmos. Environ.* 24A(6):1501–1507 (1990).
30. McCurdy, T., and R. Atherton. "Annual and Multi-Year Variability in Ozone Expected Exceedances," *J. of Air and Waste Manage. Assoc.* 41:1189–1194 (1991).
31. Hough, A. M., and R. G. Derwent. "Changes in the Global Concentration of Tropospheric Ozone Due to Human Activities," *Nature* 344:645–648 (12 April 1990).
32. Staehelin, J., and W. Schmid. "Trend Analysis of Tropospheric Ozone Concentrations Utilizing the 20-Year Data Set of Ozone Balloon Soundings Over Payerne (Switzerland)," *Atmos. Environ.* 25A:1739–1749 (1991).
33. Wunderli, S., and R. Gehrig. "Surface Ozone in Rural, Urban and Alpine Regions of Switzerland," *Atmos. Environ.* 24A:2641–2646 (1990).
34. Logan, J. A. "Tropospheric Ozone: Seasonal Behavior, Trends, and Anthropogenic Influence," *J. Geo. Res.* 90(DG):10463–10482 (October 20, 1985).
35. Logan, J. A. "Ozone in Rural Areas of the United States," *J. Geo. Res.* 94(DG):8511–8532 (June 20, 1989).
36. Bohm, M., B. McCune, and T. Vandetta. "Diurnal Curves of Tropospheric Ozone in the Western United States," *Atmos. Environ.* 25A:1577–1590 (1991).
37. McCurdy, T. "Descriptive Statistical Analyses of Ozone Air Quality Indicators in Rural/Remote Areas," unpublished U.S. EPA Report (1987).

38. Johnson, T., A. Ferdo, and J. Capel. "Relationships Between Selected Three-Month Air Quality Indicators Used to Predict Ozone Damage to Crops and Woodland," (Durham, NC: PEI Associates, 1987).
39. Rombout, P. J. A., et al. "The Need for an Eight Hour Ozone Standard," in *Atmospheric Ozone Research and Its Policy Implications,* T. Schneider, et al., Eds. (Amsterdam: Elsevier, 1989) pp. 167–176.
40. DeLeeuw, F. A., H. Leyssius, and P. Builtjes. "Calculations of Long Termed Averaged Ground Level Ozone Concentrations," *Atmos. Environ.* 24A:185–193 (1990).
41. DeLeew, F. A., and H. J. V. R. Leyssius. "Calculation of Long-Term Averaged and Episodic Oxidant Concentrations for the Netherlands," *Atmos. Environ.* 25A:1809–1818 (1991).
42. Lee, E. H., D. T. Tingey, and W. E. Hogsett. "Evaluation of Ozone Exposure Indices in Exposure-Response Modeling," *Environ. Pollut.* 53:43–62 (1988).
43. Lefohn, A. S., J. A. Lawrence, and R. J. Kohut. "A Comparison of Indices that Describe the Relationship Between Exposure to Ozone and Reduction in the Yield of Agricultural Crops," *Atmos. Environ.* 22:1229–1240 (1988).
44. Lefohn, A. S., S. V. Krupa, and D. Winstanley. "Surface Ozone Measured at Clean Locations Around the World," *Environ. Pollut.* 63:189–224 (1990).
45. Lefohn, A. S., D. S. Shadwick, and V. A. Mohnen. "The Characterization of Ozone Concentrations at a Select Set of High-Elevation Sites in the Eastern United States," *Environ. Pollut.* 67:147–178 (1990).
46. Clarke, B. B., B. Greenhalgh-Weidman, and E. G. Brennan. "An Assessment of the Impact of Ambient Ozone on Field-Grown Crops in New Jersey Using the EDU Method: Part 1. White Potato (*Solanvm tuberosvm*)," *Environ. Pollut.* 66:351–360 (1990).
47. Brennan, E. G., et al. "An Assessment of the Impact of Ambient of the Impact of Ambient Ozone on Field-Grown Crops in New Jersey Using the EDU Method: Part 2. Soybean (*Glycine max.* (L.) Merr.)," *Environ. Pollut.* 66:361–373 (1990).
48. Lefohn, A. S., and A. A. Lucier. "Spatial and Temporal Variability of Ozone Exposure in Forested Areas of the United States and Canada:1978–1988," *J. Air Waste Manage. Assoc.* 41:694–701 (1991).
49. Lefohn, A. S., and D. S. Shadwick. "Ozone, Sulfur Dioxide, and Nitrogen Dioxide Trends at Rural Sites Located in the United States," *Atmos. Environ.* 25A:491–501 (1991).
50. Tiedemann, A. V., et al. "Ozone Episodes in Southern Lower Saxony (FRG) and Their Impact on the Susceptibility of Cereals to Fungal Pathogens," *Environ. Poll.* 67:43–59 (1990).
51. Low, P. S., et al. "Uncertainties in Surface Ozone Trend at Hohenpeissenberg," *Atmos. Environ.* 25A:511–515 (1991).
52. Hakola, H., et al. "Transport, Formation, and Sink Processes Behind Surface Ozone Variability in North European Conditions," *Atmos. Environ.* 25A:1437–1447 (1991).
53. Meagher, J. F., K. J. Olszyna, and F. P. Watherford. "The Availability of H_2O_2 and O_3 for Aqueous Phase Oxidation of SO_2. The Question of Linearity," *Atmos. Environ.* 24A:1825–1829 (1990).
54. Aneja, V. P., et al. "Exceedances of the National Ambient Air Quality Standards for Ozone Occurring at a 'Pristine' Area Site," *J. Air Waste Manage. Assoc.* 40(2): 217–220 (1990).

55. Pinkerton, J. E., and A. S. Lefohn. "The Characterization of Ozone Data for Sites Located in Forested Areas of the Eastern United States," *J. Air Pol. Control Assoc.* 37(9):1005–1010 (1987).
56. Angle, P., and H. S. Sandhu. "Rural Ozone Concentrations in Alberta, Canada," *Atmos. Environ.* 20:1221–1228 (1986).
57. Pratt, G. C., et al. "Ozone and Oxides of Nitrogen in the Rural Upper-Midwestern U.S.A.," *Atmos. Environ.* 17:2013–2023 (1983).
58. Galbally, I. E., et al. "Surface Ozone at Rural Sites in the Latrobe Valley and Cape Grim, Tasmania," *Atmos. Environ.* 20:2403–2422 (1986).
59. Singh, H. B., F. L. Ludwig, and W. B. Johnson. "Tropospheric Ozone: Concentrations and Variabilities in Clean Remote Atmospheres," *Atmos. Environ.* 12:2185–2196 (1978).
60. Girgzdiene, R. "Surface Ozone Measurements in Lithuania," *Atmos. Environ.* 25A:1791–1794 (1991).
61. Gay, M. J. "Meteorological and Altitudinal Influences on the Concentration of Ozone at Great Dunn Fell," *Atmos. Environ.* 25A:1767–1779 (1991).
62. Puxbaum, H., et al. "A One-Year Record of Ozone Profiles in an Alpine Valley (Zillertal/Tyrol, Austria, 600–200m a.s.l.)," *Atmos. Environ.* 25A:1759–1765 (1991).
63. Wunderli, S., and R. Gehrig. "Influence of Temperature on Formation and Stability of Surface PAN and Ozone," *Atmos. Environ.* 25A:1599–1608 (1991).
64. Altshuller, A. P. "Review Paper: The Role of Nitrogen Oxides in Non-urban Ozone Formation in the Planetary Boundary Over N. America, W. Europe and Adjacent Areas of Ocean," *Atmos. Environ.* 20:245–268 (1986).

CHAPTER 3

ACUTE AND CHRONIC HEALTH EFFECTS OF OZONE

Steven M. Horvath and David J. McKee

INTRODUCTION

The focus of this chapter is on those health effects associated with levels of O_3 that have been monitored in many of the largest cities of the world. Of the major photochemical oxidants in ambient air, only O_3 has been reported to exist at sufficiently high concentrations to be of significant concern for human health. Since other photochemical oxidants such as hydrogen peroxide (H_2O_2) and peroxyacetyl nitrate (PAN) are known to produce health effects only at concentrations much higher than those found in most cities, this chapter will focus only on documented evidence of pulmonary and extrapulmonary effects of O_3 reported in human and animal studies. Animal toxicological studies have been used primarily to support the possibility of human health effects caused by O_3 where there is an absence of adequate human data, although differences in dosimetry and species sensitivity limit quantitative extrapolation of animal data to human health effects.

O_3 ABSORPTION AND MECHANISMS OF EFFECT

O_3 enters the human body through the respiratory system. During inhalation and exhalation, O_3 is removed from the air by airway surfaces through a process

0-87371-475-X/94/$0.00+$.50
© 1994 by Lewis Publishers

of dissolution and chemical reaction. The rate, amount, and sites of O_3 uptake and removal largely determine the extent to which exposure of an individual to a particular concentration of O_3 will evoke adverse health effects.

Numerous studies help to explain the quantity and sites of O_3 uptake in mammalian respiratory tracts. Nasopharyngeal removal studies suggest that (1) the fraction of O_3 uptake depends inversely on the flow rate, (2) tracheal and exposure chamber concentrations of O_3 are positively correlated,[1-3] and (3) approximately 50% of inhaled O_3 (> 72% for dogs and > 50% for rabbits) is removed in the nasopharyngeal region of animals exposed to between 0.1 and 0.2 ppm O_3, thereby indicating the role of the nasopharynx in removing O_3 before it reaches the more sensitive lung tissues. Only one study measured O_3 uptake in the lower respiratory tract and reported 80 to 87% uptake by the lower respiratory tract of dogs.[1] There are, however, no published data currently available for human nasopharyngeal or lower respiratory tract absorption.

Inhaled O_3 not absorbed in the nasopharyngeal region can be deposited along the entire respiratory tract. Penetration into the lower respiratory tract increases in rats as inhaled O_3 concentrations increase from 0.2 to 0.8 ppm.[4] Models have been developed to estimate lower respiratory uptake.[5-7] One O_3 deposition model has predicted that similar patterns of O_3 deposition occur in humans, rabbits, and guinea pigs and that the junction of the conducting airways and gas exchange region receives the maximal dose.[5,6] This prediction is consistent with location of O_3-induced lesions as indicated by pathology data from several species of O_3-exposed animals. The Miller dosimetry model has been used to estimate effects of lower respiratory tract secretions and exercise on the uptake of O_3 in the human lung.[6]

Biologically important functional groups which react relatively rapidly with O_3 include the alkenic (carbon-carbon double bonds), amino, and sulfhydryl groups. Alkenic bonds are found in essential fatty acids and polyunsaturated fatty acids (PUFA) which are important components of the lipids in cell membranes. Oxidation of amino and sulfhydryl groups can result in protein denaturation with concomitant loss of structural and functional integrity for the affected protein.

Although there is general agreement that the oxidative properties of O_3 cause toxic effects, the precise molecular mechanism of toxicity remains unclear. Several theories have been proposed.

1. Oxidation of polyunsaturated lipids contained mainly in cell membranes
2. Oxidation of sulfhydryl, alcohol, aldehyde, or amine groups in low molecular weight compounds or proteins
3. Formation of toxic compounds (ozonides and peroxides) through reaction with polyunsaturated lipids
4. Formation of free radicals, either directly or indirectly, through lipid peroxidation
5. Injury mediated by some pharmacologic action, such as via a neurohormonal mechanism or release of histamine

Molecular targets (e.g., carbon-carbon double bond, sulfhydryl groups) are shared across all species. Therefore, if O_3 initiates mechanisms such as those listed above in an animal which ultimately result in lung structural damage, there is reason for concern that the same mechanisms may be activated in humans with similar outcomes. This assumption is limited by differences in human and animal dosimetry which produce different doses of O_3 for equivalent exposures. Also, defense and repair mechanisms are likely to provide quantitative differences in the display of toxicity resulting from equivalent doses. Nonetheless, mechanistic information strongly supports the hypothesis that equivalent effects may occur in humans and animals, albeit not necessarily at the same concentrations.

FACTORS AFFECTING SUSCEPTIBILITY TO O_3

There are numerous factors which could affect susceptibility to O_3 exposure and alter physiological responsiveness. These factors include such individual characteristics as age, sex, smoking status, and nutritional status. In addition, environmental stresses and exercise level during exposure can influence the extent and level of an individual's response to O_3 by increasing the volume of inhaled O_3 and by promoting penetration deeper into the lungs.

Age and Gender

There is a growing body of evidence which suggests that age and gender may be factors in determining responsiveness to O_3. Older men and women (ages 51–76 years), all of whom were nonsmokers, were exposed to 0.45 ppm O_3 during intermittent moderate exercise without any significant response; this led to the suggestion by the authors that older persons may be less responsive than younger persons.[8] These results were extended in a subsequent study by Bedi et al.[9] who exposed 16 older subjects (ages 60–89 years) to O_3 on 2 to 3 consecutive days and reported no significant response, compared to filtered air exposure, on any of the 3 days. This contrasts to earlier studies (e.g., see Reference 10) which indicated significant response for similar first-day exposures followed by attenuation of response on subsequent days. Other studies[11-13] from the same laboratories in Santa Barbara have reported that older individuals tend to have smaller pulmonary responses than younger subjects. Reisenauer et al.[14] report that older women but not older men have an increase in respiratory resistance of 13% when exposed to 0.30 ppm O_3; however, subjects were younger and had lower total O_3 exposures compared to studies cited above.

Numerous clinical, field, and epidemiology studies have provided evidence of pulmonary function and/or symptomatic response in younger persons exposed to less than 0.15 ppm O_3.[15-21] A series of camp studies[22-26] was conducted with young children who were in an outdoor environment for several weeks but were, in general, exposed to relatively low concentrations of ambient O_3 (i.e., below the NAAQS). These studies have raised some concern as to potential ill effects in this young population. Children in these studies have experienced decreased lung function with increases in ambient O_3, but effects appear more causally related to levels on a day preceding the effects observation day. In some instances, the increased sensitivity persisted for a period of days.[23] The above findings are based primarily on regression analysis of pulmonary parameters against ambient O_3 concentrations and testing the derived slopes for differences from zero. There are, however, some confounding conclusions reported. Bock et al.[24] reported significant negative slopes for girls but not for boys, and Spektor et al.[25] found greater declines in lung function if O_3 concentrations above 0.06 parts per million (ppm) were excluded from analysis. Higgins et al.[27] reported in a separate camp study that lung function decrements occurred only if the O_3 levels exceeded 0.12 ppm. In all these camp studies, there are some confounding factors which could not be controlled or precisely measured. For example, activity levels of the children could not be precisely determined, times at which the tests were performed varied, and interaction with other pollutants introduced uncertainty in the analyses. Nevertheless, these findings, in concert with the converse findings in older adults, suggest the need to reevaluate age implications of the earlier studies, performed mostly on young adults, with regard to susceptibility.

Studies conducted in countries other than the U.S. provide support for the contention that children may be more susceptible to the adverse effects of O_3. Symptoms have been reported for children living in Japan who were exposed to ambient oxidant concentrations of 0.10 ppm and higher.[28] The Castillejos et al. study[29] of children living in Mexico City raises some important new issues. This study found that some, but not all, lung function parameters decreased in a manner which led the authors to speculate that lung inflammation was occurring in the children. Each child was tested following treadmill exercise sessions in a school yard during O_3 exposures up to 0.32 ppm. Results showed that half of the children had chronic respiratory symptoms. Follow-up studies are now being conducted in Mexico City on younger children. In Basel and Zurich, which frequently have high O_3 levels, Braun-Fahrlander et al.[30] investigated respiratory symptoms in children up to 49 months of age. They found a likely association between incidence and duration of symptoms and particulate matter concentrations, but did not evaluate similar associations with O_3 concentrations.

Although it has been reported that older individuals may be less likely than younger persons to experience reductions in lung function and symptoms when exposed to O_3, this may in fact put them at higher risk for developing lung damage because they may continue to carry on with normal activities even when ambient O_3 levels are elevated. On the other hand, because children tend to exercise outdoors more often than most adults, they are more likely to experience acute and chronic health effects of O_3 and, therefore, should also be considered at higher risk.

Due to the small number of female subjects included in most controlled human exposure studies, no definitive conclusions have been drawn regarding gender differences in response to O_3 exposure. Four studies which provide enough information for limited comparative evaluation[31-34] suggest that lung function of women may be affected more than that of men for similar O_3 exposures and exercise levels. More recent studies[8,11,12,14,36] have recruited female subjects, but thus far there has been no organized effort to assess gender differences.

Most O_3 research on animals also has been performed using only males. When female animals were used, gender comparisons usually have not been made. An exception is the study by Graham et al.[37] which reports increased pentobarbital-induced sleeping time in all females, but not in male mice or rats after 5-h exposures to 1.0 ppm O_3. Although reasons for these gender differences are not yet well understood, the differences reported in several research studies provide support for the need of further investigation in both humans and animals.

Smoking Status

Results of studies comparing the effects of O_3 on smokers vs. nonsmokers remain somewhat inconclusive, although below 0.5 ppm O_3, smokers appear to be somewhat less responsive. Two studies reported greater pulmonary function changes for nonsmokers than for smokers at rest during exposure to O_3 levels of 0.37 to 0.5 ppm; this finding was reversed for higher O_3 levels (0.75 ppm), with smokers showing greater response.[38,39] Greater pulmonary function response was also reported in nonsmokers than in smokers at 0.5 ppm[40a] and at 0.3 ppm.[33] Effects of O_3 were greater for nonsmokers compared to smokers at 0.5 and at 0.3 ppm;[40] in a subsequent study, exercising nonsmokers showed a greater response to specofic airway conductance (SG_{aw}) than exercising smokers at 0.15 ppm O_3.[41] Exercising smokers also showed[42] slower and smaller spirometric variable changes than nonsmokers exposed 2 h at 0.5 and 0.75 ppm O_3. Although none of these studies has examined the effects of different amounts of smoking, available data supports the contention that smokers are less responsive to O_3 than nonsmokers, at least for lower O_3 exposures. Why smokers appear to be less responsive to O_3 than nonsmokers is not yet clear, though it has been suggested that altered lung function and increased mucus production caused by smoking could influence O_3 deposition in the lungs.

Nutritional Status

Results of studies investigating nutritional status as a factor affecting susceptibility to O_3 are inconclusive. Some inconsistencies between human and animal data are apparent, but might be explained based on different experimental approaches. Human subjects receiving 4 to 8 times the recommended daily allowance of vitamin E showed no statistically significant differences in blood biochemistry or agglutination compared to unsupplemented subjects after 2-h exposures to 0.5 ppm O_3.[43-45] Animal studies show that vitamin E deficiency makes rats more susceptible to O_3-induced enzymatic changes[46-48] and that vitamin E alters the rate and extent of O_3 toxicity but not the lesion in the centriacinar region of the lungs.[49,50] Lesions have been reported to be worse in vitamin E-deficient rats or in rats marginally supplemented with vitamin E when compared to highly supplemented rats;[46,47] this supports results of mortality[51] and biochemical studies suggesting that vitamin E is protective in rats.

Since redistribution of vitamin E from extrapulmonary stores in humans to the lungs is slow, it is possible that such a protective effect does not occur over short-term exposures to O_3 in human studies. Another possible explanation is that an effect is only seen when subjects are vitamin E deficient. Animal studies in which effects have been reported were conducted with a vitamin E-deficient group for comparison and for long-term exposures,[52-54] neither of which is possible in human studies. Although laboratory animal studies of vitamin E support the hypothesis that lipid peroxidation is involved in O_3 toxicity, benefits to be derived from dietary vitamin E supplementation in connection with O_3 exposure have not yet been demonstrated in humans.

Environmental Stress

Subjective symptoms and physiological impairment caused by O_3 exposure can be substantially worsened by environmental stresses such as heat and high relative humidity (rh). High temperatures (31 to 40° C) and/or humid conditions (85% rh), when combined with exercise during O_3 exposure, have been shown to reduce FEV_1 more than similar O_3 exposures at a more moderate temperature (25 °C) and humidity (50% rh).[55-57] Enhanced effects of O_3 caused by high temperatures and humidity may be the result of the higher ventilation caused by these conditions, which increases the volume of inhaled O_3 and promotes deeper penetration into the lungs. There may also be an independent effect of elevated body temperature on pulmonary function. Since many urban areas around the U.S. tend to experience the highest O_3 levels during periods of high temperature and humidity, environmental stress should be considered a factor of concern in assessing potential effects from O_3 exposure.

Exercise

Exercise stresses respiratory, cardiovascular, and musculoskeletal systems and increases ventilation and, therefore, total dose to the lungs. Exercise thus should be considered an important susceptibility factor when O_3 exposure occurs concurrently. The most apparent and well-studied effects of O_3 during exercise occur in the respiratory system. In particular, pulmonary function decrements and respiratory symptoms caused by O_3 exposure are increased by a greater work load, which is characterized by increased frequency and depth of breathing.[15-18,58-62] Representative activities and associated ventilation rates are summarized in Table 1. Higher ventilation increases total volume of O_3 breathed and could cause deeper penetration of O_3 into peripheral areas of the lung which tend to be most sensitive to injury. Since most individuals engage in some type of outdoor exercise where O_3 exposure is possible during summer months, exercise should be considered a susceptibility factor in assessing health effects of O_3.

POTENTIALLY SENSITIVE POPULATION SUBGROUPS

This section identifies potentially susceptible groups or subpopulations and provides a rationale for identifying these groups. Susceptibility to any pollutant may depend upon many factors such as those previously discussed (age, sex, smoking, nutrition, environment, and exercise), but is also influenced by individual sensitivity. Greater than normal sensitivity to a particular pollutant may be conferred by numerous individual characteristics, including (1) predisposition to pulmonary infection, (2) preexisting disease or nutritional deficiency, (3) some aspect of growth or decline in lung development, (4) a prior infection or immunological problem, (5) sensitization caused by prior pollutant exposure or challenge with respiratory irritants, and (6) genetic variability in the population. These sensitivity and susceptibility factors should be considered in identifying groups or subpopulations which may be susceptible to O_3 exposure.

Individuals Having Preexisting Respiratory Disease

The first major group discussed here which might be at increased risk to O_3 exposure is those individuals with preexisting respiratory disease. This includes individuals with asthma, allergy, and chronic obstructive lung disease (COLD) such as emphysema. Asthmatics, who experience variable airway obstruction and/or reactivity and who may have altered immunological states (e.g., atopy) or cellular function (e.g., eosinophilia), may be expected to be more sensitive to O_3 than nonasthmatics even though they have not been shown to be more sensitive in studies to date. Asthma, however, is not a specific homogeneous disease and efforts to precisely define asthma have been unsuccessful. Susceptibility to O_3 of patients with COLD remains somewhat uncertain, depending on their clinical and functional state.

Although several controlled human exposure studies suggest that adults with preexisting respiratory disease may not be more responsive to O_3 than healthy adults, two considerations place these subgroups at potentially higher risk to O_3 exposure. First, due to concern for the health of these persons, higher concentrations of O_3 and higher exercise levels have not been used during controlled exposure studies, thus making comparisons with healthy subjects difficult. A second consideration is that for individuals with already compromised respiratory systems, any further decrement in pulmonary function may be expected to impair the ability to perform normal activities. The same decrement in a healthy individual may go unnoticed. Increases in symptoms reported for asthmatics and allergic individuals exposed to O_3 may also be expected to interfere with normal function.

Exercising Individuals

A second group potentially at higher risk to O_3 exposure is that group of individuals including both normal, healthy persons and those with preexisting

Table 1. Estimated Values of Oxygen Consumption and Minute Ventilation Associated with Representative Types of Exercise[a]

	Work performed		O_2 consumption,	Minute ventilation	
Level of work	Watts	kg-m/min[b]	L/min	L/min	Representative activities[c]
Light	25	150	0.65	12–16	Level walking at 2 mph; washing clothes
Light	50	300	0.96	17–23	Level walking at 3 mph; bowling; scrubbing floors
Moderate	75	450	1.25	23–30	Dancing; pushing wheelbarrow with 15-kg load; simple construction; stacking firewood
Moderate	100	600	1.54	29–38	Easy cycling; pushing wheelbarrow with 75-kg load; using sledgehammer
Moderate	125	750	1.83	35–46	Climbing stairs; playing tennis; digging with spade
Heavy trenches	150	900	2.12	42–55	Cycling at 13 mph; walking on snow; digging
Heavy	175	1050	2.47	52–57	Cross-country skiing; rock climbing; stair climbing with load; playing squash and handball; chopping with axe
Very heavy	200	1200	2.83	62–79	
Very heavy	225	1350	3.19	73–93	
Very heavy	250	1500	3.55	89–110	Level running at 10 mph; competitive cycling
Severe	300	1800	4.27	107–132 skiing	Competitive long distance running; cross-country

[a] See text of Criteria Document (pp. 10-13 to 10-15) for discussion.
[b] kg-m/min = work performed each minute to move a mass of 1 kg through a vertical distance of 1 m against the force of gravity.
[c] Adapted from Astrand and Rodahl (1977).

respiratory disease whose regular outside activities cause increased minute ventilation. The most prominent modifier of response to O_3 in the general population is ventilatory volume, which increases proportionately with increasing exercise workload. The increased proportion of oral breathing with increased exercise also contributes to the increase in effects since more O_3 is delivered to the lungs, due to reduced efficiency of O_3 scrubbing in the mouth. Although individuals with preexisting respiratory or cardiovascular disease may not exercise as heavily as healthy persons, any increases in activity over resting levels will increase O_3 exposure and resultant effects.

Exercise has become recognized in numerous recent studies as perhaps the major factor which can predispose all individuals to O_3 health effects.[15-18,60-66] Thus, activities which increase minute ventilation ($\dot{V}_E$) out-of-doors will also increase the health risks associated with exposure to O_3.

Individual responsiveness to O_3 of apparently healthy, homogeneous groups has been shown to vary widely. For example, during heavy exercise ($\dot{V}_E = 45–51$ l/min) and exposure to 0.4 ppm O_3, subjects showed FEV_1 decrements ranging from below 10% to as much as 40% of control values for those subjects, with an average of 26%.[67,68] Mode (oral vs. nasal) and pattern (rapid shallow vs. slow deep) of breathing may contribute to intersubject variability, but cannot fully explain it. Unsuccessful attempts have been made to determine which factors are responsible for modifying individual responsiveness.[69] Factors such as prior exposure to O_3 or other pollutants, latent infections, and nutritional deficiencies may also contribute to differential response. The potential for these contributions is biologically plausible, but these have yet to be demonstrated in human subjects.

Unusual responsiveness to O_3 has been typically observed only in exercising individuals. These individuals, often referred to as "responders", have not been characterized as having any particular medical problem, but experience significantly greater pulmonary function decrements than the average response of the groups concurrently studied during O_3 exposure. It is not known if these individuals are a population subgroup with a specific risk factor or simply represent the upper 5 to 20% of the O_3 response distribution. As yet, there is no means of identifying these highly responsive individuals prior to O_3 exposure.

Although the specific factors which modify individual response to O_3 are not yet identified, it has been suggested that intersubject variability in the magnitude of effects induced by O_3 is caused by large differences in intrinsic responsiveness to O_3.[61,66] As measured by FEV_1 and FVC, individual responses to O_3 are highly reproducible for at least the 10-month period of observation and, for given individual differences, tend to be much smaller than differences between subjects.[32,61] Hence, some intrinsic factor or factors appear to be responsible for individual responsiveness.

Changes in FEV_1 of exercising subjects exposed to clean air are small and normally distributed in the subject population. As O_3 concentrations increase, this distribution widens and becomes skewed toward larger FEV_1 decrements, the largest changes representing the most responsive subjects.[60,66] Classification of subjects into "responders" and "nonresponders" has been somewhat arbitrary.

Examples of retrospective classification include identifying responders by using medical history and exposure test results[70] and selecting those with greater than 10% post-exposure decrements[10] or with FEV_1 decrements greater than 2 standard deviations above control.[67] While there are no clearly established criteria for identifying or selecting "responders", it has been suggested that between 5 and 20% of the healthy population may represent a subgroup of the population which is more responsive and, therefore, at higher risk to O_3 exposures.

HEALTH EFFECTS ASSOCIATED WITH EXPOSURE TO O_3

There exists a wide variety of health effects which have been linked to O_3 exposure. While the databases which establish concentration-response relationships for each of these effects are highly variable, there are several effects which are of greatest concern to health scientists due to their more serious or potentially incapacitating nature. Those health effects of greatest concern include (1) acute alterations in pulmonary function, (2) respiratory symptoms, (3) decrements in exercise performance, (4) increased airway reactivity, (5) aggravation of existing respiratory disease, (6) pulmonary inflammation, (7) morphological effects, (8) host defense system alterations, (9) morbidity and mortality, and (10) extrapulmonary effects.

Respiratory effects provide by far the largest, most quantitative database for considering adverse health effects of O_3. Discussion of these effects will be the primary focus of this and a subsequent chapter on risk assessment (see Chapter 5). Extrapulmonary effects such as behavioral, blood chemistry, chromosomal, cardiovascular, reproductive, teratological, liver metabolism, and endocrine system modifications have also been associated with O_3 exposure, but most are considered to be of uncertain functional importance at this time and thus do not provide very strong evidence of adverse effects associated with O_3 exposure.

Evidence for the above effects has come from both human and animal studies. The most quantifiable data are provided by controlled human exposure studies, but these studies are limited to acute (short-term) exposures. Field studies are similarly limited but permit investigation of the effects of oxidants in ambient air and allow for better characterization of exposure than epidemiological studies. Although use of most community studies has been hampered by the difficulty in adequately characterizing exposure and numerous confounding environmental variables, these investigations provide important supporting evidence for effects occurring in populations. Animal toxicology studies have provided evidence of many acute and chronic (long-term) exposure effects which can be detected only with invasive procedures; however, uncertainties inherent in dosimetry and species sensitivity differences have limited quantitative extrapolation of effects reported in animals to human health effects.

Assessment of health effects attributed to O_3 requires consideration of the database in each of the above areas of study. This chapter integrates research in each of these areas to provide an indication of the strength of the database for each effect. A more thorough review and evaluation of individual studies can be found in Chapters 9 to 12 of Reference 71, its Addendum,[72] and Reference 73.

Acute Alterations in Pulmonary Function

The best documented evidence of human health effects of O_3 exposure are pulmonary function decrements. Controlled human exposure, field, epidemiology, and animal toxicology studies have provided evidence that exposure to O_3 can modify such pulmonary measurements as 1-s forced expiratory volume (FEV_1), forced expiratory flow (FEF), forced vital capacity (FVC), vital capacity (VC), tidal volume (V_T), peak expiratory flow rate (PEFR), inspiratory capacity (IC), total lung capacity (TLC), airway resistance (R_{aw}), and breathing frequency (f_B).

Early controlled experimental studies of resting human subjects exposed to O_3 levels up to 0.75 ppm for 2 h demonstrated little or no change in FVC,[68,74] FEV_1, and FRC.[68] Flow rate variables such as FEF 25% and FEF 50% showed up to 30% decreases in some subjects exposed at rest to 0.75 ppm O_3,[68,75] while only small increases in R_{aw} (<17%) were reported for >0.5-ppm O_3 exposures.[75,76] Other studies have reported the occurrence of FEV and FEF decrements during resting exposures to ≥ 0.5 ppm O_3;[31,64] however, no statistically significant changes in R_{aw} and only suggestive changes in residual volume (RV) and TLC have been reported for similar exposures.[42] R_{aw} is not generally affected in resting subjects at these O_3 levels. Changes in pulmonary function have not been observed in resting subjects exposed to O_3 levels between 0.12 ppm[77] and 0.3 ppm,[64] though some subjects exhibit O_3-induced pulmonary symptoms during resting exposures.[76,78] In general, however, because subjects were at rest in most of the older studies, significant respiratory effects were not reported even for higher O_3 exposures.

Exercise, which causes increased $\dot{V}_E$, enhances individual and group mean response to O_3 exposure. As discussed earlier in this chapter, exercise increases f_B and depth of breathing, resulting in greater total dose of O_3 inhaled and increased penetration to the most sensitive lung tissue. As exercise levels increase to the point where $\dot{V}_E$ exceeds approximately 35 l/min, oronasal or oral breathing tends to predominate;[79] thus, at higher exercise levels, a greater portion of the inhaled O_3 will bypass the nose and nasopharynx.[80] Individual variability will affect the $\dot{V}_E$ at which oral or oronasal breathing predominates. The relationship between exercise and magnitude of response is illustrated quite well by Figure 1, showing group mean decrements in FEV_1 caused by exposure of exercising subjects to various O_3 levels based on results from 25 different studies. The curves clearly demonstrate that as exercise levels increase for a given O_3 exposure, there is a resultant increase in the group mean FEV_1 decrements.

Pulmonary function decrements have been reported in controlled exposure studies of healthy exercising human adults exposed to O_3 levels between 0.12 and 0.24 ppm, a range for which a concentration-response relationship has been established. During 2 h of intermittent, very heavy exercise ($\dot{V}_E$ = 64 l/min), healthy subjects experienced group mean decrements in FEV_1 of 4.5, 6.2, and 14.5% for O_3 exposures of 0.12, 0.18, and 0.24 ppm, respectively.[60] Positive associations between group mean magnitude of FVC (3, 4, and 12%) and FEF (7.2, 12, and 23%) decrements and O_3 exposures of (0.12, 0.18, and 0.24 ppm) were also reported in McDonnell's study;[60] statistically significant increases in specific airway resistance (SR_{aw}) and f_B did not occur until O_3 was ≥0.24 ppm. In a separate study, McDonnell et al.[15] reported a statistically significant but small decline (3.4%) in group mean FEV_1 of children (8–11 years) after a 2-h exposure to 0.12 ppm O_3 during intermittent heavy exercise ($\dot{V}_E$ = 39 l/min). Furthermore, there was a suggestion that the small decrements in FEV_1 persisted for 16 to 20 h after O_3 exposure ended.

Support for O_3-induced pulmonary function changes also comes from other controlled exposure studies. Avol et al.[62] report statistically significant decreases (6.1%) in FEV_1 at 0.16 ppm O_3, with larger decrements (19.1%) occurring at ≥0.24 ppm O_3. Decrements have been reported in FEV_1, FVC, and FEF for distance runners and distance cyclists exposed to 0.20 and 0.21 ppm O_3, respectively, during 1 h of continuous, heavy exercise ($\dot{V}_E$ = 77.5 and 81 l/min).[65,81] Separate studies of continuously exercising males ($\dot{V}_E$ = 61.8 l/min) and females ($\dot{V}_E$ = 46 l/min) exposed to 0.2 ppm O_3 for about an hour showed no statistically significant FEV_1 decrements, but these negative results can be attributed in part to the small number of subjects (i.e., inadequate power), 8 male and 6 female, respectively.[34,82] In another study involving O_3 concentrations ranging between 0.10 and 0.25 ppm, exponential decreases in FVC, FEV_1, FEF_{25-75}, specific airway conductance (SG_{aw}), IC, and TLC have been reported with exposure to increasing O_3 concentration during very heavy exercise ($\dot{V}_E$ = 68 l/min); time of exposure was related to linear decreases in FVC and FEV_1.[66]

Field studies, which contain elements of both controlled human exposure and epidemiologic studies, provide the most quantitatively useful human exposure data available for ambient photochemical oxidants. Results from field studies are consistent with pulmonary function decrements reported in controlled O_3 exposure studies. For a l-h exposure to mean O_3 concentrations of 0.144 ppm, small but significant group mean decreases in FVC (3.3%), $FEV_{0.75}$ (4.0%), FEV_1 (4.2%), maximum mid-expiratory flow rate (MMFR) (3.2%), and PEFR (3.9%) relative to preexposure levels were reported in 59 healthy, continuously exercising ($\dot{V}_E$= 32 l/min) adolescents.[17,18] Based on the comparison of ambient air and control exposures in these studies, it appears that O_3 was the causative agent of the pulmonary function effects.

In a separate study of 50 healthy, adult, continuously exercising ($\dot{V}_E$ = 53 l/min) bicyclists, mean O_3 concentrations of 0.153 ppm for 1 h produced statistically significant group mean decreases in FEV_1 (5.3%) compared to preexposure.[62] This study showed that similar effects result in subjects exposed to comparable O_3 levels in ambient and controlled O_3 exposures. Small but statistically significant decreases in FEV_1 and FVC were also reported in exercising healthy and asthmatic adults during exposure to mean O_3 concentrations of 0.165 and 0.174 ppm.[63,83,84]

Although the group mean changes in lung function in the above studies are small, considerable intersubject variability in the magnitude of individual pulmo-

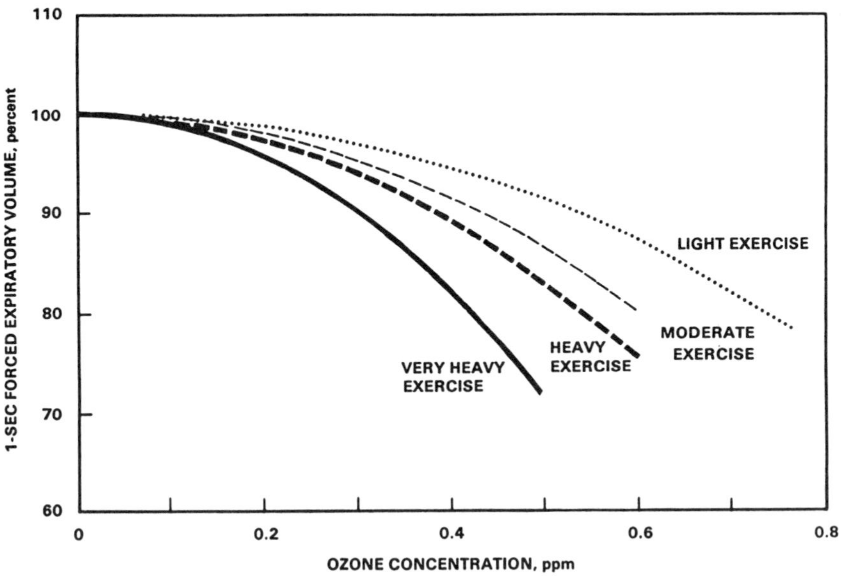

Figure 1. Group mean decrements in 1-s forced expiratory volume during 2-h O_3 exposures with different levels of intermittent exercise: light ($\dot{V}_E \leq 23$ l/min); moderate ($\dot{V}_E = 24–43$ l/min); heavy ($\dot{V}_E = 44–63$ l/min); and very heavy ($\dot{V}_E \geq 64$ l/min). From O_3 Criteria Document, U.S. EPA, 1986.

nary response exists, and some subjects experienced responses which were quite large. Controlled exposures to 0.12 ppm O_3 during very heavy exercise have resulted in individual pulmonary function decrements up to 16% for adults[60] and up to 22% for children.[15] Individual subject data from several studies[60,62,66] have been used to estimate the fraction of population which experiences ≥10 and ≥20% FEV_1 decrements due to O_3 exposure. The exposure-response relationships suggest that between 2 and 20% of the heavily exercising population might exhibit ≥10% decreases in FEV_1 when exposed to O_3 levels of approximately 0.12 ppm. It is estimated that between 0 and 5% of the heavily exercising population might experience FEV_1 decrements of ≥20% when O_3 exposures are 0.12 ppm. These estimates must be considered approximate considering: (1) the amount of variability of response between the three studies analyzed, (2) the small number of subjects tested in each study (i.e., between 20 and 50 subjects at any given exposure level), (3) the fact that the subjects in these three studies were not derived from population-based sampling, and (4) the use of a fitted function based on empirical rather than biological grounds. Further discussion of this analysis can be found in Chapter 5 of this book and in References 85 and 86.

Recent controlled human exposure, epidemiology, and animal toxicology studies have increased concern about enhanced respiratory effects associated with multi-hour (6–8 h) exposures to O_3. Folinsbee et al.[87] published data indicating that 10 nonsmoking males (age 18–33 years) exposed to 0.12 ppm O_3 during intermittent moderate exercise for 6.6 h experienced progressively decreasing

FEV_1 over the exposure period. At the end of the exposure, group mean FEV_1 had decreased by 13.0%, FVC by 8.3%, and FEV_{25-75} by 17.4% when compared to clean-air exposures. One individual experienced a 47.6% decrease in FEV_1. There were also progressive increases in symptom ratings of cough and pain on deep inspiration, as well as a marked increase in methacholine airway responsiveness.

In a continuation of the Folinsbee et al. research,[87] Horstman et al.[88] exposed each of 21 nonsmoking, healthy, male subjects to 0.0, 0.08, 0.10, and 0.12 ppm O_3 on separate days. Exposure protocol was similar to that described above for the Folinsbee study,[87] and results were consistent. Substantial pulmonary function decrements, respiratory symptoms, and increases in nonspecific airway reactivity were reported for all three O_3 exposures when compared to filtered air exposure and became progressively greater after each hour of exposure. Group mean decreases in FEV_1 ranged from 7.0% at 0.08 ppm to 12.3% at 0.12 ppm, but all group means were statistically significant. An additional follow-up by McDonnell et al.[89] reported a group mean decrease in FEV_1 of 8.4% at 0.08 ppm using a similar protocol with 38 healthy young men.

Several analyses have been performed on data from the USEPA Clinical Research Branch to assess the relationships between exposure duration, exposure concentration, lung function decrements, and rate of decay of effects. In a multiple linear regression analysis of the Horstman et al.[88] data, Larsen et al.[90] found that when concentrations approximate normal peak ambient O_3 levels, exposure time is almost as important as exposure concentration in estimating FEV_1 decrements. Analyzing additional data from the same laboratory, Folinsbee et al.[91] found that 3- and 4-h sliding averages of O_3 concentration provide the best fit for FEV_1 decrements and that all data fit a curvilinear response relationship if only exposure during exercise is considered. Folinsbee and Hazucha[92] analyzed data of moderately exercising young adults exposed to 0.35 ppm O_3 and found that mean FEV_1 decrements were 21% immediately after exposure, and had decreased to only 4% after 18 h and 2% after 42 h. Finally, comparison of a constant exposure of 0.12 ppm O_3 vs. a variable exposure (ranging from 0 to 0.24 ppm O_3) protocol for a group of healthy young adults exposed to O_3 for 8 h led Hazucha et al.[93] to conclude that for the same cumulative exposure, greater decrements in FEV_1 resulted from variable exposures.

Further support of the relationship between acute O_3 exposure and pulmonary function decrements is provided by the first of several epidemiological studies of children engaged in normal activity at an eastern summer camp.[22,24] These studies reported lung function impairment following acute exposures to ambient O_3 concentrations during the study period. Lioy et al.[23] further reported that a persistent decrement in lung function of children at a New Jersey camp lasted as much as a week after the end of a smog period of 4 days during which peak 1-h O_3 levels were in the range of 0.135 to 0.186 ppm. Other persistent decrements reported by Lioy et al.[23] (i.e., altered epithelial permeability and changes in airway secretion) represent a potentially more important response than the more transient effects reported in most controlled exposure studies; however, the authors recognized that elevated concentrations of inhalable particulate matter associated with a large-scale photochemical smog episode were also associated with the effects reported. Subsequently, Lioy and Dyba[94] proposed that the PEFR decrements reported by Lioy et al.[23] resulted from total O_3 dose rather than persistence of the effect from one day to the next.

Spektor et al.[25] have followed up the work of Lioy et al.[23] with a summer camp study in New Jersey of children exposed to ambient O_3 levels which did not exceed the current standard level of 0.12 ppm during the study. Multiple regression analyses of lung function measurements indicate that the most explanatory environmental variables for daily change in lung function were: (1) previous-hour O_3 levels, (2) cumulative daily O_3 exposure between 9 a.m. and the time of measurement of lung function, (3) ambient temperature, and (4) humidity. A subsequent study[26] conducted at the same summer camp provided confirmatory evidence of significant associations between O_3 levels and lung function declines. Persistence of measurable lung function decrements reported for at least a day after exposures to $\geq$0.15 ppm O_3 is consistent with Kinney et al.[95] and Lioy et al.,[23] but not with controlled human exposure studies for similar exposures, thus suggesting that other factors (e.g., acid aerosols) may be involved.

Several other summer camp studies have reported similar results. Significant reductions in late-afternoon peak expiratory flow associated with increasing peak O_3 concentrations and increased frequency of respiratory symptoms as the daily peak concentration rose above 0.12 ppm O_3 were reported in a New Jersey study conducted at a different camp site.[96] Coefficients of response were similar to those reported by Spektor et al.,[25,26] but differed from results of a camp study in the San Bernadino mountains of southern California.[27] In the latter study, children also experienced lung function decrements which declined in proportion to O_3 levels; however, coefficients were smaller than those in the New Jersey studies cited above, possibly due to the higher acid aerosol levels in the eastern camp sites. One study[97] which specifically looked at the acid aerosol contribution was conducted at an eastern Connecticut camp for healthy and asthmatic children. Thurston et al.[97] reported significant associations between medication use by asthmatics and peak O_3 concentration; however, the greatest increases in use occurred when both O_3 and hydrogen ion levels peaked.

Studies of healthy adults engaged in outdoor activity also confirm the association between ambient O_3 levels and lung function decrements. Although ventilatory rates and period of exercise varied substantially between subjects, and O_3 measurements ranged from only 0.021 to 0.124 ppm, Spektor et al.[98] still found significant associations between O_3 levels and functional decrements. Results, when compared to chamber studies, suggest that comparable decrements occurred after shorter exposures in the field study, thus leading the authors to conclude that cofactors in the environment may exacerbate effects of O_3. A study by Korrick et al.[99] of 103 males and females (ages 18–53 years) who hiked on Mount Washington, NH reported decrements in FVC, FEV_1, and FEF associated with maximum hourly O_3 levels ranging from 0.04 to 0.08 ppm. This study further showed that decrements were not as pronounced in older hikers and that longer hikes resulted in greater decrements, both results consistent with chamber studies discussed earlier in the chapter. Another study of healthy adults exercising outdoors was conducted by Kinney et al.[100] at the U.S. Military Academy at West Point, NY, before and after training exercises at bases in New Jersey, Georgia, Missouri, and Oklahoma. Cadets trained in New Jersey, where O_3 levels exceeded 0.10 ppm for 23 h, showed an average decline in FEV_1 of 78 ml. This is in contrast to those cadets trained at the other sites, where O_3 levels never exceeded 0.10 ppm, who had an average FEV_1 decline of only 31 ml. In a separate study conducted at West Point, Kinney et al.[101] examined baseline lung function of cadets during their first week on campus and found that those coming from communities with high oxidant levels had mean FVC measurements which were 2% lower than those from communities with lower oxidant levels.

While it is clear that no single epidemiological study provides definitive quantitative data by itself, usually for a variety of reasons (e.g., methodological problems, lack of exposure data, and confounding variables), the aggregation of studies provides reasonably good evidence of association between ambient photochemical oxidant exposure and acute pulmonary effects. The association is strengthened by the consistency of these epidemiological results with the findings in numerous clinical studies reporting decrements in pulmonary function for individuals exercising intermittently during exposures to O_3 ranging from 0.08 to 0.12 ppm, depending on length of exposure.

In conclusion, the weight of currently available evidence indicates that healthy, heavily exercising, young subjects can experience pulmonary function decrements during controlled exposures of $\geq$0.12 ppm O_3 for 1 or 2 h. It is clear that level of exercise and individual responsiveness play a major role in determining the extent of pulmonary function loss caused by exposure to any level of O_3. More recent clinical studies have reported lung function decrements for exposures as low as 0.08 ppm when exposures lasted for 6.7 h. Epidemiology and field studies provide supporting evidence of measurable functional decrements for exposures below 0.12 ppm, perhaps in part due to pollutant interactions and other environmental variables. Further discussion and a different perspective on these results can be found in Chapter 9 of this book.

Respiratory Symptoms

Respiratory symptoms have been associated with group mean pulmonary function changes in adults acutely exposed in controlled exposures to O_3 and in ambient air containing O_3 as the predominant pollutant. Despite a close association observed between changes in group mean FEV_1 and group mean respiratory symptoms for O_3 exposures, Hayes et al.[86] reports only a weak-to-moderate correlation between FEV_1 changes and symptoms severity when the analysis is conducted using individual data. It should be noted that symptoms reported are inherently more subjective than FEV_1 decrements measured.

In controlled O_3 exposures, some heavily exercising ($\dot{V}_E \geq 65$ l/min) adult subjects have experienced cough, shortness of breath, and pain on deep inspiration at 0.12 ppm O_3, although the group mean response was statistically significant for cough only.[60] Above 0.12 ppm O_3, respiratory and nonrespiratory symptoms which have been reported include throat dryness, chest tightness, substernal pain, cough, wheeze, pain on deep inspiration, shortness of breath, dyspnea, lassitude, malaise, headache, and nausea.[40,57,60,62,65,66,81,83,102-104] At 0.2 ppm O_3 and higher, controlled exposure studies of exercising subjects have reported positive "symptom scores", with elimination of symptoms for most subjects within 24 h post-exposure.

Data reported in three controlled human exposure studies[60,62,66] have been analyzed to estimate the fraction of the heavily exercising population that might experience mild and moderate respiratory symptoms during exposure to various O_3 levels. Figures 2 and 3 provide the estimated exposure-response relationships. As O_3 exposure levels rise in these studies, the fraction of subjects experiencing respiratory symptoms generally increases. For example, it is estimated that for 0.12 ppm O_3 exposures, between 0 and 15% of the heavily exercising population might experience moderate cough or chest discomfort. As the O_3 exposure level rises to 0.2 ppm, the estimated fraction of the population that might experience moderate cough or chest discomfort increases to a range between 3 and 40%. As with the lung function analysis discussed in the previous section, these estimates must be considered approximate considering (1) the amount of variability of response between the three studies analyzed, (2) the small number of subjects tested in each study (i.e., between 20 and 50 subjects at any given exposure level), (3) the fact that the subjects in the three studies were not derived from population-based sampling, and (4) the use of a fitted function based on empirical rather than biological grounds. Chapter 5 of this book and References 85 and 86 present a more detailed discussion of the respiratory symptom risk analysis.

Comparisons of symptoms reported in controlled O_3 exposure studies have been made with those reported in field studies. One study, which provided a direct comparison of symptoms in exercising ($\dot{V}_E = 57$ l/min) adults caused by exposure to either purified air containing 0.16 ppm O_3 or oxidant-polluted ambient air which contained 0.15 ppm O_3, showed no statistically significant differences, suggesting that increased symptoms associated with lung function impairment were caused by O_3 alone.[62] Several epidemiology studies have provided evidence of qualitative associations between ambient oxidant levels greater than 0.10 ppm and symptoms in children and young adults such as throat irritation, chest discomfort, cough, and headache.[25,105] Thus, it can be concluded that most symptoms reported in individuals exposed to O_3 in purified air are similar, but not identical, to those found for ambient air exposures.

An exception is eye irritation, a common symptom associated with exposure to photochemical oxidants, which has not been reported for controlled exposures to O_3 alone. This appears to hold even at O_3 concentrations much higher than would be found in the ambient air. It is widely accepted that other oxidants such as aldehydes and PAN are primarily responsible for eye irritation and are generally found in atmospheres containing higher ambient O_3 levels.[106–109]

Pulmonary function decrements have been reported in studies which did not report increases in symptoms. Children (age 8–11 years) intermittently exercising

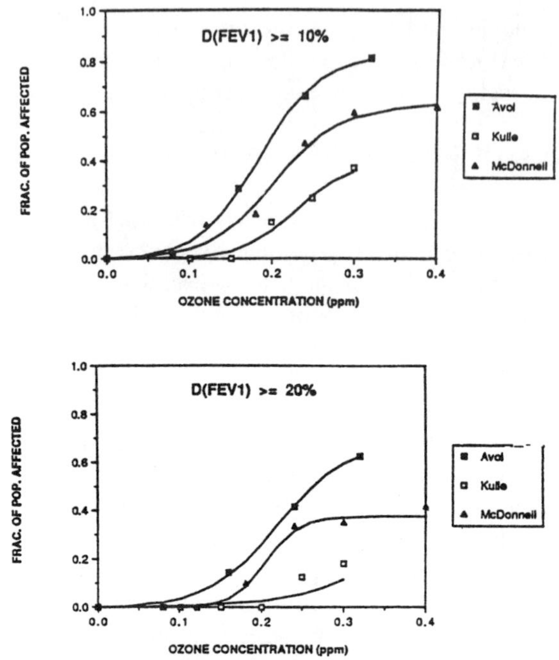

Figure 2. Fraction of heavily exercising population experiencing ≥10% and ≥20% change in FEV_1 due to various O_3 levels. Exposures were for 1-h continuous exercise for Avol et al. Data for 2-h intermittent exercise for McDonnell et al. and Kulle et al. data.

($\dot{V}_E = 39$ l/min) for 2.5 h at 0.12 ppm O_3 showed small but statistically significant decreases in FEV_1, but showed no changes in frequency or severity of cough compared to control.[15,61] Similarly, adolescents (age 12–15 years), continuously exercising ($\dot{V}_E = 31$–33 l/min) during exposure to 0.144 ppm mean O_3 in ambient air, showed no changes in symptoms despite statistically significant decrements in group mean FEV_1 (4%) which persisted at least 1 h during resting post-exposure.[17,18] Since symptoms can be viewed as an early warning of related lung function impairment or possibly other more significant health effects, the lack of symptoms in children and adolescents during exposures which induce functional decrements may be of concern. It is possible that children may be at higher risk since, with no warning signal, they may not exhibit avoidance behavior which might otherwise limit exposure to elevated O_3 concentrations.

Symptoms, when combined with objective measures of lung function, are considered useful adjuncts in assessing health effects caused by O_3 and photochemical oxidants. Group mean symptoms have been shown to be closely associated with the time course and magnitude of group mean pulmonary function changes associated with O_3 exposures; however, only weak to moderate correlations exist when data are analyzed on an individual basis.

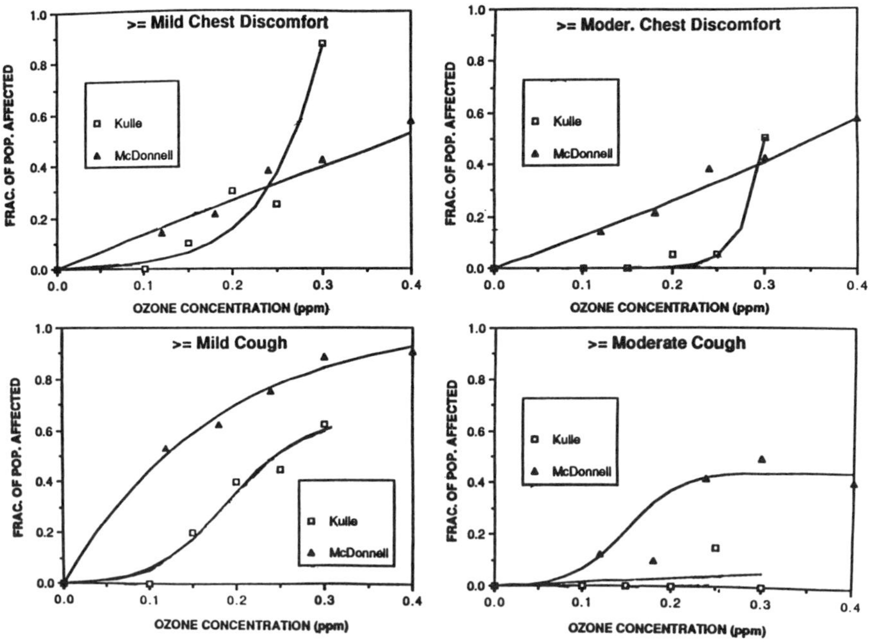

Figure 3. Fraction of heavily exercising population experiencing mild and moderate symptoms due to various O_3 levels. Exposures were for 2 h of intermittent exercise.

Decrements in Exercise Performance

A limited database provides suggestive evidence of decrements in exercise performance associated with O_3 exposure. Early epidemiological evidence on high school students suggested that the percentage of track team members failing to improve performance increased with increasing oxidant concentrations the hour before a race.[110] The authors concluded that the effects may have been related to increased R_{aw} or to associated discomfort which may have limited motivation to run at maximal levels. Controlled exposure studies of heavily exercising competitive runners have demonstrated decreased max $\dot{V}_E$ at 0.3 ppm O_3[111] and decreased FVC, FEV, and FEF at 0.20 ppm O_3.[81] At 0.21 ppm O_3, Folinsbee et al.[65] reported decreases in FVC, FEV, FEF, IC, and maximal ventilatory volume (MVV) at 75% max VO_2 as well as symptoms in seven distance cyclists exercising heavily ($\dot{V}_E$ = 81 l/min).

Too limited a database is available to draw any conclusive judgments regarding effects of O_3 on exercise performance. Although subjective statements by individuals engaged in sports indicate possible voluntary curtailing of activities during high-oxidant episodes, increased temperature and relative humidity may be involved in provoking the symptoms and lung function decrements observed in the studies cited above. Controlled studies of O_3 exposure have, however, demonstrated lung function impairment and subjective symptoms which cause individuals to reduce work load and performance at 0.12 ppm O_3 and higher.[35,62,65,81] Since O_3 has been implicated, at least in part, in reducing exercise performance during periods of high oxidants and some individuals may experience this effect at levels near the O_3 NAAQS, precautions should be taken to avoid unnecessary outdoor exercise when O_3 levels are high.

Airway Reactivity

Airway reactivity is associated with exposure to O_3 in numerous recent studies. Using airway responsiveness to various drugs (e.g., acetylcholine, methacholine, and histamine) as a measure of nonspecific airway sensitivity, researchers have observed increased airway reactivity in both healthy and asthmatic subjects exposed to O_3 levels as low as 0.12 ppm after just 1 h of exposure. It should be noted that most of these studies were conducted on resting or moderately exercising individuals.

In both atopic and non-atopic subjects exposed (using a noseclip) to 0.6 ppm O_3 during intermittent light exercise, bronchial response to histamine and methacholine was greater when subjects were exposed to O_3 than to filtered air.[112] Symptoms of bronchial irritation in the Holtzman et al.[112] study were not detectable by the next day, which contrasts with results of the Golden et al.[76] study of healthy resting subjects exposed to 0.6 ppm O_3 reporting enhanced response to histamine challenge for as much as 1 to 3 weeks after exposure.

Significant increases in airway reactivity were reported with acetylcholine challenge in healthy adult subjects exposed to either 0.32 ppm or 1.0 ppm O_3.[78] Healthy subjects showed no alterations of bronchomotor response to histamine aerosol when exposed to 0.2 ppm O_3 for 2 h, although exposure to 0.4 ppm O_3 did enhance bronchial responsiveness to inhaled histamine.[113] Kulle et al.[114] reported significantly enhanced airway reactivity to methacholine of healthy subjects exposed to 0.4 ppm O_3 as compared to filtered air. Although Folinsbee et al.[87] reported no apparent relationship between O_3-associated changes in methacholine reactivity and changes in FVC or FEV_1, subsequent research[88] in the same lab provided evidence of increased methacholine responsiveness after 6.6 h-exposures to 0.08 ppm (56%), 0.10 ppm (89%), and 0.12 ppm (121%). A closer relationship between changes in methacholine reactivity and lung function were reported in exercising healthy subjects sequentially exposed to acid fog and O_3.[115] Molfino et al.[116] have recently reported that airway reactivity may be much greater in resting asthmatics exposed for 1 h to 0.12 ppm O_3 followed by exposure to allergen (e.g., ragweed or grass) than in non-asthmatics. These results are particularly disturbing since the exposures were conducted while subjects were at rest at levels of O_3 which could commonly be encountered in the summer. This work is currently being followed up in other laboratories.

In conclusion, there is increasing evidence to indicate that O_3 enhances airway reactivity in both healthy and asthmatic subjects when they are exposed to O_3 levels which are seen in the most highly polluted cities. However, emerging evidence seems to indicate that asthmatics may be more greatly affected than healthy individuals at O_3 levels seen in most large cities if acid aerosols are also present.

Aggravation of Existing Respiratory Disease

Only limited evidence is available from controlled exposure to suggest that individuals with asthma, chronic bronchitis, or emphysema are more significantly affected by O_3 or other photochemical oxidants than healthy persons. However, several epidemiological studies have suggested an association between photochemical smog and aggravation of existing respiratory disease. Individuals with preexisting respiratory disease are considered to be especially "at risk" to O_3 exposure due to their already compromised respiratory systems, and there is reason for concern that increased symptoms or pulmonary function decrements may interfere with normal activity.

In controlled human exposure studies, statistically significant group mean decrements in pulmonary function were not reported for adult asthmatics exposed for 2 h at rest[117] or with intermittent light exercise[118] to O_3 concentrations of 0.25 ppm or lower. Similarly, no statistically significant group mean changes in pulmonary function or symptoms were found in adolescent asthmatics exposed to 0.12 ppm O_3 for 1-h at rest.[77] Subjects with COLD performing light to moderate intermittent exercise show no statistically significant group mean decrements in pulmonary function during 1- and 2-h exposures to $\leq$0.30 ppm O_3,[84,119-122] and only small, statistically significant group mean decrements in FEV_1 are observed for 3-h exposures of chronic bronchitics to 0.41 ppm O_3.[123]

While the above controlled exposure results suggested that individuals with preexisting respiratory disease may not be more sensitive to O_3 than healthy subjects, a pilot study by Molfino et al.[116] provided evidence that asthmatics are more affected by allergens than healthy individuals when preexposed to O_3. Seven asthmatics (age 27 to 53 years) identified as reactive to ragweed or grass were exposed to 0.12 ppm O_3 or air, while at rest, followed by exposure to an allergen. Although the O_3 exposures resulted in no changes in baseline FEV_1, airway reactivity was significantly increased for 6 out of 7 subjects and for the whole group only when exposure to allergen was preceded by exposure to O_3. In a separate study, Frampton et al.[124] investigated the impact of sulfuric acid and sham sodium chloride aerosols followed by 0.08, 0.10, and 0.12 ppm O_3 on the lung function of 30 healthy and 30 allergic, asthmatic, young adults. Results of the Frampton et al.[124] study indicate that even though healthy subjects were not affected by preexposure to either of the aerosols, preexposure of asthmatics to 100 $\mu g/m^3$ sulfuric acid enhanced functional response to O_3 exposures as low as 0.08 ppm. Preexposure to sodium chloride aerosol did enhance functional response at 0.12 and 0.18 ppm O_3 in the asthmatic subjects, but not as significantly as the acid aerosol.

Although epidemiological studies do not provide a clear concentration-response relationship between O_3 and aggravation of disease, they do provide evidence of how "real-world" exposures affect individuals with respiratory disease. One study[125] conducted in Los Angeles reported increased daily asthma attack rates on cool days and days with high oxidants and particulates when median daily maximum hourly oxidant levels were ≤0.15 ppm. Questionable exposure assessment, lack of control for medication, pollen, respiratory infections, and other pollutants limit the use of this study for developing quantitative dose-response relationships.

Despite several exposure and variable control limitations, the Houston Area Oxidants Study[126,127] concluded that for the study period in which the daily maximum hourly O_3 concentrations were below 0.21 ppm O_3 near the subjects' residence, (1) there was increased incidence of nasal and respiratory symptoms and increased frequency of medication use for asthmatics with increasing O_3 levels; (2) FEV_1 and FVC decreased with increasing O_3 and total oxidants; and (3) increased incidence of chest discomfort, eye irritation, and malaise occurred at high PAN concentrations. In a subsequent related study,[128-130] increased probability of an asthma attack was associated with the occurrence of a previous attack and with exposure to increased O_3 concentrations and temperature when maximum 1-h averages for O_3 were between 0.001 and 0.127 ppm; however, other pollutants such as SO_2 and particulates may have been involved.

In a series of studies conducted in a Tucson community, adults with asthma, allergies, or airway obstructive disease (AOD) were observed during an 11-month period in which 1-h daily maximum O_3 concentrations were ≤0.12 ppm.[19-21] After adjusting for covariables, O_3 and TSP levels were significantly associated with decreases in peak expiratory flow rate in adults with AOD, and the interaction between O_3 and temperature was significantly associated with alterations in peak flow and symptoms in asthmatics. While these results suggest an effect of O_3 in individuals with preexisting respiratory disease, interpretation is difficult due to the small sample size in relation to the number of covariates and the fact that individual exposure data were not available.

Several studies have reported an association between ambient O_3 levels and admissions to hospitals for acute respiratory effects. One of the earliest studies reporting such a relationship was conducted by Bates and Sizto,[130] who investigated respiratory admissions to 79 acute care hospitals in southern Ontario, Canada. Subsequent analysis by Bates and Sizto[131] confirmed the association between respiratory hospital admissions and O_3 levels, but also indicated that as strong an association existed for sulfates as for O_3. Establishing clear concentration-response relationships has proven difficult due to the close association between ambient O_3 and sulfate levels.[132,133] In a similar study of hospital admissions, Ozkaynak et al.[134] reported positive associations between daily hospital admissions caused by pneumonia and influenza and the maximum hourly O_3 levels monitored in cities in Massachusetts. In a multi-year study of the respiratory hospital admissions in several New York cities, Thurston et al.[135,136] analyzed 1988 and 1989 admissions data and revealed strong associations between total respiratory and asthma admissions on same or subsequent days and summer haze pollutants (i.e., H^+, O_3, SO_4^{2-}). Although significant associations were reported for each of the three pollutants, O_3 consistently had the highest mean effects estimates. In a retrospective study conducted in New Jersey, Cody et al.[137] analyzed the relationship between ambient O_3 levels and emergency department visits for asthma, bronchitis, and finger wounds (a nonrespiratory control) for May through August of 1988 and 1989. A highly significant relationship between asthma visits and O_3 levels was reported, with 13 to 15% of the variability of asthma visits explained in the regression model by temperature and O_3 concentrations. Hospital admissions studies are discussed further in Chapter 9.

Several limitations should be recognized in drawing conclusions about the above-cited studies. None of the studies cited above individually or as a group has yet demonstrated a concentration-response relationship between O_3 exposure and aggravation of preexisting respiratory disease. In general, the epidemiology studies report effects which may also be related to inhalable particle exposure, temperature, or other environmental variables. Furthermore, most offer limited characterization of exposure. However, the epidemiology studies taken as a whole support the contention that exposures to ambient levels of O_3, at levels commonly monitored in many cities (see Chapter 2), appear to increase responsiveness of asthmatics and others with respiratory disease. In combination with other pollutants, O_3 seems to be significantly contributing to adverse respiratory effects. The limited evidence provided by controlled exposure studies of asthmatics and individuals with other respiratory disease further reinforces the contention that O_3 exposure adversely affects the health of individuals with preexisting respiratory disease.

Pulmonary Inflammation

Until the last few years, pulmonary inflammation was a health effect which could only be studied by conducting animal toxicology studies since most techniques for obtaining lung tissue samples could not ethically be used on human subjects. This resulted in obvious limitations to quantitative extrapolation due to major differences in dosimetry and species sensitivity. A new technique called "bronchoalveolar lavage" (BAL), however, has permitted ethical removal of small quantities of cellular material which can then be analyzed to assess the presence of biochemical indicators of inflammation (e.g., increases in polymorphonuclear leukocytes, PMNs).

Several studies have provided important new information on biochemical changes caused by exposure to O_3 which are related to pulmonary inflammation. An early study by Seltzer et al.[138] indicated that the inflammatory process, as indicated by an increase in PMNs in BAL fluid, is initiated soon after exposure of subjects to 0.40 or 0.60 ppm O_3 while undergoing moderate intermittent exercise. Studies conducted by Koren et al.[139-142] provided evidence that the inflammatory response (i.e., an eightfold increase in PMNs) persisted for at least 18 h after exposure of subjects for 2 h to 0.40 ppm O_3 during intermittent heavy exercise. Schelegle et al.[143] further studied the time course of inflammation and found that PMNs in BAL fluid peaked after 6 h and continued at elevated levels after 24 h. Biomarkers of fibrosis and fibrinolysis were also elevated following exposure to O_3.

Increased epithelial permeability was reported, as indicated by doubling of protein, albumin, and immunoglobulin G, thus confirming results of an earlier study by Kehrl et al.[144] Biomarkers of tissue injury and repair (e.g., a sixfold increase in fibronectin) and of coagulation (e.g., increases in tissue factor and urokinase plasminogen activator) were also reported by Koren et al.[142] Significant increases in neutrophils, prostaglandins, interleukin-6, and lactate dehydrogenase in BAL fluid, indicative of pulmonary inflammation, were reported by Devlin et al.[145] and Koren et al.[147] following exposure of subjects of 0.08 and 0.10 ppm O_3 for 6.6 h during intermittent moderate exercise, thus showing that the inflammatory process occurs even at relatively low O_3 levels.

These studies as a group provide clear evidence that even relatively short-term exposures to low levels of O_3 can induce the inflammatory process in the lungs. One or a few such exposures probably would not be harmful to most individuals, just as a suntan or single sunburn would probably not cause permanent damage to the skin. However, many animal studies have shown that repeated exposure to O_3 with resultant repeated inflammation may eventually cause changes in lung structure, reduced elasticity of the lungs, and accelerated permanent loss of lung function. These studies are discussed in the next section and, in conjunction with the studies discussed above, provide a basis of concern for many scientists who believe that aging individuals living in areas with elevated O_3 levels will experience a reduced quality of life resulting from O_3 exposure.

Morphological Effects

Morphological effects (i.e., structural changes) in the lungs caused by exposure to O_3 have thus far been reported almost exclusively in animal toxicology studies. For this reason, it is important to consider differences in dosimetry and sensitivity between humans and laboratory animals. The following discussion provides an indication of some of the structural changes which might occur in human lungs as a result of repeated acute (short-term) and chronic (long-term) exposures to O_3.

Despite the differences in lung structure between humans, dogs, monkeys, mice, rats, and guinea pigs, a characteristic lesion occurs at the junction of the conducting airways and the gaseous exchange tissues. In all species examined, the typical effect is damage to the ciliated and Type 1 cells and hyperplasia of non-ciliated bronchiolar and Type 2 cells; an increase in inflammatory cells is also observed. These effects were reported after 7 days, 8-h/day exposures to 0.2 ppm O_3 in monkeys,[4,146a] and in rats exposed for 7 days, 8 and 24 h/day.[50] Similar effects with different exposures were reported in rats (0.26 ppm, 6 h, endotracheal tube),[147] mice (0.5 ppm, 35 days),[148] and guinea pigs (0.5 ppm, 6 months).[149]

Other controlled-exposure studies of adult and neonate rats[150-154] provide evidence that prolonged, repeated exposures to ≥0.25 ppm O_3 can affect the junctional airways of the distal bronchioles and the proximal alveoli. Barry et al.[150] exposed rats to 0.25 ppm O_3 for 6 weeks and found alterations in surface characteristics of ciliated and non-ciliated (Clara) cells and significant changes in the normal structure of terminal bronchiolar epithelial cells. Huang et al.[152] investigated effects of variable exposures (0.06 to 0.25 ppm O_3) over varying time periods (3 or 12 weeks) and concluded that in all groups of rats tested, hyperplasia of type I alveolar cells in the proximal alveoli was linearly related to cumulative O_3 exposure. In a related study, Pinkerton et al.[157] reported that rats exposed to O_3 for 6 weeks and once to asbestos aerosol fibers for 5 h have three times the fiber count in their lungs compared to rats exposed to clean air and the asbestos aerosol. Implications of this study point to the possible cocarcinogenic effects of O_3 when the lung tissue is exposed to insoluble particles of fibers which may have carcinogenic potential. Further investigation of this effect is currently being conducted by the National Institutes of Health. Small, but significant decrements in lung function consistent with early signs of focal fibrogenesis in the proximal bronchoalveolar region were observed by Grose et al.[151] who exposed rats to O_3 for 22 h per day for 12 months with a variable pattern of exposure (i.e., 0.06 ppm background with peaks of 0.25 ppm, equivalent to a square wave pattern of 0.19 ppm averaged over 9 h). Residual and total lung capacity were reduced, suggesting a stiffening of the lungs; biochemical changes indicative of the continued oxidant exposure were present; however, no immunological changes were observed. Tepper et al.[158] followed up the Grose study[151] by using the same exposure pattern for as much as 78 weeks, and reported small but statistically significant changes in breathing patterns and mechanisms after 1, 3, 13, 52, and 78 weeks during post-exposure challenge with 0, 4, and 8% carbon dioxide. The prolonged exposure to O_3 also caused an increase in expiratory resistance. Chang et al.[154] used the electron microscope to study the development of lung injury, both during and after the 78-week exposure period. They found that 1 week of exposure resulted in epithelial inflammation, interstitial edema, interstitial cell hypertrophy, and influx of macrophages. Although these responses were reduced 3 weeks post-exposure, prolonged exposure resulted in progressive epithelial and interstitial tissue response, including epithelial hyperplasia, fibroplast proliferation, and interstitial matrix accumulation, involving both type I and type II epithelial cells. These changes persisted throughout exposure and continued even during post-exposure recovery. Evidence of injury to ciliated and Clara cells was apparent in the terminal bronchioles. These and other results led Chang et al.[154] to the conclusion that chronic exposure to a pattern of O_3 similar to that in some of our more polluted cities can cause epithelial inflammation and interstitial fibrosis in the proximal alveolar regions of the lungs and bronchiolar epithelial cell injury.

Structural changes in the centriacinar region have been reported after long-term exposures of rats,[159-162] monkeys,[155,156,163–165] and dogs.[166] While cell repair begins within 18 h of exposure,[167–170] cell damage continues throughout long-term exposures but at a slower rate. In a study by Tyler et al.,[155] three groups of 7-month-old male monkeys were exposed to 0.25 ppm O_3 for 8 h per day, either every day or on the days of alternate months, or to filtered air over an 18-month period. Although all monkeys exposed to O_3 developed respiratory bronchiolitis with increases in morphometric indicators of response, only those monkeys exposed on alternate months significantly increased lung collagen content, chest wall compliance, and inspiratory capacity. Furthermore, Tyler et al.[155] reported that those monkeys exposed on alternate months had larger biochemical and physiological alterations and equivalent morphometric changes than those exposed daily throughout the 18-month period, even though they had received only half the dose of O_3.

Increases of lung collagen content in the centriacinar interalveolar septa may be indicative of lung fibrosis and structural damage. This type of morphological effect has been associated with biochemical changes in activity of prolyl hydroxylase and in hydroxyproline content, both related to lung collagen content and fibrosis.[171,172] Lung collagen content increased after exposures to O_3 concentrations as low as 0.5 ppm[171] and continued to increase during long-term exposures.[173] Weanling and adult rats exposed for 6 and 13 weeks, respectively, and young monkeys for 1 year to <1.0 ppm O_3 also showed increased collagen content in the lungs.[173] In the Last study,[173] examination of some of the exposed weanlings and controls at 6 weeks post-exposure indicated a continued increase in lung collagen content, a result demonstrating that damage continued to occur during the post-exposure period. Also, there was little apparent difference in collagen levels of rat lungs for those animals exposed continuously vs. intermittently. This would suggest that intermittent periods of clean air were insufficient for recovery.

The centriacinar inflammatory process also continues during long-term O_3 exposures and appears to be related to remodeling of the centriacinar airways[159,160] and to increased lung collagen which appears mainly in the centriacinar regions.[159,160,171,173] In addition, there is morphometric,[165] morphologic,[166] and functional[174,175] evidence of distal airway narrowing.

While the distal airway narrowing and lesions at the junction of the conducting airways and gaseous exchange region are similar to the changes which have been found in lungs of cigarette smokers,[176-179] there is no evidence of emphysema in the lungs of animals exposed to only O_3 based on the currently accepted definition for emphysema.

In the only autopsy study which has focused on the chronic effects of air pollution on human lungs, Sherwin and Richters[180] studied 107 lungs of accident victims (ages 14 to 25 years) who lived in Los Angeles County. They found that 27% of the lungs studied had severe structural abnormalities and bronchiolitis, far greater than would normally be expected for such a young group. Similar abnormalities, though less severe, were reported in 48% of the lungs investigated. While it is possible that some fraction of the lung tissue damage was caused by other factors such as smoking or use of drugs, a cause-and-effect relationship will be difficult to establish until future studies can be conducted in other cities with much cleaner air. These studies are currently underway.

Many of the lung structural changes that have been reported for long-term exposures to less than 1.0 ppm O_3 in several different species would be considered adverse and relevant if they were demonstrated in humans at ambient exposure concentrations of O_3. Although the lowest O_3 concentrations and shortest averaging times which could produce structural changes in human lungs is uncertain at this time, there is a need to protect the public from O_3 exposures which may induce such adverse effects. Further analysis is needed to clarify how animal data should be used to estimate lung structural changes in humans caused by O_3 exposure. However, if new autopsy studies continue to provide evidence of structural abnormalities in the lungs of long-term residents of highly polluted cities such as Los Angeles and possibly even in cleaner cities, a much stronger case can be made for using the animal data as confirmatory evidence.

Effects on Host Defense Mechanisms

Respiratory systems of mammals are protected from bacterial and viral infections by the closely interrelated particle removal (both mucociliary and phagocytic) and immunological defense systems. Numerous factors such as poor nutrition, preexisting disease, and environmental stress may influence or sufficiently alter these host defense systems of individuals to permit development of respiratory infections. Animal studies have indicated that exposure to O_3 is one of those factors.

Both *in vivo* (live animal) and *in vitro* (isolated cell) studies have demonstrated that O_3 can affect the ability of the clearance and immune systems to defend against infection. Increased susceptibility to bacterial infection has been reported in mice at 0.08 to 0.10 ppm O_3 for a single 3-h exposure[181-183] and at 0.10 ppm O_3 for subchronic exposures.[184] Several related alterations of the pulmonary defenses caused by short-term and subchronic exposures to O_3 include (1) impaired ability to inactivate bacteria in rabbits and mice;[185-188] (2) delayed mucociliary clearance;[189-192] (3) immunosuppression;[184,193,194] (4) significantly reduced number of pulmonary defense cells in rabbits;[185,195] and (5) impaired macrophage phagocytic activity, less macrophage mobility, more fragility and membrane alterations, and reduced lysosomal enzymatic activity.[187,196-204] Some of these effects have been shown to occur in a variety of species, including mice, rats, rabbits, guinea pigs, dogs, sheep, and monkeys.

Other studies indicate similar effects for short-term and subchronic exposures of mice to O_3 combined with pollutants such as SO_2, NO_2, H_2SO_4, and particles.[184,189,205-209] Similar to human pulmonary function response to O_3, activity levels of mice exposed to O_3 has been shown to play a role in determining the lowest effective concentration which alters the immune defenses.[210]

Although this large body of evidence clearly demonstrates that short-term and subchronic exposures to O_3 can impair the immune defense systems of animals, technical and ethical considerations have limited similar research on human subjects. Thus, inferences have been drawn and models developed to assess the relevance of animal data to humans. For example, animal endpoints other than increased mortality should be appropriately compared to increased morbidity in humans, such as the increased prevalence of respiratory illness in the community.

Based on current understanding of physiology, metabolism, and immune defenses, it is generally accepted that the basic mechanisms of action associated with defense against infectious agents are similar in humans and animals. Similarities between human and rodent antibacterial systems have been discussed in detail,[211] but differences in lung structure and biological response will inevitably cause differences in dosimetry, sensitivity, and endpoints of response. In addition, other factors such as preexisting disease, nutrition, presence of other pollutants, and environmental stresses (e.g., NO_2, SO_2, PM, high temperature or humidity) can influence effects of exposure to O_3 and infectious agents. Despite the differences in related factors which make precise estimation of human response from animal data difficult, it is reasonable to suggest that humans exposed to O_3 could experience impairment of host defenses.

Morbidity and Mortality

Relatively few studies of the effects of O_3 on morbidity and mortality have been conducted. One such study by Ostro and Rothschild[212] examined the association between air pollution concentrations and respiratory morbidity. Using the National Center for Health Statistics database collected with the Health Interview Survey, Ostro and Rothschild found associations between fine particulate matter and both minor and major restricted activity days, but only found that minor restricted activity days were associated with O_3. Controlling for temperature, Kinney and Ozkaynak[213] analyzed mortality records of Los Angeles County over a 10-year period and revealed associations between pollution levels (excluding sulfate and sulfuric acid levels) and short-term variations in both total mortality and cardiovascular mortality. The strongest associations with mortality were with O_3 and temperature. A more recent analysis by Kinney and Ozkaynak[214] indicated a statistically significant association between O_3 levels and daily mortality in New York City between 1971 and 1976. The magnitude of the effect for a given level of O_3 appeared to be quite similar to the Los Angeles results. While the limited data associating O_3 with mortality and morbidity do not permit drawing clear conclusions, this information does offer preliminary evidence of significant adverse effects which may be occurring as a result of both acute and chronic exposures to ambient O_3.

Extrapulmonary Effects

Extrapulmonary effects which have been demonstrated in humans or laboratory animals following exposure to O_3 include alterations in red blood cell morphology and enzyme activity, cytogenetic effects in circulating lymphocytes, and subjective limitations in vigilance tasks. Additionally, animal toxicology studies provide limited evidence for cardiovascular, reproductive, teratological, endocrine system, and liver metabolism effects. This wide variety of extrapulmonary effects is probably caused by oxidative reaction products of O_3. Due to the high reactivity of O_3 with biological tissue, mathematical models predict that only a small fraction of O_3 actually reaches the circulatory system.[215]

Of the extrapulmonary effects reported, cytogenetic and mutational effects are probably the most controversial. Statistically significant increases in frequency of sister chromatid exchanges (chromosomal alterations) have been caused by *in vitro* exposures to 0.25 ppm for 1 h,[216] suggesting a mutagenic response. However, *in vitro* responses are not generally extrapolatable to *in vivo* responses, and homeostatic mechanisms are not represented. Therefore, isolated *in vitro* exposure studies have not been used to provide estimations of risk for exposure to O_3. *In vivo* animal studies have shown significant increases in the number of chromosomal and chromatid aberrations following 4- and 5-h exposures to 0.2 and 0.43 ppm O_3, respectively.[217-220] However, other animal studies[220] and controlled human exposures to O_3 levels as high as 0.5 ppm have shown no significant cytogenetic effects attributable to O_3,[216,221-223] and epidemiology studies provide no evidence of chromosomal changes induced by ambient O_3.[224,225] Thus, while the animal studies are suggestive of possible cytogenetic effects from O_3, human studies have not demonstrated such effects for realistic ambient exposures to O_3.

Limited hematological and serum chemistry effects data indicate that O_3 or one of its reactive products can cross the blood gas barrier and interfere with biochemical mechanisms in human blood erythrocytes and sera. Much of the data is *in vitro* and primarily useful in studying mechanisms of action. Ozone-induced effects such as pentobarbital-induced sleep time alterations and hormone level changes have been reported[37] and cause-effect hypotheses made. With regard to other extrapulmonary effects, exact mechanisms remain to be elucidated and physiological significance for humans remains uncertain. Nonetheless, the body of this evidence suggests that O_3 can cause effects distant from the lungs in animals, and hence possibly in humans.

Summary

The body of evidence contained in this chapter makes it clear that O_3 is a highly toxic chemical which can adversely affect the respiratory systems of any individual depending on the conditions of exposure. Acute effects such as decrements in lung function and respiratory symptoms (e.g., cough, chest pain, or throat irritation) have been shown to become more severe as exposure concentrations of O_3 increase, as exercise levels increase, and as exposure duration increases; however, severity of response is highly dependent on the sensitivity of each individual's respiratory system and often on the state of health of the person exposed. The association between O_3 exposure and acute effects such as increased susceptibility to asthma attack and pulmonary infection or increased hospital admissions has become better understood as a result of recent studies conducted in the U.S. and Canada.

Chronic health effects of O_3 present a potentially far more serious problem for those populations who reside in cities with elevated oxidant levels. Although use of the extensive animal toxicology literature on chronic effects — such as inflammation, structural changes in the lung, and accelerated decline in lung function — is limited by uncertainties of dosimetry and species sensitivity differences, there is a consistency of these effects across species. Recent controlled-exposure human studies support the contention that ambient levels of O_3 do induce inflammation in human lungs, generally accepted as a precursor to irreversible lung damage. The autopsy studies of accident victims living in Los Angeles provide additional support for the involvement of oxidants in causing chronic lung damage. As animal extrapolation models improve and new studies provide answers to the unresolved questions about health effects, exposure, and risk, the need for controlling O_3 in the ambient air should become clearer in the near future.

REFERENCES

1. Yokoyama, E., and R. Frank. "Respiratory Uptake of Ozone in Dogs," *Arch. Environ. Health* 25:132–138 (1972).
2. Moorman, W. J., J. J. Chmiel, J. F. Stara, and T. R. Lewis. "Comparative Decomposition of Ozone in the Nasopharynx of Beagles: Acute vs. Chronic Exposure," *Arch. Environ. Health* 26:153–155 (1973).
3. Miller, F. J., C. A. McNeal, J. M. Kirtz, D. E. Gardner, D. L. Coffin, and D. B. Menzel. "Nasopharyngeal Removal of Ozone in Rabbits and Guinea Pigs," *Toxicology* 14:273–281 (1979).
4. Dungworth, D. L., W. L. Castleman, C. K. Chow, P. W. Mellick, M. G. Mustafa, B. Tarkington, and W. S. Tyler. "Effect of Ambient Levels of Ozone on Monkeys," *Fed. Proc. Fed. Am. Soc. Exp. Biol.* 34:1670–1674 (1975).
5. Miller, F. J., D. B. Menzel, and D. L. Coffin. "Similarity Between Man and Laboratory Animals in Regional Deposition of Ozone," *Environ. Res.* 17:84–101 (1978).
6. Miller, F. J., J. H. Overton, Jr., R. H. Jaskot, and D. B. Menzel. "A Model of the Regional Uptake of Gaseous Pollutants in the Lung. I. The Sensitivity of the Uptake of Ozone in the Human Lung to Lower Respiratory Tract Secretions and to Exercise," *Toxicol. Appl. Pharmacol.* 79:11–27 (1985).
7. McJilton, C., J. Thielke, and R. Frank. "Ozone Uptake Model for the Respiratory System," in *Abstr. Technical Papers: American Industrial Hygiene Conf.* San Francisco, CA, *Am. Ind. Hyg. Assoc. J.* 33:Paper No. 45 (1972).
8. Drechsler-Parks, D. M., J. F. Bedi, and S. M. Horvath. "Pulmonary Function Responses of Older Men and Women to Ozone Exposure," *Exp. Gerontol.* 42:91–101 (1987).
9. Bedi, J. F., S. M. Horvath, and D. M. Drechsler-Parks. "Adaptation of Older Individuals Exposed to 0.45 ppm Ozone for Two Hours," *J. Air Pollut. Assoc.* 39:194–199 (1989).
10. Horvath, S. M., J. A. Gliner, and L. J. Folinsbee. "Adaptation to Ozone: Duration of Effect," *Am. Rev. Respir. Dis.* 123:496–499 (1981).
11. Drechsler-Parks, D. M., J. F. Bedi, and S. M. Horvath. "Pulmonary Function Response of Young and Older Adults to Mixtures of O_3, NO_2, and PAN," *Tox. Ind. Health.* 5:505–517 (1989).
12. Drechsler-Parks, D. M., S. M. Horvath, and J. F.Bedi. "Environmental Pollutants and Age," *Progr. Biometerol.* 7:11–19 (1989).
13. Drechsler-Parks, D. M., J. F. Bedi, and S. M. Horvath. "Evaluation of the Effective Dose Concept in Older Subjects Exposed to Ozone," *Exp. Gerontol.* 25:107–115 (1990).
14. Reisenauer, C. S., J. Q. Koenig, M. S. McManus, M. S. Smith, G. Kusic, and W. E. Pierson. "Pulmonary Response to Ozone Exposures in Healthy Individuals Aged 55 Years or Greater," *JAPCA* 38:51–53 (1988).

15. McDonnell, W. F. III, R. S. Champan, M. W. Leigh, G. L. Strope, and A. M. Collier. "Respiratory Responses of Vigorously Exercising Children to 0.12 ppm Ozone Exposure," *Am. Rev. Respir. Dis.* 132:875–879 (1985).
16. McDonnell, W. F., R. S. Champan, D. H. Horstman, M. W. Leigh, and S. Abdul-Salaam. "A Comparison of the Responses of Children and Adults to Acute Ozone Exposure," in *Proc. Int. Specialty Conf. Evaluation of the Scientific Basis for an Ozone/Oxidant Standard*, APCA Transaction Vol. 4, November 1984, Houston, TX (Pittsburgh, PA: Air Pollution Control Association, 1985).
17. Avol, E. L., W. S. Linn, D. A. Shamoo, L. M. Valencia, U. T. Anzar, and J. D. Hackney. "Respiratory Effects of Photochemical Oxidant Air Pollution in Exercising Adolescents," *Am. Rev. Respir. Dis.* 132:619–622 (1985).
18. Avol, E. L., W. S. Linn, D. A. Shamoo, L. M. Valencia, U. T. Anzar, and J. D. Hackney. "Short-Term Health Effects of Ambient Air Pollution in Adolescents," in *Proc. Int. Specialty Conf. Evaluation of the Scientific Basis for an Ozone/Oxidant Standard*; APCA International Specialty Conference Transactions, Vol. 4 (Pittsburgh. PA: Air Pollution Control Association), pp. 329–336.
19. Lebowitz, M. D., M. K. O'Rourke, R. Dodge, C. J. Holberg, G. Corman, R. W. Hoshaw, J. L. Pinnas, R. A. Barbee, and M. R. Sneller. "The Adverse Health Effects of Biological Aerosols, Other Aerosols, and Indoor Microclimate on Asthmatics and Nonasthmatics," *Environ. Int.* 8:375–380 (1982).
20. Lebowitz, M. D., C. J. Holberg, and R. R. Dodge. "Respiratory Effects on Populations from Low Level Exposures to Ozone," Presented at 34th Annual Meeting of the Air Pollution Control Association, Atlanta, GA (Pittsburgh, PA: Air Pollution Control Association, 1983) Paper No. 83-12.5.
21. Lebowitz, M. D. "The Effects of Environmental Tobacco Smoke Exposure and Gas Stoves on Daily Peak Flow Rates in Asthmatic and Non-Asthmatic Families," *Eur. J. Respir. Dis.* 65 (Suppl. 133):90–97 (1984).
22. Lippmann, M., P. J. Lioy, G. Leikauf, K. B. Green, D. Baxter, M. Morandi, and B. S. Pasternack. "Effects of Ozone on the Pulmonary Function of Children," in *Int. Symp. Biomedical Effects of Ozone and Related Photochemical Oxidants. Advances in Modern Technology, Vol. 5*, S. D. Lee, M. G. Mustafa, and M. A. Mehlman, Eds. (Princeton, NJ: Princeton Scientific Publishers, 1983), pp. 423–446.
23. Lioy, P. J., T. A. Vollmuth, and M. Lippmann. "Persistence of Peak Flow Decrement in Children Following Ozone Exposures Exceeding the National Ambient Air Quality Standard," *J. Air Pollut. Control Assoc.* 35:1068–1071 (1985).
24. Bock, N., M. Lippmann, P. Lioy, A. Munoz, and F. Speizer. "The Effects of Ozone on the Pulmonary Function of Children," in *Proc. Int. Specialty Conference on the Evaluation of the Scientific Basis for an Ozone/Oxidant Standard*, November 1984, Houston, TX, APCA International Specialty Conference Transactions, Vol. 4 (Pittsburgh, PA: Air Pollution Control Association), pp. 297–308.
25. Spektor, D. M., M. Lippmann, and P. J. Lioy. "Effects of Ambient Ozone on Respiratory Function in Active Normal Children," *Am. Rev. Respir. Dis.* 137:313–320 (1988).
26. Spektor, D. M., G. D. Thurston, J. Mao, D. He, C. Hayes, and M. Lippmann. "Effects of Single- and Multi-Day Ozone Exposures on Respiratory Function in Active Normal Children," *Environ. Res.* 55:107–122 (1991).
27. Higgins, I. T. T., J. B. D'Arcy, and D. I. Gibbons. "Effects of Exposure to Ambient Ozone on Ventilatory Function in Children," *Am. Rev. Respir. Dis.* 141:1136–1146 (1990).

28. Makino, K., and I. Mizoguchi. "Symptoms Caused by Photochemical Smog," *Nippon Koshu Eisei Zasshi* 22:421–430 (1975).
29. Castillejos, M., D. R. Gold, and D. Dockery. "Effects of Ambient Ozone on Respiratory Function and Symptoms in Mexico City Children," *Am. Rev. Respir. Dis.* 145:276–282 (1992).
30. Braun-Fahrlander, C. U., E. Ackermann-Liebrich, and J. Schwartz. "Air Pollution and Respiratory Symptoms in Preschool Children," *Am. Rev. Respir. Dis.* 145:42–47 (1992).
31. Horvath, S. M., J. A. Gliner, and J. A. Matsen-Twisdale. "Pulmonary Function and Maximum Exercise Responses Following Acute Ozone Exposure," *Aviat. Space Environ. Med.* 50:901–905 (1979).
32. Gliner, J. A., S. M. Horvath, and L. J. Folinsbee. "Pre-exposure to Low Ozone Concentrations Does Not Diminish the Pulmonary Function Response on Exposure to Higher Ozone Concentration," *Am. Rev. Respir. Dis.* 127:51–55 (1983).
33. DeLucia, A. J., J. A. Whitaker, and L. R. Bryant. "Effects of Combined Exposure to Ozone and Carbon Monoxide in Humans," in *Int. Symp. Biomedical Effects of Ozone and Related Photochemical Oxidants. Advances in Modern Environmental Technology, Vol. 5*, M. A. Mehlman, S. D. Lee, and M. G. Mustafa, Eds. Pinehurst, NC, March 1982 (Princeton, NJ: Princeton Scientific Publishers, 1983), pp. 145–149.
34. Lauritzen, S. K., and W. C. Adams. "Ozone Inhalation Effects Consequent to Continuous Exercise in Females: Comparison to Males," *J. Appl. Physiol.* 59:1601–1606 (1985).
36. Horvath, S. M., J. F. Bedi, D. M. Drechsler-Parks, and R. E. Williams. "Alterations in Pulmonary Function Parameters During Exposure to 80 ppb Ozone for 6.6 Hours in Healthy Middle Aged Individuals," in *Tropospheric Ozone and the Environment: Papers from an International Conference*; R. L. Bergland, D. R. Lawson, and D. J. McKee, Eds., AWMA Transaction Series No. TR-19, March 1990, Los Angeles, CA (Air and Waste Management Association, 1991), pp. 59–70.
37. Graham, J. A., D. B. Menzel, F. Miller, J. W. Illing, and D. E. Garner. "Influence of Ozone on Pentobarbital-Induced Sleeping Time in Mice, Rats, and Hamsters," *Toxicol. Appl. Pharmacol.* 61:64–73 (1981).
38. Hazucha, M., F. Silverman, C. Parent, S. Field, and D. V. Bates. "Pulmonary Function in Man After Short-term Exposure to Ozone," *Arch. Environ. Health* 27:183–188 (1973).
39. Bates, D. V., and M. Hazucha. "The Short-Term Effects of Ozone on the Human Lung," in *Proc. Conf. Health Effects of Air Pollutants*; Serial No. 93-15, October (Washington, DC: U.S. Senate, Committee on Public Works), pp. 507–540.
40. Kagawa, J., and K. Tsuru. "Effects of Ozone and Smoking Alone and in Combination on Bronchial Reactivity to Inhaled Acetylcholine," *Nippon Kyobu Shikkan Gakkai Zasshi* 17:703–709 (1979).
40[a]. Kerr, H. D., T. J. Kulle, M. L. McIlhany, and P. Swidersky . "Effects of Ozone on Pulmonary Function in Normal Subjects," *Am. Respir. Dis.* 111:763–773 (1975).
41. Kagawa, J. "Effects of Ozone and Other Pollutants on Pulmonary Function in Man," in *Int. Symp. Biomedical Effects of Ozone and Related Photochemical Oxidants. Advances in Modern Environmental technology, Vol. 5*, S. D. Lee, M. G. Mustafa, and M. A. Mehlman, Eds. March 1982, Pinehurst, NC (Princeton, NJ: Princeton Scientific Publishers, 1983), pp. 441–422.
42. Shephard, R. J., B. Urch, F. Silverman, and P. N. Corey. "Interaction of Ozone and Cigarette Smoke Exposure," *Environ. Res.* 31:125–137 (1983).

43. Posin, C. I., K. W. Clark, M. P. Jones, R. D. Buckley, and J. D. Hackney. "Human Biochemical Response to Ozone and Vitamin E," *J. Toxicol. Environ. Health* 5:1049–1058 (1979).
44. Hamburger, S. J., and B. D. Goldstein. "Effect of Ozone on the Agglutination of Erythrocytes by Concanavalin A. I: Studies in Rats," *Environ. Res.* 19:292–298 (1979).
45. Hamburger, S. J., B. D. Goldstein, R. D. Buckley, J. D. Hackney, and M. A. Amoruso. "Effect of Ozone on the Agglutination of Erythrocytes by Concanavalin A. II. Study of Human Subjects Receiving Supplemental Vitamin E," *Environ. Res.* 19:299–305 (1979).
46. Chow, C. K., C. G. Plopper, M. Chiu, and D. L. Dungworth. "Dietary Vitamin E and Pulmonary Biochemical and Morphological Alterations of Rats Exposed to 0.1 ppm ozone," *Environ. Res.* 24:315–324 (1981).
47. Plopper, C. G., C. K. Chow, D. L. Dungworth, and W. S. Tyler. "Pulmonary Alterations in Rats Exposed to 0.2 and 0.1 ppm Ozone: A correlated Morphological and Biochemical Study," *Arch. Environ. Health* 34:390–395 (1979).
48. Chow, C. K., and A. L. Tappel. "An Enzymatic Protective Mechanism Against Lipid Peroxidation Damage to Lungs of Ozone-Exposed Rats," *Lipids* 7:518–524 (1972).
49. Stephens, R. J., M. F. Sloan, M. J. Evans, and G. Freeman. "Early Response of Lung to Low Levels of Ozone," *Am. J. Pathol.* 74:31–58 (1974).
50. Schwartz, L. W., D. L. Dungworth, M. G. Mustafa, B. K. Tarkington, and W. S. Tyler. "Pulmonary Responses of Rats to Ambient Levels of Ozone: Effects of 7-Day Intermittant or Continuous Exposure," *Lab. Invest.* 34:565–578 (1976).
51. Donovan, D. H., S. J. Williams, J. M. Charles, and D.B. Menzel. "Ozone Toxicity: Effect of Dietary Vitamin E and Polyunsaturated Fatty Acids," *Toxicol. Lett.* 1:135–139 (1977).
52. Sato, S., M. Kawakami, S. Maeda, and T. Takishima. "Scanning Electron Microscopy of the Lungs of Vitamin E-Deficient Rats Exposed to a Low Concentration of Ozone," *Am. Rev. Respir. Dis.* 113:809–821 (1976).
53. Sato, S., S. Shimura, M. Kawakami, T. Hirose, S. Maeda, T. Takishima, S. Kimura, M. Yashiro, S. Okazaki, and M. Ito. "Biochemical and Ultrastructural Studies on the Effects of Long-Term Exposure of Ozone on Vitamin E-Depleted Rats," *Nippon Kyobu Shikkan Gakkai Zasshi* 16:260–268 (1978).
54. Sato, S., S. Shimura, T. Hirose, S. Maeda, M. Kawakami, T. Takashima, and S. Kimura. "Effects of Long-Term Ozone Exposure and Dietary Vitamin E in Rats," *Tohoku J. Exp. Med.* 130:117–128 (1980).
55. Folinsbee, L. J., F. Silverman, and R. J. Shephard. Decrease of Maximum Work Performance Following Ozone Exposure," *J. Appl. Physiol.: Respir. Environ. Exercise Physiol.* 42:531–536 (1977).
56. Folinsbee, L. J., S. M. Horvath, P. B. Raven, J. F. Bedi, A. R. Morton, B. L. Drinkwater, N. W. Bolduan, and J. A. Gliner. "Influence of Exercise and Heat Stress on Pulmonary Function During Ozone Exposure," *J. Appl. Physiol.: Respir. Environ. Exercise Physiol.* 43:409–413 (1977).
57. Gibbons, S. I., and W. C. Adams. "Combined Effects of Ozone Exposure and Ambient Heat on Exercising Females," *J. Appl. Physiol.* 57:450-456 (1984).
58. Folinsbee, L. J., B. L. Drinkwater, J. F. Bedi, and S. M. Horvath. "The Influence of Exercise on the Pulmonary Changes Due to Exposure to Low Concentrations of Ozone," in *Environmental Stress: Individual Human Adaptations*, L. J. Folinsbee, J. A. Wagner, J. F. Borgia, B. L. Drinkwater, J. A. Gliner, and J. F. Bedi, Eds. (New York: Academic Press, 1978), pp. 125–145.

59. Folinsbee, L. J., J. F. Bedi, and S. M. Horvath. "Pulmonary Function Changes Following 1-Hour Continuous Heavy Exercise While Breathing 0.21 ppm Ozone," *J. Appl. Physiol.: Respir. Environ. Exercise Physiol.* 57:984–988 (1984).
60. McDonnell, W. F., D. H. Horstman, M. J. Hazucha, E. Seal, Jr., E. D. Haak, S. Salaam, and D. E. House. "Pulmonary Effects of Ozone Exposure During Exercise: Dose-Response Characteristics," *J. Appl. Physiol.: Respir. Environ. Exercise Physiol.* 54:1345–1352 (1983).
61. McDonnell, W. F., III, D. H. Horstman, S. Abdul-Salaam, and D. E. House. "Reproducibility of Individual Responses to Ozone Exposure," *Am. Rev. Respir. Dis.* 131:36–40 (1985).
62. Avol, E. L., W. S. Linn, T. G. Venet, D. A. Shamoo, and J. D. Hackney. "Comparative Respiratory Effects of Ozone and Ambient Oxidant Pollution Exposure During Heavy Exercise," *J. Air Pollut. Control Assoc.* 34:804–809 (1984).
63. Avol, E. L., W. S. Linn, D. A. Shamoo, T. G. Venet, and J. D. Hackney. "Acute Respiratory Effects of Los Angeles Smog in Continuously Exercising Adults," *J. Air Pollut. Control Assoc.* 33:1055–1060 (1983).
64. Folinsbee, L. J., B. L. Drinkwater, J. F. Bedi, and S. M. Horvath. "The Influence of Exercise on the Pulmonary Changes Due to Exposure to Low Concentrations of Ozone," in *Environmental Stress: Individual Human Adaptations*, Folinsbee, L. J., J. A. Wagner, J. F. Borgia, B. L. Drinkwater, J. A. Gliner, and J. F., Bedi, Eds. (New York: Academic Press, 1978), pp. 125–145.
65. Folinsbee, L. J., J. F. Bedi, and S. M. Horvath. "Pulmonary Function Changes Following 1-Hour Continuous Heavy Exercise While Breathing 0.21 ppm Ozone," *J. Appl. Physiol.*: *Respir. Environ. Exercise Physiol.* 57:984–988 (1984).
66. Kulle, T. J., L. R. Sauder, J. R. Hebel, and M. D. Chatham. "Ozone Response Relationships in Healthy Nonsmokers," *Am. Rev. Respir. Dis.* 132:36–41 (1985).
67. Haak, E. D., M. J. Hazucha, R. W. Stacy, D. E. House, B. T. Ketcham, E. Seal, Jr., L. J. Roger, and J. R. Knelson. "Pulmonary Effects in Healthy Young Men of Four Sequential Exposures to Ozone," U.S. EPA, Research Triangle Park, NC, Report No. EPA-600/1-84-033. Available from NTIS, Springfield, VA.
68. Silverman, F., L. J. Folinsbee, J. Barnard, and R. J. Shephard. "Pulmonary Function Changes in Ozone-Interaction of Concentration and Ventilation," *J. Appl. Physiol.* 41:839–864 (1976).
69. Hazucha, M. "Ozone-Induced Hyperreactivity as Measured by Histamine in Normal Healthy Subjects," in *Proc. Research Planning Workshop on Health Effects of Oxidants;* January 1980, EPA Report No. EPA-600/9-81-001 (Raleigh, NC. Research Triangle Park, NC: U.S. Environmental Protection Agency, 1981), pp. 314–327. Available from NTIS, Springfield, VA; PB81-178832.
70. Hackney, J. D., W. S. Linn, D. C. Law, S. K. Karuza, H. Greenberg, R. D. Buckley, and E. E. Pedersen. "Experimental Studies on Human Health Effects of Air Pollutants. III. Two-hour Exposure to Ozone Alone and in Combination with Other Pollutant Gases," *Arch. Environ. Health* 30:385–390 (1975).
71. U.S. Environmental Protection Agency. Air Quality Criteria for Ozone and other Photochemical Oxidants. Environmental Criteria and Assessment Office, Research Triangle Park, NC. EPA Report No. EPA-600/8-84/020a to 020e (1986). Available from NTIS, Springfield, VA, PB 87-142949.
72. U.S. Environmental Protection Agency. Summary of Selected New Information on Effects of Ozone on Health and Vegetation: Draft Supplement to Air Quality Criteria for Ozone and Other Photochemical Oxidants. Environmental Criteria and Assessment Office, Research Triangle Park, NC. EPA Report No. EPA/600/8-88/105F (1992). Available from NTIS, Springfield, VA.

73. Lippmann, M. "Health Effects of Tropospheric Ozone: Implications of Recent Research Findings to Ambient Air Quality Standards," *J. Exp. Anal. Env. Epid.* 3:103–129 (1993).
74. Folinsbee, L. J., F. Silverman, and R. J. Shephard. "Exercise Responses Following Ozone Exposure," *J. Appl. Physiol.* 38:996–1001 (1975).
75. Bates, D. V., G. M. Bell, C. D. Burnham, M. Hazucha, J. Mantha, L. D. Pengelly, and F. Silverman. "Short-Term Effects of Ozone on the Lung," *J. Appl. Physiol.* 32:176–181 (1972).
76. Golden, J. A., J. A. Nadel, and H. A. Boushey. "Bronchial Hyperirritability in Healthy Subjects After Exposure to Ozone," *Am. Rev. Respir. Dis.* 118:287–294 (1978).
77. Koenig, J. Q., D. S. Covert, M. S. Morgan, N. Horike, S. G. Marshall, and W. E. Pierson. "Acute Effects of 0.12 ppm Ozone or 0.12 ppm Nitrogen Dioxide on Pulmonary Function in Healthy and Asthmatic Adolescents," *Am. Rev. Respir. Dis.* 132:648–651 (1985).
78. Konig, G., H. Rommelt, H. Kienele, K. Dirnagl, H. Polke, and G. Fruhmann. "Changes in the Bronchial Reactivity of Humans Caused by the Influence of Ozone," *Arbeitsmed. Sozialmed. Praeventivmed.* 151:261–263 (1980).
79. Niinimaa, V., P. Cole, S. Mintz, and R. J. Shepherd. "The Switching Point from Nasal to Oronasal Breathing," *Respir. Physiol.* 42:61–71 (1980).
80. Niinimaa, V., P. Cole, S. Mintz, and R. J. Shepherd. "Oronasal Distribution of Respiratory Airflow," *Respir. Physiol.* 43:69–75 (1981).
81. Adams, W. C., and E. S. Schelegle. "Ozone and High Ventilation Effects on Pulmonary Function and Endurance Performance," *J. Appl. Physiol.: Respir. Environ. Exercise Physiol.* 55:805–812 (1983).
82. Adams, W. C., W. M. Savin, and A. E. Christo. "Detection of Ozone Toxicity during Continuous Exercise via the Effective Dose Concept," *J. Appl. Physiol: Respir. Environ. Exercise Physiol.* 51:415–422 (1981).
83. Linn, W. S., M. P. Jones, E. A. Bachmayer, C. E. Spier, S. F. Mazur, E. L. Avol, and J. D. Hackney. "Short-term Respiratory Effects of Polluted Air: A Laboratory Study of Volunteers in a High-Oxidant Community," *Am. Rev. Respir. Dis.* 121:243–252 (1980).
84. Linn, W. S., D. A. Shamoo, T. G. Venet, C. E. Spier, L. M. Valencia, U. T. Anzar, and J. D. Hackney. "Response to Ozone in Volunteers with Chronic Obstructive Pulmonary Disease," *Arch. Environ. Health* 38:278–283 (1983).
85. Hayes, S. R., A. S. Rosenbaum, T. S. Wallsten, R. G. Whitfield, and R. L. Winkler. "Assessment of Lung Function and Symptom Health Risks Associated with Attainment of Alternative Ozone NAAQS," (San Rafael, CA: Systems Applications, 1987).
86. Hayes, S. R., M. Moezzi, T. S. Wallsten, and R. L. Winkler. "An Analysis of Symptom and Lung Function Data from Several Human Controlled Ozone Exposure Studies," (San Rafael, CA: Systems Applications, 1987).
87. Folinsbee, L. J., W. F. McDonnell, and D. H. Horstman. "Pulmonary Function and Symptom Responses After 6.6 Hour Exposure to 0.12 ppm Ozone with Moderate Exercise," *J. Air Pollut. Control Assoc.* 38:28–35 (1988).
88. Horstman, D. H., L. J. Folinsbee, P. J. Ives, S. Abdul-Salaam, and W. F. McDonnell. "Ozone Concentration and Pulmonary Response Relationships for 6.6 Hour Exposures with Five Hours of Moderate Exercise to 0.08, 0.10, and 0.12 ppm," *Am. Rev. Respir. Dis.* 142:1158–1163 (1990).
89. McDonnell, W. F., H. R. Kehrl, S. Abdul-Salaam, P. J. Ives, L. J. Folinsbee, R. B. Devlin, J. J. O'Neil, and D. H. Horstman. "Respiratory Response of Humans Exposed to Low Levels of Ozone for 6.6 Hours," *Arch. Environ. Health* 46:145–150 (1991).

90. Larsen, R. I., W. F. McDonnell, D. H. Horstman, and L. J. Folinsbee. "An Air Quality Data Analysis System for Interrelating Effects, Standards, and Needed Source Reductions. II. A Lognormal Model Relating Human Lung Function Decrease to Ozone Exposure," *J. Air Waste Manage. Assoc.* 41:455–459 (1991).
91. Folinsbee, L. J., M. J. Hazucha, E. Seal, H. R. Kehrl, and D. H. Horstman. "Cumulative Ozone Exposure Dose Explains Responses to Different Exposure Regimens," *Am. Rev. Respir. Dis.* 145(4:2):195 (1992).
92. Folinsbee, L. J., and M. J. Hazucha. "Persistence of Ozone-Induced Changes in Lung Function and Airway Responsiveness," in *Atmospheric Ozone Research and Its Policy Implications,* T. Schneider, S. D. Lee, G. J. R. Wolters, and L. D. Grant, Eds. (Nijmegen, The Netherlands: Elsevier, 1989).
93. Hazucha, M. J., E. Seal, and L. J. Folinsbee. "Effects of Steady-State and Variable Ozone Concentration Profiles on Pulmonary Function of Man," *Am. Rev. Respir. Dis.* 141(4:2):A71 (1990).
94. Lioy, P. J. and R. V. Duba. "The Dynamics of Human Exposure to Tropospheric Ozone," in *Atomspheric Ozone Research and its Policy Implications: Proceedings of the Third U.S.-Dutch International Symposium,* May 1988, Nigmegen, The Netherlands, Studies in Environmental Science 35 (Amsterdam: Elsevier Science, 1989), pp. 711–721.
95. Kinney, P. L., J. H. Ware, J. D. Spengler, Dockery, F. E. Speizer, and B. G. Ferris, Jr. "Short-term Pulmonary Change in Association with Ozone Levels," *Am. Rev. Respir. Dis.* 139:56–61 (1989).
96. Berry, M., P. J. Lioy, K. Gelperin, G. Buckler, and J. Klotz. "Ozone Accumulated Exposure to Ozone and Measurement of Health Effects in Children and Counselors at Two Summer Camps," *Environ. Res.* 54:135–150 (1991).
97. Thurston, G. D., N. D'Souza, M. Lippmann, M. Bartoszek, and J. Fine, "Associations Between Summer Haze Air Pollution and Asthma Exacerbations," *Am. Rev. Respir. Dis.* 145(4:2):A429 (1992).
98. Spektor, D. M., M. Lippmann, G. D. Thurston, P. J. Lioy, J. Stecko, G. O'Connor, E. Garshick, F. E. Speizer, and C. Hayes. "Effects of Ambient Ozone on Respiratory Function in Healthy Adults Exercising Outdoors," *Am. Rev. Respir. Dis.* 138:821–828 (1988).
99. Korrick, S. A., D. R. Gold, D. W. Dockery, B. A. Rosner, F. E. Speizer, G. A. Allen, K. D. Kimball, and C. G. Wager. "Pulmonary Effects of Ozone Exposures in Hikers on Mount Washington," *Am. Rev. Resp. Dis.* 145(4:2):A94 (1992).
100. Kinney, P. L., M. Skorkinski, D. K. He, J . Gorczynski, C. Hayes, and M. Lippmann. "Subchronic Change in FEB_1 Associated with Ambient Ozone Exposures," *Am. Rev. Respir. Dis.* 143(4:2):A96 (1992).
101. Kinney, P. L., R. Hayes, R. Butler, and M. Lippmann. "Pulmonary Function of Military Academy Freshmen from Areas of High and Low Ozone Levels: Preliminary Results," presented at the *1991 Int. Speciality Conf. Tropospheric Ozone, Air Waste Management Assoc.,* Atlanta, GA (1992).
102. DeLucia, A. J., and W. C. Adams. "Effects of Ozone Inhalation During Exercise on Pulmonary Function and Blood Biochemistry," *J. Appl. Physiol.: Respir. Environ. Exercise Physiol.* 43:75–81 (1977).
103. Kagawa, J., and K. Tsuru. "Respiratory Effects of 2-Hour Exposure to Ozone and Nitrogen Dioxide Alone and in Combination in Normal Subjects Performing Intermittent Exercise," *Nippon Kyobu Shikkan Gakkai Zasshi* 17:765–774 (1979).

104. Kagawa, J., and K. Tsuru. "Respiratory Effect of 2-Hour Exposure with Intermittent Exercise to Ozone and Sulfur Dioxide Alone and in Combination in Normal Subjects," *Nippon Eiseigaku Zasshi* 34:690–696 (1979).
105. Hammer, D. I., V. Hasselblad, B. Portnoy, and P. F. Wehrle. "Los Angeles Student Nurse Study: Daily Symptom Reporting and Photochemical Oxidants," *Arch. Environ. Health* 28:255–260 (1974).
106. Altshuller, A. P. "Eye Irritation As an Effect of Photochemical Air Pollution," *J. Air Pollut. Control Assoc.* 27:1125–1126 (1977).
107. National Research Council. "Toxicology," in *Ozone and Other Photochemical Oxidants*. (Washington, DC: National Academy of Sciences, Committee on Medical and Biologic Effects of Environmental Pollutants, 1977), pp. 323-387.
108. U. S. Environmental Protection Agency. Air Quality Criteria for Ozone and Photochemical Oxidants. Environmental Criteria and Assessment Office, Research Triangle Park, NC. EPA Report No. EPA-600/8-78-004 (1978). Available from NTIS, Springfield, VA, PB80-124753.
109. Okawada, N., I. Mizoguchi, and T. Ishiguro. "Effects of Photochemical Air Pollution on the Human Eye — Concerning Eye Irritation, Tear Lysome and Tear pH," *Nagoya J. Med. Sci.* 41:9–20 (1979).
110. Wayne, W.S., P. F. Wehrle, and R. E. Carroll. "Oxidant Air Pollution and Athletic Performance," *J. Am. Med. Assoc.* 199:151–154 (1967).
111. Savin, W. M., and W. C. Adams. "Effects of Ozone Inhalation on Work Performance and $V_{o2max,}$" *J. Appl. Physiol.: Respir. Environ. Exercise Physiol.* 46:309–314 (1979).
112. Holtzman, M. I., J. H. Cunningham, J. R. Sheller, G. B. Irsigler, J. A. Nadel, and H. A. Boushey. "Effect of Ozone on Bronchial Reactivity in Atopic and Nonatopic Subjects," *Am. Rev. Respir. Dis.* 120:1059–1067 (1979).
113. Dimeo, M. J., M. G. Glenn, M. J. Holtzman, J. R. Sheller, J. A. Nadel, and H. A. Boushey. "Threshold Concentration of Ozone Causing an Increase in Bronchial Reactivity in Humans and Adaptation with Repeated Exposure," *Am. Rev. Respir. Dis.* 124:245–248 (1981).
114. Kulle, T. J., L. R. Sauder, H. D. Kerr, B. P. Farrell, M. S. Bermel, and D. M. Smith. "Duration of Pulmonary Function Adaptation to Ozone in Humans," *Am. Ind. Hyg. Assoc. J.* 43:832–837 (1982).
115. Aris, R., D. Christian, D. Sheppard, and J. R. Balmes. "The Effects of Sequential Exposure to Acidic Fog and Ozone on Pulmonary Function in Exercising Subjects," *Am. Rev. Respir. Dis.* 143:85–91 (1991).
116. Molfino, N. A., S. C. Wright, I. Katz, S. Tarlo, F. Silverman, P. A. McClean, J. P. Szalai, M. Raizenne, A. S. Slutsky, and N. Zamel. "Effect of Low Concentrations of Ozone on Inhaled Allergen Responses in Asthmatic Subjects," *Lancet* 338:199–203 (1991).
117. Silverman, F. "Asthma and Respiratory Irritants (Ozone)," *EHP Environ. Health Perspect.* 29:131–136 (1979).
118. Linn, W. S., R. D. Buckley, C. E. Spier, R. L. Blessey, M. P. Jones, D. A. Fischer, and J. D. Hackney. "Health Effects of Ozone Exposure in Asthmatics," *Am. Rev. Respir. Dis.* 117:835–843 (1978).
119. Linn, W. S., D. A. Medway, U. T. Anzar, L. M. Valencia, C. E. Spier, F. S.-O. Tsao, D. A. Fischer, and J. D. Hackney. "Persistence of Adaptation to Ozone in Volunteers Exposed Repeatedly Over Six Weeks."*Am. Rev. Respir. Dis.* 125:491–495 (1982).

120. Solic, J. J., M. J. Hazucha, and P. A. Bromberg. "The Acute Effects of 0.2 ppm Ozone in Patients with Chronic Obstructive Pulmonary Disease," *Am. Rev. Respir. Dis.* 125:664–669 (1982).
121. Kehrl, H. R., M. J. Hazucha, J. J. Solic, and P. A. Bromberg. "The Acute Effects of 0.2 and 0.3 ppm Ozone in Persons with Chronic Obstructive Lung Disease (COLD)," in *Int. Symp. Biomedical Effects of Ozone and Related Photochemical Oxidants. Advances in Modern Environmental Toxicology, Vol. 5,* S. D. Lee, M. G. Mustafa, and M. A. Mehlman, Eds. (Princeton, NJ: Princeton Scientific Publishers, 1983), pp. 213–225.
122. Kehrl, H. R., M. J. Hazucha, J. J. Solic, and P. A. Bromberg. "Responses of Subjects with Chronic Pulmonary Disease after Exposures to 0.3 ppm Ozone," *Am. Rev. Respir. Dis.* 131:719–724 (1985).
123. Kulle, T. J., J. H. Milman, L. R. Sauder, H. D. Kerr, B. P. Farrell, and W. R. Miller. "Pulmonary Function Adaptation to Ozone in Subjects with Chronic Bronchitis," *Environ. Res.* 34:55–63 (1984).
124. Frampton, M. W., P. E. Morrow, C. Cox, P. C. Levy, D. M. Speer, F. R. Gibb, J. J. Condemi, and M. J. Utell. "Does Preexposure to Acidic Aerosols Alter Airway Responses to Ozone in Humans?" *Am. Rev. Respir. Dis.* 145(4:2):A428 (1992).
125. Whittemore, A. S., and E. L. Korn. "Asthma and Air Pollution in the Los Angeles Area," *Am. J. Public Health* 70:687–696 (1980).
126. Johnson, D. E., R. J. Prevost, K. T. Kimball, D. E. Jenkins, and E. Bourhofer. "Study of Health-Related Responses to Air Pollution of Persons with Chronic Obstructive Pulmonary Disease in Houston, Texas," SWRI Project No. 01-4902 (Houston, TX: Southwest Research Institute, 1979).
127. Javitz, H. S., R. Kransnow, C. Thompson, K. M. Patton, D. E. Berthiaume, and A. Palmer. "Ambient Oxidant Concentrations in Houston and Acute Health Symptoms in Subjects with Chronic Obstructive Pulmonary Disease: A Reanalysis of the HAOS Health Study," in *Int. Symp. Biomedical Effects of Ozone and Related Photochemical Oxidants. Advances in Modern Environmental Toxicology, Vol. 5,* S. D. Lee, M. G. Mustafa, and M. A. Mehlman, Eds. (Princeton, NJ: Princeton Scientific Publishers, 1983), pp. 227–256.
128. Stock, T. H., A. H. Holguin, B. J. Selwyn, B. P. Hsi, C. F. Contant, P. A. Buffler, and D. J. Kotchmar. "Exposure Estimates for the Houston Area Asthma and Runners Studies," in *Int. Symp. Biomedical Effects of Ozone and Related Photochemical Oxidants. Advances in Modern Environmental Toxicology, Vol. 5,* S. D. Lee, M. G. Mustafa, and M. A. Mehlman, Eds. (Princeton, NJ: Princeton Scientific, 1983).
129. Holguin, A. H., P. A. Buffler, C. F. Contant, Jr., T. H. Stock, D. J. Kotchmar, B. P. Hsi, D. E. Jenkins, B. M. Gehan, L. M. Noel, and M. Mei. "The Effects of Ozone on Asthmatics in the Houston Area," in *Proc. Int. Specialty Conf. Evaluation of the Scientific Basis for an Ozone/Oxidant Standard,* APCA International Specialty Conference Transaction, Vol. 4 (Pittsburgh, PA: Air Pollution Control Association, 1985), pp. 262–280.
130. Contant, C. F., Jr., B. M. Gehan, T. H. Stock, A. H. Holguin, and P. A. Buffler. "Estimation of Individual Ozone Exposures Using Microenvironment Measures," in *Proc. Int. Specialty Conf. Evaluation of the Scientific Basis for an Ozone/Oxidant Standard*; APCA Specialty Conference Transactions, Vol. 4 (Pittsburgh, PA: Air Pollution Control Association, 1985), pp. 250–261.
131. Bates, D. V., and R. Sizto. "Air Pollution and Hospital Admissions in Southern Ontario: The Acid Summer Haze Effect," *Environ. Res.* 43:317–331 (1987).
132. Bates, D. V. "Ozone — Myth and Reality," *Environ. Res.* 50:230–237 (1989).

133. Bates, D. V., and R. Sizto. "The Ontario Air Pollution Study: Identification of the Causative Agent," *Environ. Health Perspect.* 79:69–72 (1989).
134. Ozkaynak, H., P. L. Kinney, and B. Burbank. "Recent Epidemiological Findings on Morbidity and Mortality Effects of Ozone," Preprint 90–150.6. Presented at the 1990 Annual Meeting of the Air Waste Management Assoc., Pittsburgh, PA.
135. Thurston, G. D., P. Kinney, K. Ito, and M. Lippmann. "Daily Respiratory Hospital Admissions and Summer Haze Air Pollution in Several New York Metropolitan Areas," *Am. Rev. Respir. Dis.* 145(4:2):A429 (1992).
136. Thurston, G. D., K. Ito, P. L. Kinney, and M. Lippmann. "A Multi-year Study of Air Pollution and Respiratory Hospital Admissions in Three New York State Metropolitan Areas: Results for 1988 and 1989 Summers," *J. Exp. Anal. Environ. Med.* (accepted for publication).
137. Cody, R. P., C. P. Weisel, G. Birnbaum, and P. J. Lioy. "The Effects of Ozone Associated with Summertime Photochemical Smog on the Frequency of Asthma Visits to Hospital Emergency Departments," *Environ. Res.* 58:184–194 (1992).
138. Seltzer, J., B. G. Bigby, M. Stulbarg, M. J. Holtzman, J. A. Nadel, I. F. Ueki, G. D. Leikauf, E. J. Goetzl, and H. A. Boushey. "Ozone-Induced Change in Bronchial Reactivity to Methacholine and Airway Inflammation in Humans," *J. Appl. Physiol.* 60:1321–1326 (1986).
139. Koren, H. S., R. B. Devlin, D. E. Graham, R. Mann, D. H. Horstman, W. J. Kozumbo, S. Becker, and W. F. McDonnell. "Cellular and Biochemical Changes in the Lower Airways of Subjects Exposed to Ozone," (Research Triangle Park, NC: U.S. Environmental Protection Agency, Health Effects Research Laboratory, 1988), Report No. EPA/600/D-88/031. Available from NTIS, Springfield, VA, PB88-170048.
140. Koren, H. S., R. B. Devlin, D. E. Graham, R. Mann, D. H. Horstman, W. J. Kozumbo, S. Becker, and W. F. McDonnell. "Cellular and Biochemical Changes in the Lower Airways of Subjects Exposed to Ozone," in *The Alveolar Macrophage,* C. Sorg, Ed. Local Immunity, Vol. 4 (Werne, Federal Republic of Germany: Stiftung Immunologische Tage, 1988), pp. 36–49.
141. Koren, H. S., R. B. Devlin, D. E. Graham, R. Mann, M. P. McGee, D. H. Horstman, W. J. Kozumbo, S. Becker, D. E. House, W. F. McDonnell, and P. A. Bromberg. "Ozone-Induced Inflammation in the Lower Airways of Human Subjects," *Am. Rev. Respir. Dis.* 139:407–415 (1989).
142. Koren, H. S., R. B. Devlin, D. E. Graham, R. Mann, and W. F. McDonnell. "The Inflammatory Response in Human Lung Exposed to Ambient Levels of Ozone," in *Atmospheric Ozone Research and Its Policy Implications: Proc. 3rd U.S.-Dutch International Symposium*, T. Schneider, S. D. Lee, G. J. R. Wolters, and L. D. Grant, Eds. Studies in Environmental Science, Vol. 35 (Amsterdam: Elsevier Science, 1989), pp. 745–753.
143. Schelegle, E. S., A. D. Siefkin, and R. J. MacDonald. "Time Course of Ozone-Induced Neurtrophilia in Normal Humans," *Am. Rev. Respir. Dis.* 143:1353–1358 (1991).
144. Kehrl, H. R., L. W. Vincent, R. J. Kowalsky, D. H. Horstman, J. J. O'Neil, W. H. McCartney, and P. A. Bromberg. "Ozone Exposure Increases Respiratory Epithelial Permeability in Humans," *Am. Rev. Respir. Dis.* 135:1124–1128 (1987).
145. Devlin, R. B., W. F. McDonnell, R. Mann, S. Becker, D. E. House, D. Schreinemachers, and H. S. Koren. "Exposure of Humans to Ambient Levels of Ozone for 6.6 Hours Causes Cellular and Biochemical Changes in the Lung," *Am. J. Respir. Cell. Mol. Biol.* 4:72–81 (1991).

146. Castleman, W. L., W. S. Tyler, and D. L. Dungworth. "Lesions in Respiratory Bronchioles and Conducting Airways of Monkeys Exposed to Ambient Levels of Ozone," *Exp. Mol. Pathol.* 26:384–400 (1977).
147. Koren, H. S., W. F. McDonnell, S. Becker, and R. B. Devlin. "Inflammation of the Lung Induced by Extended Exposure of Humans to 0.08 and 0.10 ppm Ozone," *Tropospheric Ozone and the Environment: Papers from The First Int. Symp. Tropospheric Ozone*, R. L. Berglund, D. R. Lawson, and D. J. McKee, Eds. AWMA Transaction Series No. TR-19 (Pittsburgh, PA: Air and Waste Management Association, 1991), pp. 71–82.
148. Zitnik, L. A., L. W. Schwartz, N. K. McQuillen, Y. C. Zee, and J. W. Osebold. "Pulmonary Changes Induced by Low-Level Ozone: Morphological Observations," *J. Environ. Pathol. Toxicol.* 1:365–376 (1978).
149. Cavender, F. L., B. Singh, and B. Y. Cockrell. "Effects in Rats and Guinea Pigs of Six-Month Exposures to Sulfuric Acid Mist, Ozone, and Their Combination," *J. Toxicol. Environ. Health* 4:845–852 (1978).
150. Barry, B. E., R. R. Mercer, F. J. Miller, and J. D. Crapo. "Effects of Inhalation of 0.25 ppm Ozone on the Terminal Bronchioles of Juvenile and Adult Rats," *Exp. Lung Res.* 14:225–245 (1988).
151. Grose, E. C., M. A. Stevens, G. E. Hatch, R. H. Jaskot, M. J. K. Selgrade, A. G. Stead, D. L. Costa, and J. A. Graham. "The Impact of a 12 Month Exposure to a Diurnal Pattern of Ozone Function, Antioxidant Biochemistry and Immunology," in *Atmospheric Ozone Research and Its Policy Implications: Proc.3rd U.S.-Dutch International Symposium,* T. Schneider, S. D. Lee, G. J. R. Wolters, and L. D. Grant, Eds. Studies in Environmental Science, Vol. 35 (Amsterdam: Elsevier Science Publishers,1989), pp. 755–762.
152. Huang, Y., L. Y. Chang, F. J. Miller, J. A. Graham, J. J. Ospital, and J. D. Crapo. "Lung Injury Caused by Ambient Levels of Oxidant Air Pollutants: Extrapolation from Animal to Man," *Am. J. Aerosol Med.* 1:180–183 (1988).
153. Chang, L., F. J. Miller, J. Ultman, Y. Huang, B. L. Stockstill, E. Grose, J. A. Graham, J. J. Ospital, and J. P. Crapo. "Alveolar Epithelial Cell Injuries by Subchronic Exposure to Low Concentrations of Ozone Correlate with Cumulative Exposure," *Toxicol. Appl. Pharmacol.* 109:219–234 (1991).
154. Chang, L., Y. Huang, B. L. Stockstill, J. A. Graham, E. C. Grose, M. G. Menache, F. J. Miller, D. L. Costa, and J. D. Crapo. "Epithelial Injury and Interstitial Fibrosis in the Proximal Alveolar Regions of Rats Chronically Exposed to a Simulated Pattern of Urban Ambient Ozone," *Toxicol. Appl. Pharmacol.* 115:241–252 (1992).
155. Tyler, W. S., N. K. Tyler, J. A. Last, M. J. Gillespie, and T. J. Barstow. "Comparison of Daily and Seasonal Exposures of Young Monkeys to Ozone," *Toxicology* 50:131–144 (1988).
156. Hyde, D. M., C. G. Plopper, J. R. Harkema, J. A. St. George, W. S. Tyler, and D. L. Dungworth. "Ozone-Induced Changes in Monkey Respiratory System," in *Atmospheric Ozone Research and Its Policy Implications*, T. Schneider, S. D. Lee, G. J. R. Wolters, and L. D. Grant, Eds. (Amsterdam: Elsevier, 1989), pp. 525–532.
157. Pinkerton, K. E., A. R. Brody, F. J. Miller, and J. D. Crapo. "Exposure to Low Levels of Ozone Results in Enhanced Pulmonary Retention of Inhaled Asbestos Fibers," *Am. Rev. Respir. Dis.* 140:1075–1081 (1989).
158. Tepper, J. S., M. J. Wiester, M. F. Weber, S. Fitzgerald, and D. L. Costa. "Chronic Exposure to a Simulated Urban Profile of Ozone Alters Ventilatory Responses to Carbon Dioxide Challenge in Rats," *Fund. Appl. Toxicol.* 17:52–60 (1991).

159. Boorman, G. A., L. W. Schwartz, and D. L. Dungworth. "Pulmonary Effects of Prolonged Ozone Insult in Rats: Morphometric Evaluation of the Central Acinus," *Lab. Invest.* 43:108–115 (1980).
160. Moore, P. F., and L. W. Schwartz. "Morphological Effects of Prolonged Exposure to Ozone and Sulfuric Acid Aerosol on the Rat Lung," *Exp. Mol. Pathol.* 35:108–123 (1981).
161. Barry, B. E., F. J. Miller, and J. D. Crapo. "Alveolar Epithelial Injury Caused by Inhalation of 0.25 ppm of Ozone," in *Int. Symp. Biomedical Effects of Ozone and Related Photochemical Oxidants. Advances in Modern Environmental Toxicology, Vol. 5,* S. D. Lee, M. G. Mustafa, and M. A. Mehlman, Eds. (Princeton, NJ: Princeton Scientific Publishers), pp. 299–309.
162. Crapo, J. D., B. E. Barry, L.-Y. Chang, and R. R. Mercer. "Alterations in Lung Structure Caused by Inhalation of Oxidants," *J. Toxicol. Environ. Health* 13:301–321 (1984).
163. Eustis, S. L., L. W. Schwartz, P. C. Kosch, and D. L. Dungworth. "Chronic Bronchiolitis in Nonhuman Primates After Prolonged Ozone Exposure," *Am. J. Pathol.* 105:121–137 (1981).
164. Fujinaka, L. E. "Respiratory Bronchiolitis Following Long-term Ozone Exposure in Bonnet Monkeys: A Morphometric Study," Master's Thesis. University of California, Davis (1984).
165. Fujinaka, L. E., D. M. Hyde, C. G. Plopper, W. S. Tyler, D. L. Dungworth, and L. O. Lollini. Respiratory Bronchiolitis Following Long-term Ozone Exposure in Bonnet Monkeys: A Morphometric Study," *Exp. Lung Res.* 8:167–190 (1985).
166. Freeman, G., R. G. Stephens, D. L. Coffin, and J. F. Stara. "Changes in Dogs' Lungs after Long-term Exposure to Ozone: Light and Electron Microscopy," *Arch. Environ. Health* 26:209–216 (1973).
167. Evans, M. J., L. V. Johnson, R. J. Stephens, and G. Freeman. "Renewal of the Terminal Bronchiolar Epithelium in the Rat Following Exposure to NO_2 or O_3,"*Lab. Invest.* 35:246–257 (1976).
168. Evans, M. J., L. V. Johnson, R. J. Stephens, and G. Freeman. "Cell Renewal in the Lungs of Rats Exposed to Low Levels of Ozone,"*Exp. Mol. Pathol.* 24:70–83 (1976).
169. Evans, M. J., R. J. Stephens, and G. Freeman. "Renewal of Pulmonary Epithelium Following Oxidant Injury," in *Lung Cells in Disease: Proceedings of a Brook Lodge Conference*, A. Bouhuys, Ed. (New York: Elsevier/North-Holland Biomedical Press, 1976), pp. 165–178.
170. Lum, H., L. W. Schwartz, D. L. Dungworth, and W. S. Tyler. "A Comparative Study of Cell Renewal after Exposure to Ozone or Oxygen: Response of Terminal Bronchiolar Epithelium in the Rat," *Am. Rev. Respir. Dis.* 118:335–345 (1978).
171. Last, J. A., D. B. Greenberg, and W. L. Castleman. "Ozone-Induced Alterations in Collagen Metabolism of Rat Lungs," *Toxicol. Appl. Pharmacol.* 51:247–258 (1979).
172. Bhatnagar, R. S., M. Z. Hussain, K. R. Sorensen, M. G. Mustafa, F. M. von Dohlen, and S. D. Lee. "Effect of Ozone on Lung Collagen Biosynthesis," in *Int. Symp. Biomedical Effects of Ozone and Related Photochemical Oxidants. Advances in Modern Environmental Toxicology, Vol. 5,* S. D. Lee, M. G. Mustafa, and M. A. Mehlman, Eds. (Princeton, NJ: Princeton Scientific Publishers), pp. 311–321.
173. Last, J. A., K. M. Reiser, W. S. Tyler, and R. B. Rucker. "Long-term Consequences of Exposure to Ozone. I. Lung Collagen Content," *Toxicol. Appl. Pharmacol.* 72:111–118 (1984).

174. Costa, D. L., R. S. Kutzman, J. R. Lehmann, E. A. Popenoe, and R. T. Drew. "A Subchronic Multi-Dose Ozone Study in Rats," in *Int. Symp. Biomedical Effects of Ozone and Related Photochemical Oxidants. Advances in Modern Environmental Toxicology, Vol. 5,* S. D. Lee, M. G. Mustafa, and M. A. Mehlman, Eds. (Princeton, NJ: Princeton Scientific Publishers, 1983), pp. 369–393.
175. Wegner, C. D. "Characterization of Dynamic Respiratory Mechanics by Measuring Pulmonary and Respiratory System Impedances in Adult Bonnet Monkeys (*Macaca radiata*): Including the Effects of Long-term Exposure to Low-Level Ozone," Dissertation. University of California, Davis. Available from University Microfilms, Ann Arbor, MI, Publication No. 82-27900.
176. Niewoehner, D. E., J. Kleinerman, and D. B. Rice. "Pathologic Changes in the Peripheral Airways of Young Cigarette Smokers," *N. Engl. J. Med.* 291:755–758 (1974).
177. Cosio, M. G., K. A. Hale, and D. E. Niewoehner. "Morphologic and Morphometric Effects of Prolonged Cigarette Smoking on the Small Airways," *Am. Rev. Respir. Dis.* 122:265–271 (1980).
178. Hale, K. A., D. E. Niewoehner, and M. G. Cosio. "Morphologic Changes in Muscular Pulmonary Arteries: Relationship to Cigarette Smoking, Airway Disease, and Emphysema," *Am. Rev. Respir. Dis.* 122:273–280 (1980).
179. Wright, J. L., L. M. Lawson, P. D. Pare, B. J. Wiggs, S. Kennedy, and J. C. Hogg. "Morphology of Peripheral Airways in Current Smokers and Exsmokers," *Am. Rev. Respir. Dis.* 127:474–477 (1983).
180. Sherwin, R. P., and V. Richters. "Centriacinar Region Disease in the Lungs of Young Adults: A Preliminary Report," in *Tropospheric Ozone and the Environment (TR-19): Proc. Int. Symp. Tropospheric Ozone*, R. L. Berglund, D. R. Lawson, and D. J. Mckee, Eds. (Pittsburgh, PA: Air and Waste Management Association, 1991), pp. 178–196.
181. Coffin, D. L., E. J. Blommer, D. E. Gardner, and R. Holzman. "Effect of Air Pollution on Alteration of Susceptibility to Pulmonary Infection," in *Proc. Third Annual Conf. Atmospheric Contamination in Confined Spaces*, Report No. AMRL-TR-67-200 (Wright-Patterson Air Force Base, OH: Aerospace Medical Research Laboratories), pp. 71–80. Available from NTIS, Springfield, VA, AD-835008.
182. Ehrlich, R., J. C. Findlay, J. D. Fenters, and D. E. Gardner. "Health Effects of Short-term Inhalation of Nitrogen Dioxide and Ozone Mixtures," *Environ. Res.* 14:223–231 (1977).
183. Miller, F. J., J. W. Illing, and D. E. Gardner. "Effect of Urban Ozone Levels on Laboratory-Induced Respiratory Infections," *Toxicol. Lett.* 2:163–169 (1978).
184. Aranyi, C., S. C. Vana, P. T. Thomas, J. N. Bradof, J. D. Fenters, J. A. Graham, and F. J. Miller. "Effects of Subchronic Exposure to a Mixture of O_3, SO_2, and $(NH_4)_2$-SO_4 on Host Defenses of Mice," *J. Toxicol. Environ. Health* 12:55–71 (1983).
185. Coffin, D. L., D. E. Gardner, R. S. Holzman, and F. J. Wolock. "Influence of Ozone on Pulmonary Cells," *Arch. Environ. Health* 16:633–636 (1968).
186. Coffin, D. L., and D. E. Gardner. "Interaction of Biological Agents and Chemical Air Pollutants," *Ann. Occup. Hyg.* 15:219–235 (1972).
187. Goldstein, B. D., S. J. Hamburger, G. W. Falk, and M. A. Amoruso. "Effect of Ozone and Nitrogen Dioxide on the Agglutination of Rat Alveolar Macrophages by Concanavalin A," *Life Sci.* 21:1637–1644 (1977).
188. Ehrlich, R., J. C. Findlay, and D. E. Gardner. "Effects of Repeated Exposures to Peak Concentrations of Nitrogen Dioxide and Ozone on Resistance to Streptococcal Pneumonia," *J. Toxicol. Environ. Health* 5:631–642 (1979).

189. Phalen, R. F., J. L. Kenoyer, T. T. Crocker, and T. R. McClure, "Effects of Sulfate Aerosols in Combination with Ozone on Elimination of Tracer Particles Inhaled by Rats," *J. Toxicol. Environ. Health* 6:797–810 (1980).
190. Frager, N. B., R. F. Phalen, and J. L. Kenoyer. "Adaptations to Ozone in Reference to Mucociliary Clearance," *Arch. Environ. Health* 34:51–57 (1979).
191. Kenoyer, J. L., R. F. Phalen, and J. R. Davis. "Particle Clearance from the Respiratory Tract as a Test of Toxicity: Effect of Ozone on Short and Long Term Clearance," *Exp. Lung Respir.* 2:111–120 (1981).
192. Abraham, W. M., A. J. Januszkiewicz, M. Mingle, M. Welker, A. Wanner, and M. A. Sackner. "Sensitivity of Bronchoprovocation and Tracheal Mucous Velocity in Detecting Airway Responses to O_3," *J. Appl. Physiol. Respir. Environ. Exercise Physiol.* 48:789–793 (1980).
193. Campbell, K. I., and R. H. Hilsenroth. "Impaired Resistance to Toxin in Toxoid-immunized Mice Exposed to Ozone or Nitrogen Dioxide," *Clin. Toxicol.* 9:943–954 (1976).
194. Fujimaki, H., M. Ozawa, T. Imai, and F. Shimizu. "Effect of Short-term Exposure to O_3 on Antibody Response in Mice," *Environ. Res.* 35:490–496 (1984).
195. Alpert, S. M., D. E. Gardner, D. J. Hurst, T. R. Lewis, and D. L. Coffin. "Effects of Exposure to Ozone on Defensive Mechanisms of the Lung," *J. Appl. Physiol.* 31:247–252 (1971).
196. Dowell, A. R., L. A. Lohrbauer, D. Hurst, and S. D. Lee. "Rabbit Alveolar Macrophage Damage Caused by *in vivo* Ozone Inhalation," *Arch. Environ. Health* 21:121–127 (1970).
197. Hurst, D. J., D. E. Gardner, and D. L. Coffin. "Effect of Ozone on Acid Hydrolases of the Polmonary Alveolar Macrophage," *J. Reticuloendothel. Soc.* 8:288–300 (1970).
198. Hurst, D. J., and D. L. Coffin. "Ozone Effect on Lysosomal Hydrolases of Alveolar Macrophages *in vitro*," *Arch. Intern. Med.* 127:1059–1063 (1971).
199. Goldstein, E., W. S. Tyler, P. D. Hoeprich, and C. Eagle. "Ozone and the Antibacterial Defense Mechanisms of the Murine Lung," *Arch. Intern. Med.* 128:1099–1102 (1971).
200. Goldstein, E., W. S. Tyler, P. D. Hoeprich, and C. Eagle. "Adverse Influence of Ozone on Pulmonary Bactericidal Activity of Murine Lungs," *Nature (London)* 229:262–263 (1971).
201. Hadley, J. G., D. E. Gardner, D. L. Coffin, and D. B. Menzel. "Enhanced Binding of Autologous Cells to the Macrophage Plasma Membrane as a Sensitive Indicator of Pollutant Damage," in *Pulmonary Macrophage and Epithelial Cells: Proc. 16th Annual Hanford Biology Symposium,* C. L. Sanders, R. P. Schneider, G. E. Dagle, and H. A. Ragan, Eds. ERDA Symposium Series, Vol. 43, September 1976, (Washington, DC: Energy Research and Development Administration), pp. 1–21. Available from NTIS, Springfield, VA, CONF-760927.
202. McAllen, S. J., S. P. Chiu, R. F. Phalen, and R. E. Rasmussen. "Effect of *in vivo* Ozone Exposure on *in vitro* Pulmonary Alveolar Macrophage Mobility," *J. Toxicol. Environ. Health* 7:373–381 (1981).
203. Witz, G., M. A. Amoruso, and B. D. Goldstein. "Effect of Ozone on Alveolar Macrophage Function: Membrane Dynamic Properties," in *Int. Symp. Biomedical Effects of Ozone and Related Photochemical Oxidants. Advances in Modern Environmental Toxicology, Vol. 5,* S. D. Lee, M. G. Mustafa, and M. E. Mehlman, Eds. (Princeton, NJ: Princeton Scientific), pp. 263–272.
204. Amoruso, M. A., G. Witz, and B. D. Goldstein. "Decreased Superoxide Anion Radical Production by Rat Alveolar Macrophages Following Inhalation of Ozone or Nitrogen Dioxide," *Life Sci.* 28:2215–2221 (1981).

205. Gardner, D. E., F. J. Miller, J. W. Illing, and J. M. Kirtz. "Increased Infectivity with Exposure to Ozone and Sulfuric Acid," *Toxicol. Lett.* 1:59–64 (1977).
206. Ehrlich, R. "Interaction Between Environmental Pollutants and Respiratory Infections," *EHP Environ. Health Perspect.* 35:89–100 (1980).
207. Ehrlich, P. R., and H. A. Mooney. "Extinction, Substitution, and Ecosystem Services," *BioScience* 33:248–254 (1983).
208. Grose, E. C., D. E. Gardner, and F. J. Miller. "Response of Ciliated Epithelium to Ozone and Sulfuric Acid," *Environ. Res.* 22:377–385 (1980).
209. Grose, E. C., J. H. Richards, J. W. Illing, F. J. Miller, D. W. Davies, J. A. Graham, and D. E. Gardner. "Pulmonary Host Defense Responses to Inhalation of Sulfuric Acid and Ozone," *J. Toxicol. Environ. Health* 10:351–362 (1982).
210. Illing, J. W., F. J. Miller, and D. E. Gardner. "Decreased Resistance to Infection in Exercised Mice Exposed to NO_2 and O_3," *J. Toxicol. Environ. Health* 6:843–851 (1980).
211. Green, G. M. "Similarities of Host Defense Mechanisms Against Pulmonary Disease in Animals and Man," *J. Toxicol. Environ. Health* 13:471–478 (1984).
212. Ostro, B. D., and S. Rothschild. "Air Pollution and Acute Respiratory Morbidity: An Observational Study of Multiple Pollutants," *Environ. Res.* 50:238–247 (1989).
213. Kinney, P. L., and H. Ozkaynak. "Associations of Daily Mortality and Air Pollution in Los Angeles County," *Environ. Res.* 54:99–120 (1991).
214. Kinney, P. L., and H. Ozkaynak. "Associations Between Ozone and Daily Mortality in Los Angeles and New York City," *Am. Rev. Respir. Dis.* 145(4:2):A95 (1992).
215. Miller, F. J., J. H. Overton, Jr., R. H. Jaskot, and D. B. Menzel. "A Model of the Regional Uptake of Gaseous Pollutants in the Lung. I. The Sensitivity of the Uptake of Ozone in the Human Lung to Lower Respiratory Tract Secretions and to Exercise," *Toxicol. Appl. Pharmacol.* 79:11–27 (1985).
216. Guerrero, R. R., D. E. Rounds, R. S. Olson, and J. D. Hackney. "Mutagenic Effects of Ozone on Human Cells Exposed *in vivo* and *in vitro* Based on Sister Chromatid Exchange Analysis," *Environ. Res.* 18:336–346 (1979).
217. Zelac, R. E., H. L. Cromroy, W. E. Bolch, Jr., B. G. Dunavant, and H. A. Bevis. "Inhaled Ozone as a Mutagen. I. Chromosome Aberrations Induced in Chinese Hamster Lymphocytes," *Environ. Res.* 4:262–282 (1971).
218. Zelac, R. E., H. L. Cromroy, W. E. Bolch, Jr., D. G. Dunavant, and H. A. Bevis. "Inhaled Ozone As a Mutagen. II. Effect on the Frequency of Chromosome Aberrations Observed in Irradiated Chinese Hamsters," *Environ. Res.* 4:325–342.
219. Tice, R. R., M. A. Bender, J. L. Ivett, and R. T. Drew. "Cytogenetic Effects of Inhaled Ozone," *Mutat. Res.* 58:293–304 (1978).
220. Gooch, P. C., D. A. Creasia, and J. G. Brewen. "The Cytogenetic Effect of Ozone: Inhalation and *in vitro* Exposures," *Environ. Res.* 12:188–195 (1976).
221. Merz, T., M. A. Bender, H. D. Kerr, and T. J. Kulle. "Observations of Aberrations in Chromosomes of Lymphocytes from Human Subjects Exposed to Ozone at a Concentration of 0.5 ppm for 6 and 10 Hours," *Mutat. Res.* 31:299–302 (1975).
222. McKenzie, W. H., J. H. Knelson, N. J. Rummo, and D. E. House. "Cytogenetic Effects of Inhaled Ozone in Man," *Mutat. Res.* 48:95–102 (1977).
223. McKenzie, W. H. "Controlled Human Exposure Studies: Cytogenetic Effects of Ozone Inhalation," in *Indicators of Genotoxic Exposure,* Bridges, B. A., B. E. Butterworth, and I. B. Weinstein, Eds. Banbury Report No. 13 (Spring Harbor, NY: Cold Spring Harbor Laboratory, 1982), pp. 319–324.

224. Scott, C. D. and J. A. Burkart. "Chromosomal Aberrations in Peripheral Lymphocytes of Students Exposed to Pollutants," EPA-600/1-78-054, U.S. Environmental Protection Agency, Health Effects Research Laboratory, Research Triangle Park, NC (1978). Available from NTIS, Springfield, VA, PB-285594.
225. Magie, A. R., D. E. Abbey, and W. R. Centerwall. "Effect of Photochemical Smog on the Peripheral Lymphocytes of Nonsmoking College Students," *Environ. Res.* 29:204–219 (1982).
226. Highfill, J. W., G. E. Hatch, R. Slade, K. M. Crissman, J. Norwood, R. B. Devlin, and D. L. Costa. "Concentration-Time Models for the Effects of Ozone on Bronchoalveolar Lavage Fluid Protein from Rats and Guinea Pigs," *Inhal. Toxicol.* 4:1–16 (1992).

CHAPTER 4

HUMAN EXPOSURE TO AMBIENT OZONE

Thomas R. McCurdy

INTRODUCTION

This chapter summarizes information regarding O_3 exposure data obtained with microenvironmental and personal monitoring instruments. It also presents results from numerous O_3 exposure modeling efforts undertaken within the last decade.

As discussed in Chapter 3, nonchronic human health effects associated with O_3 vary with the level, duration, and *pattern* of O_3 concentrations encountered by sensitive people. They also vary with the breathing, or ventilation, rate of exposed individuals.[1] The O_3 "effective dose" received by an exposed person is a function of O_3 air quality (concentration, time, pattern), the volume of air inhaled, and the absorption of O_3 molecules along the respiratory tract.[2,3] An O_3 exposure assessment, to be useful, must therefore contain both air quality *and* respiratory state information for exposed individuals.

O_3 is unusual in that no special functional responsiveness to O_3 exposure has been demonstrated among population groups with preexisting lung disease, except — perhaps — asthmatics.[2] Some normally healthy people respond adversely to O_3 exposures at ambient levels, while others do not. It is difficult to identify these individuals *a priori*; responders can come from all age/occupation categories of people. In USEPA's O_3 standard-setting review, emphasis is placed on

0-87371-475-X/94/$0.00+$.50
© 1994 by Lewis Publishers

people of all health statuses who exercise regularly.[4-6] These people, therefore, should also be the focus of any O_3 exposure assessment effort.

Important factors in an O_3 exposure assessment should be:

- Magnitude of the exposure (concentration level)
- Duration of the exposure (time)
- Concentration pattern of the exposure (e.g., rapid build-up followed by a slow decline; slow build-up followed by a rapid decline; steady-state pattern; double-peaked; etc.)
- Frequency of repeated peak exposures
- Ventilation state of the individual at time of exposure

As we shall see, most O_3 exposure assessments do not include observations or estimates for many of these important factors.

DIRECT AND INDIRECT EXPOSURE ASSESSMENTS

A distinction is made in the exposure literature between *direct* and *indirect* estimation approaches.[7] In the direct approach, individual exposures are measured directly by an instrument known as a personal exposure monitor (or PEM) that is carried by an individual. In the indirect approach, exposures for an individual (or a group of people) are estimated from data associated with the various components that affect human exposure to an ambient pollutant.

Usually, the indirect approach combines information from human activity surveys on how people spend time with data on pollutant concentrations monitored in specified locations (called microenvironments). Thus, if it is estimated that an individual spends time t in microenvironment m having a time-averaged pollutant concentration c, the indirect exposure estimate for that person is $c_m \times t_m$. For a more detailed discussion of exposure modeling, see the section entitled "Modeling Human Exposure to Ozone".

It should be realized from the discussion in the previous section that as t increases, a $c_m \times t_m$ O_3 exposure estimate conveys little relevant information from a health effects viewpoint. What is more meaningful is to estimate what is called the "exposure profile" for the individual.[8] See Figure 1. A number of meaningful exposure measures, when linked to an individual's ventilation rate, can be derived from this exposure profile.

An exposure profile can be obtained directly from a PEM if it has the capability of sampling and storing exposure information for relatively short averaging times (usually 1 min). Such real-time active PEMs exist for O_3. One was developed under contract to the USEPA by Tranducer Research Inc.[9] Another was developed by the Rancho Los Amigos Medical Center using a reagent that changes color.[10] Both PEMs are still being evaluated; neither of them has been used in any large-scale personal monitoring survey to date.

An hourly averaged active PEM is also under development.[11] It is a sequential O_3 sampler that can provide *semicontinuous* exposure measurements by storing 24 1-h samples on a disk for subsequent analysis.

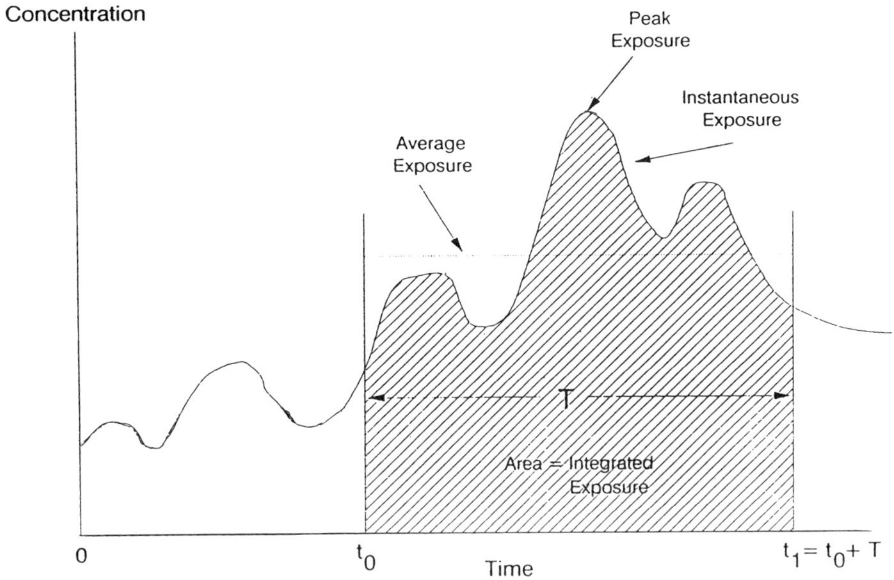

Figure 1. Hypothetical exposure profile: pollutant exposure as a function of time for a human subject. (From Duan, N., A. Dobbs, and W. Ott. *Total Exposure Assessment Methodology*, (Pittsburgh: Air & Waste Manage. Assoc., 1990), pp. 166–195.)

Two passive O_3 PEMs are under development. They both work on the principle of Brownian diffusion, where molecules of O_3 come in contact with nitrate ions.[11,12] One of the passive PEMs is based on a Palmes tube design and can be used to obtain a 24-h sample.[11] The other involves nitrate in a sodium/glycerine solution and can be used for 8-h or longer O_3 estimates.[12] It performed well (within $\pm$ 10%) when the passive PEM weekly averages were compared with those obtained using the standard reference method.[12] The 8+ hourly averaging time of these devices renders them infeasible for monitoring acute O_3 exposure profiles experienced by exercising individuals.

Non-PEM Direct Exposure Assessments

Most O_3 epidemiology studies use a fixed-site monitor to provide O_3 exposure data. Obviously, these monitors cannot provide any information on indoor O_3 exposures or ambient exposures away from the monitor location. However, there have been a number of non-PEM studies that provide a reasonable estimate of direct O_3 exposures for individuals of special interest. A mobile O_3 exposure chamber sampling ambient air was used by Rancho Los Amigos Hospital scientists in 1978–1981 to simulate exercising conditions in the Los Angeles basin.[13] The same scientists have studied O_3 exposures in a summer camp located 190 km southeast of Los Angeles.[14,15] Since the campers essentially lived in ambient air, a continuously operating O_3 monitor located in the camp provided a reasonable approximation of the camper's personal exposure to O_3 as they went through their daily activities. The current O_3 NAAQS (National Ambient Air Quality Standard) of 0.12 ppm (highest hour in a day) was exceeded on 32% of the monitored days in 1988.[14,15]

There also have been a number of O_3 camp studies undertaken by scientists at the Robert Wood Johnson Medical School in New Jersey and at the New York Medical Center.[16-20] Hourly averaged direct O_3 exposure estimates were obtained from a monitor located at the camp site. Again, this monitor probably closely approximated the camper's personal O_3 exposure. Hourly O_3 concentrations exceeded 0.12 ppm on 19% of the days monitored in the 1988 study.[18]

Probably the most comprehensive O_3 personal exposure study undertaken to date is the 1981 monitoring work done by the University of Texas School of Public Health at Houston.[21-24] A three-tiered monitoring system was operated for 6 months (May through October). Two areas of Houston were chosen; a centrally located ambient monitor was established in each subdivision. The two subdivisions are known as Sunnyside and Clear Lake. Clear Lake is a well-to-do area whose residents used central air conditioning during the entire monitoring period. Many residents of Sunnyside, on the other hand, either did not have air conditioning or did not use it during the monitoring period.

Indoor/outdoor O_3 concentrations were also monitored in a sample of homes. A mobile van was used to sequentially sample indoor air from three rooms in each monitored house: the main activity room, a bedroom, and the kitchen. Each room was sampled for 20 min per hour, throughout each hour. Another monitor in the van continuously sampled ambient O_3 outside the residence. Finally, 27 subjects were accompanied by a crew of technicians with portable — not personal — O_3 monitors. These monitors were moved around on wheeled equipment racks. In general, each participant was monitored for two 12-h daytime periods (7 a.m.–7 p.m.) as the subject went about his/her normal daily activities. A highly aggregated (1-h minimum time breakdown) daily activity diary was also kept for the participants.[21] Seven microenvironments were used:

- Indoors
    - Home
    - School *or* work
    - Other
- Outdoors
    - In the study area (within 5 mi of the fixed-site monitor)
    - Outside the study area
- Motor vehicle
    - Windows open
    - Windows closed

An example of the type of exposure data obtained from the Houston study appears as Figure 2. Personal O_3 exposures are higher than those monitored by either of the "inside home" or "outside home" monitors. In fact, personal O_3 levels exceed the current NAAQS even though outside-the-home and inside-the-home

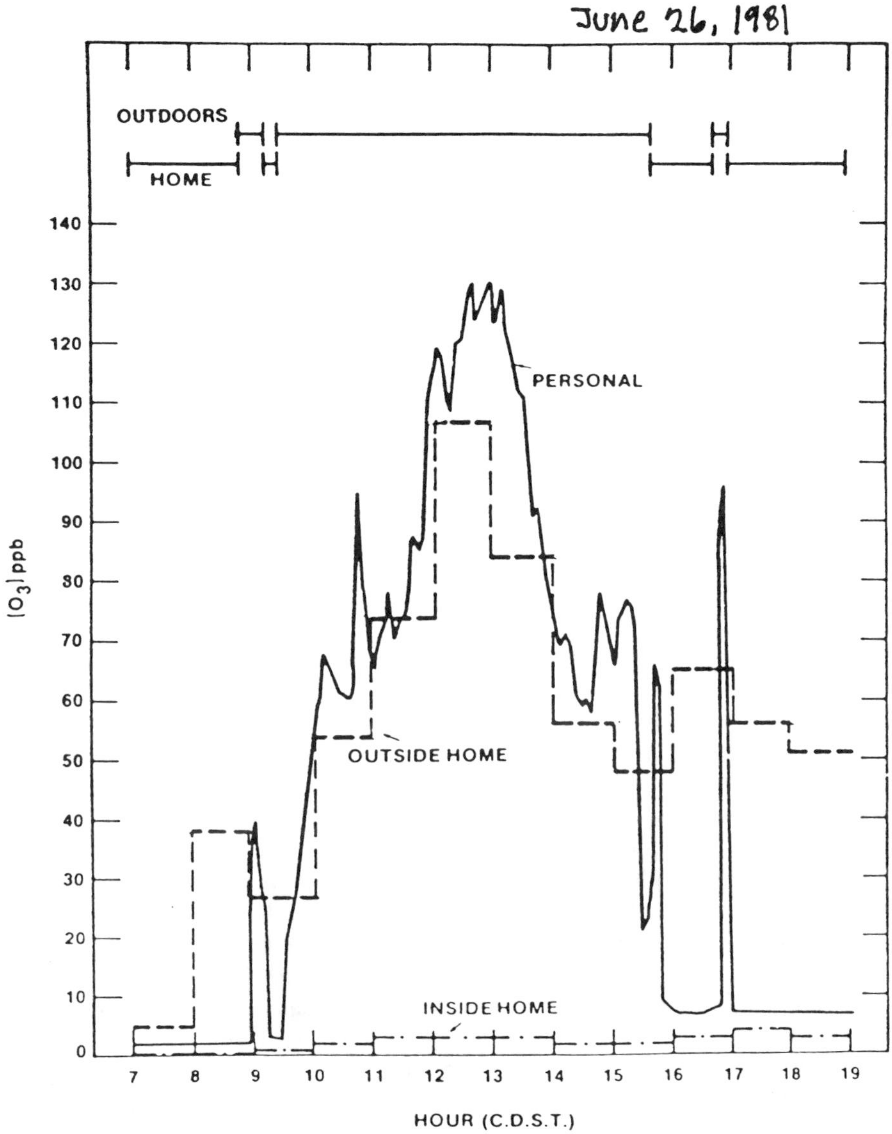

Figure 2. Example of the three-tier O_3 monitoring data available from the Houston study. (From Contant, C. F., Jr., et al. *Evaluation of the Scientific Basis for Ozone/Oxidant Standards,* S. D. Lee, Ed. (Pittsburgh: Air Pollution Control Association, 1985), pp. 250–257. With permission.)

levels do not. The "inside home" concentrations are low and quite flat, even though outdoor O_3 varies greatly. Although not shown, the fixed-site monitor recorded values very similar to those measured outside the home by the van.[21] The very low indoor O_3 concentrations are typical of those seen in closed-system, centrally air conditioned microenvironments.

Table 1. Personal O_3 Monitoring Data (in ppm) from the 1981 Houston Study

	Microenvironment		
Statistic	Indoors	Outdoors	Vehicle
Mean	0.011	0.052	0.023
Std. Dev.	—	—	—
Median	0.005	0.042	0.014
Minimum	0.001	0.001	0.001
Maximum	0.147	0.250	0.140
No. of obs.[a]	249	92	49

[a] No. of obs. = the number of valid variable-duration observations that were recorded among 27 participants.

From Contant, C. F., Jr., et al. *Evaluation of the Scientific Basis for Ozone/Oxidant Standards* (Pittsburgh: Air Pollution Control Association, 1985), pp. 250–257. With permission.

Summary personal exposure information from the Houston study for three aggregated microenvironments appear in Table 1. The summary concentration estimates are independent of time; the exposure periods can be as short as 1 min or as long as 1 h. Thus, comparisons among the microenvironments should not be made.

The Houston data have been reanalyzed by a USEPA consultant (Johnson[25]) with the original researchers' full cooperation. A summary of microenvironmental concentrations developed in the reanalysis appears as Table 2. Tables 1 and 2 are not directly comparable because of a change in how readings below the lower limit of detection (LLD) of 0.004 ppm were handled. In Table 1, any observation <0.002 ppm was set to 0.001 ppm; in Table 2, any value below the LLD was set to 0.002 ppm. There are also sample size differences between the two tables.

The impact on indoor O_3 concentrations of using an air conditioner was evaluated in the Johnson reanalysis.[25] See Table 3. The data are for the Sunnyside subdivision only. (There were very few "air conditioner off" values for Clear Lake.) In homes, the median microenvironmental (μe) O_3 concentration is 20 times higher when the air conditioner (A/C) is off than when it is on. Windows, of course, were closed when the A/C was on; however, windows could be either opened or closed when the A/C was off. (In other words, window status was not recorded, and this is a confounding effect in the data shown in Table 3.)

The problem with window position status for the "A/C off" situation probably causes the very small difference in median μe O_3 concentration for the work/school μe shown in Table 3. A two-sample t test indicates that the mean differences in me concentrations associated with A/C status are statistically significant at $p < 0.05$.[25]

Johnson computed indoor/outdoor ratios for two microenvironments in Sunnyside distinguished by A/C status. The data are shown in Table 4. The differences in mean values for the "indoors at home" μe are statistically significant.[25] He completed the same analysis for the "in-vehicle" μe, but only for Clear

Table 2. Summary Microenvironmental Statistics (in ppm) Obtained by Portable O_3 Monitors in Houston

Statistic	Indoors at home	Indoors at work or school	Indoors other	Outdoors	Inside a vehicle
Clear Lake subdivision					
Mean	0.005	0.003	0.004	0.045	0.017
Std. dev.[a]	0.003	0.002	0.003	0.032	0.019
Median	0.004	0.002	0.002	0.041	0.008
Minimum	0.002	0.002	0.002	0.002	0.002
Maximum	0.015	0.012	0.012	0.130	0.085
No. of Obs.[b]	531	599	158	338	107
Sunnyside subdivision					
Mean	0.021	0.009	0.007	0.048	0.011
Std. dev.[a]	0.032	0.014	0.005	0.029	0.012
Median	0.006	0.002	0.007	0.045	0.005
Minimum	0.002	0.002	0.002	0.002	0.002
Maximum	0.147	0.105	0.038	0.144	0.054
No. of Obs.[b]	504	482	73	172	48

Microenvironment columns: Indoors at home, Indoors at work or school, Indoors other, Outdoors, Inside a vehicle.

[a] Std. dev. = standard deviation.
[b] No. of obs. = number of observations in the microenvironment using the POM.

From Johnson, T. R . "Analysis of Second Data File from Houston Asthmatic Study," (Durham, NC: PEI Associates, July 1989). With permission.

Table 3. Differences in Microenvironmental O_3 Concentrations (in ppm) Due to the Use of Air Conditioning (A/C) in Sunnyside, Houston

	Indoors at home		Indoors at work/school	
Statistic	A/C on	A/C off	A/C on	A/C off
Mean	0.004	0.044	0.008	0.011
Std. dev.	0.003	0.040	0.016	0.011
Median	0.002	0.041	0.002	0.005
Minimum	—	—	—	—
Maximum	0.014	0.147	0.051	0.105
No. of obs.[a]	292	212	240	242

[a] No. of obs. = number of observations in the microenvironment using the POM.

From Johnson, T. R. "Analysis of Second Data File from Houston Asthmatic Study," (Durham, NC: PEI Associates, July 1989). With permission.

Lake residents. (Sunnyside residents used their vehicle A/Cs infrequently, if they even had one.) See Table 5. The data are imprecise concerning the window status of in-vehicle A/C use. When the A/C is off, the windows are open; when the A/C is on, however, no information exists as to whether or not the windows are closed.

Table 4. Differences in Indoor/Outdoor[a] Ratio of Microenvironmental O_3 Concentrations Due to the Use of Air Conditioning (A/C) in Sunnyside, Houston

	Indoors at home		Indoors at work/school	
Statistic	**A/C on**	**A/C off**	**A/C on**	**A/C off**
Mean	0.13	0.80	0.14	0.15
Std. dev.	0.14	0.38	0.16	0.15
Median	0.09	0.89	0.10	0.10
Minimum	—	—	—	—
Maximum	0.65	1.57	0.81	0.73
No. of Obs.[b]	234	173	151	185

[a] Indoor/Outdoor = (microenvironment O_3)/(central-site O_3).
[b] No. of obs. = number of observations in the microenvironment using the POM.

From Johnson, T. R. "Analysis of Second Data File from Houston Asthmatic Study," (Durham, NC: PEI Associates; July 1989). With permission.

Table 5. Differences in In-Vehicle Microenvironmental O_3 Concentrations and the Indoor/Outdoor Ratio for Clear Lake Residents

	In-vehicle μe concentration (ppm)		In-vehicle μe outdoor ratio[a]	
Statistic	**A/C on**	**A/C off**	**A/C on**	**A/C off**
Mean	0.008	0.021	0.26	0.68
Std. Dev.	0.012	0.020	0.28	0.45
Median	0.004	0.014	0.20	0.67
Minimum	—	—	—	—
Maximum	0.054	0.085	1.08	2.00
No. of Obs.[b]	33	74	13	47

[a] Ratios were calculated only when the central-site O_3 concentration was > 0.02 ppm.
[b] No. of obs. = number of observations using the POM data.

From Johnson, T. R. "Analysis of Second Data File from Houston Asthmatic Study," (Durham, NC: PEI Associates, July 1989). With permission.

Microenvironmental O_3 Monitoring Studies

As mentioned above, the indirect exposure assessment approach combines data on microenvironmental concentrations with information on human activities. This section reviews data obtained for specified microenvironments using nonportable monitors. The Texas study obtained some data of this type and it is included in summary format in certain tables that follow. The Texas data will not be discussed further, however, due to the extensive treatment already given to their data in our presentation of the USEPA-sponsored reanalysis.[25] In fact, microenvironmental data in general are presented with little commentary. The reader is referred to the original articles or reports for a detailed understanding of how the data were obtained.

A few general statements regarding indoor O_3 should suffice to set the stage for the data that follow. Unlike many air pollutants, there are few significant indoor sources of O_3. Any moving electric motor generates small amounts of O_3 due to ionization of oxygen via electrical arcing. Office equipment that uses high voltage or ultraviolet light can generate O_3, which means that photocopiers and laser printers potentially are indoor O_3 sources.[26] Indoor O_3 concentrations due to photocopiers made after ~1975 will be small, however; in a moderately ventilated room, O_3 concentrations will be 0.002 ppm or less from photocopiers.[27] O_3 may also be produced indoors from oxidant precursors being irritated with ultraviolet light from fluorescent lights. Again, subsequent O_3 concentrations are low from this photochemical process.

O_3 has a short half-life indoors. Estimates vary, but Mueller et al. calculates a 6–11 min half-life because of its high reactive rate with various building and furniture surfaces.[28] Estimates of the reactive rate also vary, with values being reported in the 0.02–0.08 cm s^{-1} range.[29] This rate, which is a normalized flux and not a physical velocity, is highly dependent upon (1) surface-to-volume geometrics, (2) the types of surface materials used, and (3) velocity of air flow inside the volume.

Generalizing to the larger indoor microenvironment, indoor O_3 generally follows the pattern of outdoor O_3. Even relatively quick, small outdoor changes are mirrored indoors.[26] How closely these patterns track largely depends upon whether or not A/C is used in the μe, the number of outdoor air exchanges that occur per unit time, and the outdoor O_3 concentration.

The number of outdoor air exchanges per hour (Ach) has a unit of h^{-1}. Information on the Ach associated with me O_3 monitoring studies appears in Tables 6 to 8, when available. Additional Ach data appear in the next section.

Peak hourly O_3 monitored indoors is presented in Table 6. The data are subdivided into three categories by A/C classification. A/C with carbon filters results in very low peak O_3 readings. Peak O_3 values are generally an order of magnitude higher for μes with A/C but no carbon filter, although some very low peaks are monitored in these μes. Most of the peak O_3 values measured in non-A/C spaces are in relatively low ambient O_3 areas, so the concentrations seen are fairly low.

Table 7 presents mean μe O_3 data for various averaging times observed during the summer O_3 season. Table 8 presents the same data for the "off season".

Outdoor Air Exchange Rate Data

Since the outdoor air exchange rate greatly affects the indoor/outdoor O_3 ratio, Ach information not obtained in conjunction with any O_3 me monitoring studies is presented below. In general, outdoor air exchange rates are very low in tight energy-conserving construction, and are quite low in typical air-conditioned buildings. The Ach rate is usually 40 to 50% higher for non-A/C buildings than

Table 6. Peak Hourly Monitored Indoor O_3 Concentrations

Microenvironment description	O_3 conc. (ppm)	Approx. outdoor air exchange rate (Ach)	Indoor/ outdoor	Ref.
Air conditioning with carbon filters				
S.W. Museum Library[a]	0.026	—	0.17	30
Huntington Air Gallery[a]	0.010	—	0.06	30
L.A. Museum of Art[a]	0.010	—	0.06	30
J.P. Getty Museum[a]	0.028	—	0.37	30
Air conditioning without carbon filters				
Scott Museum[a]	0.065	1.6	0.29	30
Montgomery Museum	0.067	1.6	0.39	30
Hospital-office	0.060	—	0.56	31
Hospital-care unit	0.055	—	0.69	31
Hospital-neonatal	0.081	—	0.70	31
House	0.105	—	—	24
Office building[b,c]	0.20	—	0.80	32
Office building[c,d]	0.18	—	0.65	32
Conference room[c,e]	0.28	11.7	0.64	33
Conference room[a,c,f]	0.16	1.2	0.45	33
Laboratory building[a,c]	0.15	2.6	0.54	34
Laboratory building[a,c]	0.13	2.6	0.46	34
Hospital-intensive[c]	0.24	—	0.64	35
Hospital-care units[c]	0.16	—	0.43	35
Laboratory[b,c,g]	0.07	12.0	0.39	36
Hospital[c,g]	0.05	—	0.26	36
Swimming pool[c,g]	0.17	—	0.63	36
School[c,g]	0.13	—	0.60	36
Department store[c,g]	0.11	—	0.52	36
No air conditioning				
Pasadena Museum[a]	0.098	0.5	0.59	30
Lang Museum	0.030	0.9	0.20	30
Villa Mont. Museum	0.022	1.8–2.6	0.49	30
Serra Museum	0.034	—	0.79	30
Southwest Museum	0.143	5.5	0.84	30
Sainbury Museum[c]	0.058	3–4	0.70	37
House in Los Angeles[c]	0.17	—	0.70	32
Hospital[c,g]	0.09	—	0.60	36
Store[c,g]	0.13	—	0.57	36

[a] The peak indoor concentration lags the peak outdoor concentration by 1 h.
[b] 100% outside make-up air.
[c] Approximate data.
[d] 70% outside make-up air.
[e] 50% outside make-up air.
[f] 10% outside make-up air.
[g] Total oxidant measurement.

for A/C structures. However, outdoor ventilation rates can increase as much as 10 times or so in residences just by opening windows. Moving vehicles have the highest Ach rates, as might be expected. Examples follow.

Tight-building Ach rates are in the range of 0.3 to 0.6 h^{-1} when fans are on, and 0.1 to 0.4 h^{-1} when fans are off. Air-conditioned spaces in typical residential and commercial construction have an Ach rate of 0.5 to 5 h^{-1} when fans are on, and 0.6 to 2 h^{-1} when fans are off. For non-A/C spaces, Ach values can range from 0.5

Table 7. Summary of Monitored Indoor O_3 Mean Concentrations Obtained During the O_3 Season

Microenvironment description	Avg. time	Mean O_3 conc. (ppm)	Std. dev. (ppm)	Outdoor air exchange rate (Ach)	Indoor/ outdoor	Ref.
Air conditioning with carbon filters	—	—	—	—	—	—
Air conditioning without carbon filters[a]						
Hospital-office	Hour	0.007	0.009	—	0.58	31
Hospital-care unit	Hour	0.021	0.021	—	0.60	31
Hospital-neonatal	Hour	0.013	0.013	—	0.72	31
All indoor micro.	Hour	0.011	—	—	0.21	24
Inside motor vehicle	Hour	0.023	—	—	0.44	24
Residence-Clear Lake	Varies	0.005	0.003	—	0.15	25
Residences-Sunnyside	Varies	0.021	0.032	—	0.41	25
Schools & offices-CL	Varies	0.003	0.002	—	0.09	25
Schools & offices-S	Varies	0.009	0.014	—	0.15	25
All other indoor-CL	Varies	0.004	0.003	—	0.14	25
All other indoor-S	Varies	0.007	0.005	—	0.17	25
Inside motor vehicles	Varies	0.008	0.012	—	0.26	25
Inside motor vehicles[b]	Varies	<0.050	—	>10	0.25	38
Residence-electric	Hour	0.017	—	—	0.20	39
Residence-gas	Hour	0.026	—	—	0.19	39
Office building	Hour	0.010	—	—	0.29	39
Residences[b]	72 hours	<.035	—	0.8–1.2	—	40
Residences (all rooms)	12 hours	0.001	0.006	—	—	23
No air conditioning						
Inside motor vehicles	Varies	0.021	0.020	—	0.68	25
Unknown air conditioning status						
Inside motor vehicles	Hour	0.015	0.017	—	0.29	41

[a] Much of these data were scaled off of graphs or were calculated from data supplied by the original researchers.
[b] These data are very approximate.

to 10 h^{-1} depending upon outside wind speed and relative direction, indoor/outdoor temperature differences, the proportion of open windows relative to the volume enclosed, and the overall building configuration. If window or door fans are used in these buildings, the Ach can go above 10 h^{-1}. The Ach for moving vehicles with windows open can go as high as 40 h^{-1}.

It should be noted that certain spaces within an air-conditioned building will have a much higher Ach value than the rest of the building. Spaces such as restrooms, conference rooms, and dining areas often have Ach rates of 6 to 25 h^{-1}.

Summary data on Ach rates found in the literature are presented in Table 9. A histogram of residential Ach values taken from a USEPA report appears as Figure 3.

Indoor/Outdoor O_3 Ratio

Indoor O_3 concentrations that are a significant fraction of monitored outdoor concentrations have been reported since the early 1970s. Usually, the fraction is reported as the ratio of indoor O_3 to outdoor O_3, or I/O. A review of O_3 I/O data

Table 8. Summary of Monitored Indoor O_3 Concentrations Obtained "Off Season"

Microenvironment description	Ave. time	Mean O_3 conc. (ppm)	Std. dev. (ppm)	Outdoor air exchange rate (Ach)	Ref.
Residences					
Kitchen A-Pre.	Week	0.004	0.003	Note 1	42
-Post	Week	0.004	0.002	Note 2	42
Kitchen B-Pre.	Week	0.005	0.005	Note 3	42
-Post	Week	0.010	0.002	Note 4	42
Fam. Room A-Pre.	Week	0.004	0.002	Note 1	42
-Post	Week	0.005	0.005	Note 2	42
Living room B-Pre.	Week	0.005	0.005	Note 3	42
-Post	Week	0.016	0.003	Note 4	42
Bedroom A-Pre.	Week	0.004	0.002	Note 1	42
-Post	Week	0.004	0.002	Note 2	42
Bedroom B-Pre.	Week	0.005	0.005	Note 3	42
-Post	Week	0.007	0.003	Note 4	42
Art gallery	2 days	0.014	—	0.3–2.0	43
School-classroom	Day	<0.005	—	Unknown	44

Note: 1. With the house fan on, the air change per hour (Ach) was 0.62 ± 0.25 (range: 0.36–0.71). With the house fan off, the Ach was 0.33 ± 0.14 (0.18–0.56).
2. The house noted above was retrofitted to save energy consumption. With the house fan on, the Ach was 0.49 ± 0.11 (0.22–0.69); with the fan off, it was 0.20 ± 0.08 (0.10–0.33).
3. This is another house. The Ach was 0.82 ± 0.07 for the fan on (range: 0.73–0.97). The Ach for the fan off was 0.33 ± 0.08 (0.24–0.40).
4. This is the post-retrofit of the house discussed in Note 3. The fan-on Ach was 0.58 ± 0.14 (0.39–0.78). The fan-off Ach was 0.23 ± 0.05 (0.16–0.39).

was undertaken in 1982 by Yocom.[52] A more recent summary review is given by Weschler et al.[26] Most of the data they report is contained in Tables 6 and 7. The majority of the reported O_3 I/O measurements are in the 0.5 to 0.8 range. Weschler et al.[46] also reported O_3 I/O ratios for 5 µes in an office complex that had been monitored intermittently between summer 1988 and winter 1990. See Table 10. As expected, I/O ratios are large for µes with a high Ach rate. Although not obvious from the table, I/O values are smaller than expected for a specific Ach rate when ambient O_3 is low.[46] Indoor O_3 scavenging is more efficient at low ambient O_3 levels.

Weschler and colleagues present a simple reformulated mass-balance model (the WSN model) to predict the I/O ratio in a µe for a steady-state condition.[26] The model is:

$$I/O = \left[E_x / \left(K_d * [A_d / V] + E_x\right)\right]$$

where E_x = air exchange rate (s^{-1})
K_d = O_3 deposition velocity ($m\ s^{-1}$)
A_d = O_3 deposition surface area (m^2)
V = volume of µe (m^3)

Table 9. Summary Data on Outdoor Air Exchange Rates Obtained from the General Literature

Microenvironment description	Average air exchange rate[a] (Ach)	Range of air exchange rate (Ach)	Ref.
Offices			
ASHRAE standard	0.84[b]	—	26
Bell Labs site #3	0.6	—	46
EPA survey	0.8	—	26
Classroom			
ASHRAE standard	4.5	—	26
EPA survey	0.9	—	26
Institutional residences			
ASHRAE standard	1.5	—	26
EPA survey	1.7	—	26
Residences			
Survey of 600 A/C homes in CA	1.5	—	26
Survey of 45 non-A/C homes in CA	—	2.2–2.4	26
Typical rate	—	0.5–1.5	48
266 low-income houses	0.75[c]	0.5–1.5[c]	48
Survey of 39 A/C houses in			
TN — HVAC on	0.90	0.36–1.44[d]	49
— HVAC off	0.56	0.19–0.93[d]	49
Experimental house/leaky			
HVAC — HVAC on	0.36	0.34–0.38[d]	49
— HVAC off	0.13	0.11–0.15[d]	49
4 houses around the country — summer	—	0.6–1.0	50
3 high-rise apartments — summer	—	0.9–1.4	50
Miscellaneous			
14 unspecified commercial bldgs.	0.94	0.29–1.73	26
Office/laboratory combination	—	4.0–8.2	26
Laboratory at site #4	10.0	—	26
Bell Labs "cleanroom"	30.0	—	26

[a] The average is given as a mean in some cases and as a median in others. If a single μe is being referenced, often only one value is given and it is unclear whether it is a single observation or a general tendency estimate.

[b] Calculated for a specified space representing a typical μe in the category listed. The ASHRAE standard is contained in Reference 45.

[c] The data shown are from a written description of a distribution:
<0.5 19%
0.5–1.0 40%
1.1–1.5 20%
>1.5 20%
The percentages do not total to 100% in the text.

[d] The calculated one standard deviation from the mean.

The term $[A_d/V]$ is obviously the surface-to-volume ratio for the μe. Wescheler et al.[46] state that a reasonable value is 2.9 m^{-1} (from the work done by Mueller et al.[28]). They also provide an estimate of average μe K_d (from Shair's measurements[33]) equal to 3.6×10^{-4} m s^{-1}. There is uncertainty concerning both terms. The variability in predicted O_3 I/O using the WSN model for an $[A_d/V]$ of 2.9 m^{-1} and different estimates of k_d is depicted in Figure 4.

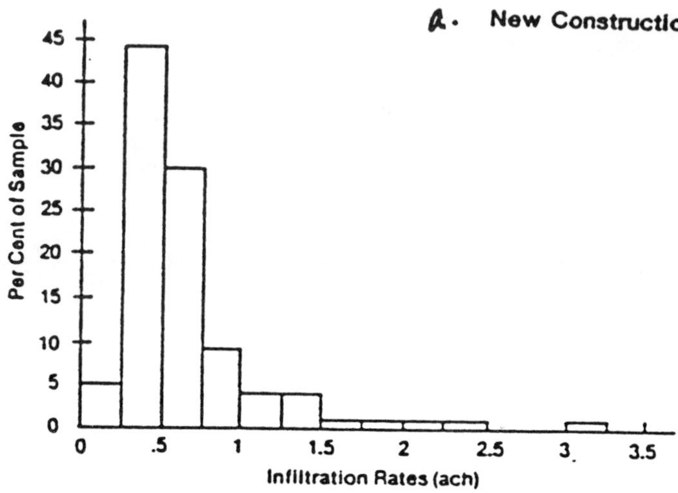

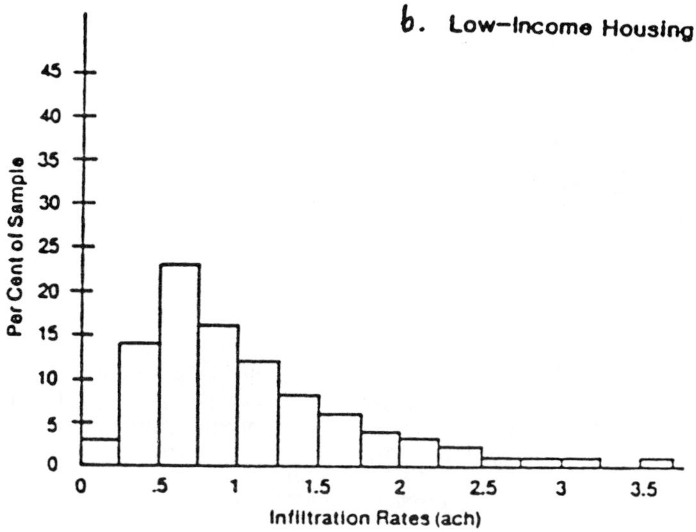

Figure 3. Natural infiltration rates in U.S. housing for new construction (A) and low-income housing (B). (From Versar, Inc. "Indoor Air Pollution, A Scoping Study," U.S. EPA Report, (1987), p. 5-2. With permission.)

A frequency distribution histogram of O_3 I/O ratios appears as Figure 5. The data are from 10-min sampling intervals in three school mes on days when school is in session.[53] Three ventilation rates were measured, as noted in the figure. The frequency distributions indicate a direct relationship between ventilation rate and O_3 I/O, as expected. The *shapes* of the I/O distributions are typical of most μes. Modeled I/O ratios are discussed below in the section on "Estimating O_3 Dose".

Table 10. O_3 Indoor/Outdoor Ratios As Measured at a Bell Labs Office Complex in New Jersey

μe Description	Approximate monitoring period	Outdoor air exchange rate (Ach)[a] (h^{-1})	Observed indoor/outdoor ratio[b]
Office/laboratory combination	Summer 1988 Summer 1989 Winter 1990	4.0	0.54
Office/laboratory combination	Summer 1988	8.2	0.71
Office	Summer 1988	0.6	0.22
Laboratory	Summer 1989	10.0	0.81
Laboratory cleanroom	Summer 1989 Fall 1989	30.0	0.89

[a] The Ach is assumed to be equal to the measured exhaust rates in the μe. The rates are relatively constant in all μes, except for the laboratory, where it varied from ~8–10 depending upon whether or not the doors were closed or open.
[b] The ratios presented are for time periods when outdoor O_3 was relatively constant, and not fluctuating rapidly.

From Weschler, C. J., H. C. Shields, and D. V. Naik. "Indoor-Outdoor Ozone Measurements: 1988–1991," Paper presented at the *84th Annu. Meet. Air and Waste Management Assoc.*, Vancouver, June 1991. With permission. See also Weschler, C. J., H. C. Shields, and D. V. Naik. "Indoor Ozone Exposures," *J. Air Pollut. Control Assoc.* 39:1562–1568 (1989).

HUMAN ACTIVITY DATABASES

As discussed, the indirect exposure approach requires an analyst to combine data on human activities with μe concentration estimates. This section reviews the human activity database appropriate for exposure modeling purposes. It is subdivided into time budget information and activity level data.

Time Budget Information

Systematic studies on how people spend their time started in the early 1970s. The first "time budget" researchers, being sociologists, were primarily interested in *what* people were doing and with *whom*. In the 1980s, exposure analysts were also interested in human activity studies in order to estimate *where* people spent time and *when* they participated in certain activities. In other words, time and space become an important desideratum of human activity information — in addition to the type of activity being undertaken. These latter data are very important to exposure analysts because they allow the analyst to match up (in a probabilistic sense) activity types with activity *levels*, or ventilation rates. Ventilation rates, as mentioned above, are important determinants of O_3 dose received by individuals going about their daily activities. In general, exposure analysts are not interested in the "with whom" datum, since the focus of exposure modeling

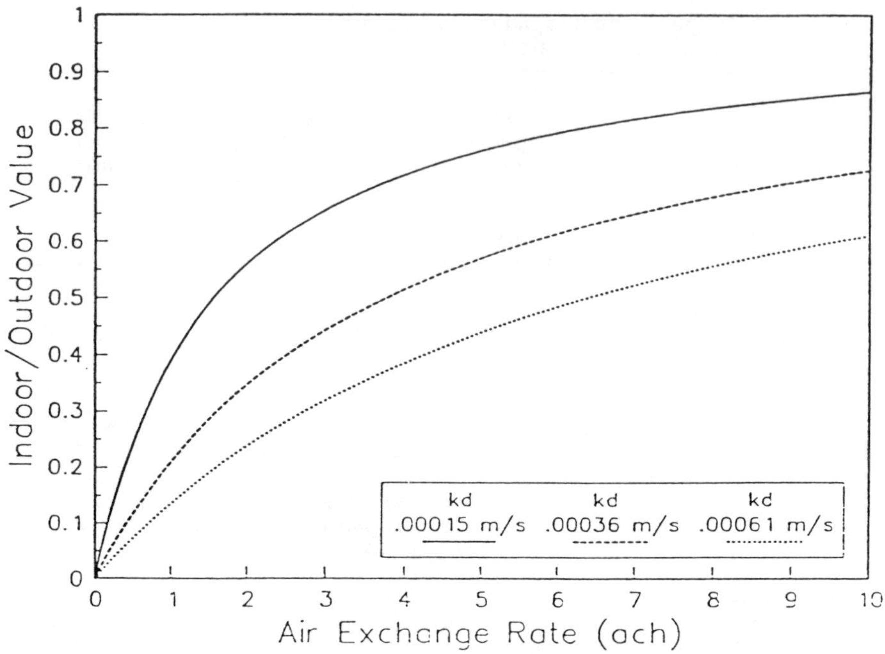

Figure 4. Predicted O_3 I/O values for a fixed [A_d/V=2.9m^{-1}] and three estimates of k_d. (From Weschler, C. J., H. C. Shields, and D. V. Naik. *Tropospheric Ozone and the Environment II*, R. L. Berglund, Ed. (Pittsburgh: Air & Waste Manage. Assoc. 1992), pp. 681–700. With permission.)

usually is on a single individual (or, a single individual representing — in a probabilistic sense — a specified group of people).

A recent National Research Council (NRC) report lists desirable properties of an activity study for exposure modeling purposes.[54] Some of these are:

1. Using a highly differentiated set of mes, especially those µes likely to lead to high exposures. Examples of the amount of differentiation are: indoor residence (kitchen, gas stove on); enclosed residential garage, car idling; gas station, pumping gas.
2. Coding of a breathing rate, or a breathing rate descriptor associated with an activity. This might include categories such as: low, medium, and high. ("Sleeping" rate can be inferred from the activity itself.)
3. Noting whether or not other people are generating indoor pollution. Examples include direct smoking by the respondent, passive smoking due to someone else's smoking, and a space heater being used at the time.
4. Coding whether or not windows in the me are open or closed. Another approach is to code what type of mechanical air handling system, if any, is in operation.

Human activity data suitable for use in an exposure assessment are obtained in a number of ways. They include:

- A telephone survey of "yesterday's" activities (also called a recall survey)

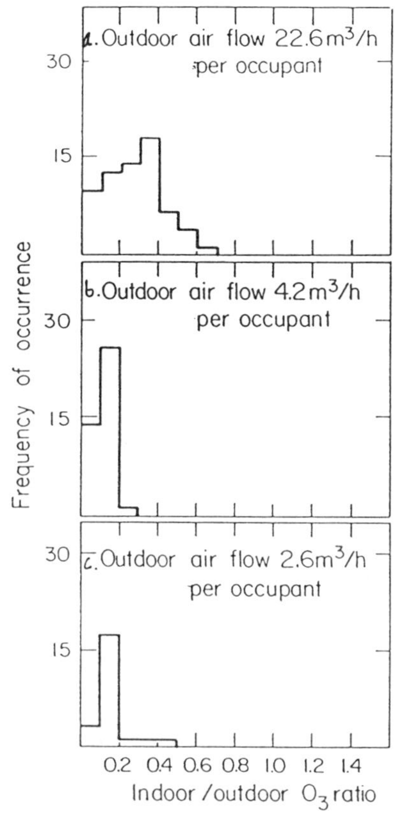

Figure 5. Frequency histogram of O_3 I/O ratios in an elementary school.

- A telephone survey of "right now" activities
- A real-time "diary" of activities that is carried by a participant (also called an open-ended diary)
- A human activity data-logger, which in essence is a computerized version of an activity diary

This typology is not standardized. Others exist; see, for instance, Reference 55.

Large-scale activity studies are usually telephone recall surveys, and the most active researcher in this field is Robinson, now of the University of Maryland.[56] He recently completed two surveys for the state of California. The first involved adults and adolescents[57] and the second focused on children under 12.[58] California is using the data to estimate population exposure to O_3 and other pollutants in the South Coast air basin.[59] The sampling frame of these surveys is a person-day of activity.

Recall telephone surveys normally cannot obtain all of the information desired for exposure assessment. Most of NRC's desirable items cannot be accurately recalled by a respondent 24 h after they occurred.[60] Extensive prompting can elicit some of the needed detail, but then the interview time stretches out to where respondents refuse to answer further questions.

Real-time telephone surveys of "what are you doing now?" usually can obtain all of NRC's desirable data. However, what is captured is a snippet of information that has no continuity with any other activity information. Estimating sequential or cumulative exposures is logically implausible when the sampling frame is a "person-activity". Because of this, real-time telephone surveys have limited use in exposure assessment data-gathering. On the other hand, they are effectively used to check activity diary data, and that seems to be their most appropriate role.

The real-time diary approach, either on paper or via an electronic logger, can theoretically provide all of the needed activity information for exposure assessment if the time basis used is small. The most valuable time basis for exposure assessment is as short as 1 min. Any activity that lasts longer than 30 s in a particular μe is recorded. When the respondent *changes* μe for >30 s, another diary entry is made. (Minute-by-minute entries are not made. Variable time-length entries are used instead.) Thus, the activity diary provides a disaggregated, *continuous* record of human activity for the day.

This type of activity study is very expensive. It requires extensive training (of both respondents and coders) and verification follow-up, in addition to those quality control procedures usually used in a self-administered survey. To maintain respondent cooperation, monetary compensation must also be given. Three open-ended studies have been completed and they are discussed below.

To reduce costs and increase cooperation, compromises often are made in the amount of data recorded in a diary. The most frequent compromise is to aggregate time into intervals much longer than a minute. Usually, a clock hour is used.[16,17,21-24, 60-64] Less frequently, the interval is 30 or 15 min.[65,66] The respondent enters one to three main activity(ies) that he or she participates in for the time interval. For the most part, transitions between mes are ignored and the sequence pattern of activities within the interval is not recorded. Other, longer intervals have been used also, but these studies are not of interest for O_3 exposure assessment.

Data loggers are still experimental in nature. At least two prototypes are in testing: the wearable PAL (personal activity logger)[67] and the "shadow sensor".[68] Neither approach has been used to date to gather human activity data suitable for use in an O_3 exposure assessment.

In this author's opinion, the best human activity surveys available for exposure assessment are the USEPA-sponsored studies in Denver[69-71] and Washington, D.C.[72-75] and the EPRI-sponsored study in Cincinnati.[76] There are a number of similarities among these studies, but significant differences also.[72] Since the two USEPA studies occurred during cold weather periods when O_3 levels are low, only the Cincinnati study will be described in any detail. All three studies used open-ended diaries. (Other studies also use them, but usually diary studies have small sample sizes, a narrow population focus, or a short duration.[77]) The two USEPA studies have 712 valid person-days of data in Washington, D.C. and 808 in Denver.[71] The EPRI study in Cincinnati captured approximately 2800 valid person-days of data.[76] People carried diaries for one day in Washington, two in

Denver, and three in Cincinnati. Thus, the sampling frame varies between 1 person-day to 3 person-days. The two USEPA studies sampled adults only, while the EPRI survey included children.

The Cincinnati study occurred in two sections. The first surveyed activities in the March to mid-April time period; the second from August to early September. Thus, it included school periods and vacation periods.

The Cincinnati diary required the respondent to describe in words what he or she was doing. Thus, there is a lot of detail concerning respondent activities, since the response is entirely open-ended. For instance, sample entries are: "walking slowly, looking at store windows", "sitting in a traffic jam", and "walking dog around the block". These descriptors can easily be used to check the other data coded by the respondent for an activity. Besides breathing rate, these other data are:

- Location of the activity (12 fixed locations and 3 fill-in blanks were used)
- Smoking status in the μe
- Use of a fireplace or woodstove (if indoors)
- Status of windows (open, closed, or uncertain)

Personal and home information were elicited from each respondent via a supplemental questionnaire. (All three cities used essentially the same questionnaire.) There were over 70 questions relating to personal characteristics (age, sex, education, health problems, and leisure and exercise patterns); work or school data (job title and work type, income, exertion level on the job, and commute time); and housing characteristics (heating and air conditioning type, value of house or rent paid, fuel used for heating and cooking, type of garage available, and fireplace use).

Most importantly for O_3 exposure assessment, the EPRI/Cincinnati diary included a question on the breathing rate associated with each activity. Four categories were mentioned in the diary

- Slow (e.g., sitting)
- Medium (e.g., brisk walk)
- Fast (e.g., running)
- Breathing problem

If the last category was chosen, the respondent was asked to provide an open-ended comment describing the problem.

A fifth breathing rate category was inferred from the activity itself. This category was sleeping, which has a very low ventilation rate associated with it (<6 $l\ min^{-1}$) for 90% of the occurrences, and <4 for the median estimate).

It is difficult to summarize the activity data available from the Cincinnati and USEPA studies. Basically, they contain activity-by-activity data for thousands of person-days. These data of course can be aggregated in an infinite number of ways. Such aggregations have been reported in the literature.[69-73,76] Aggregated data, however, generally minimize the importance of the "tails of the distribution"

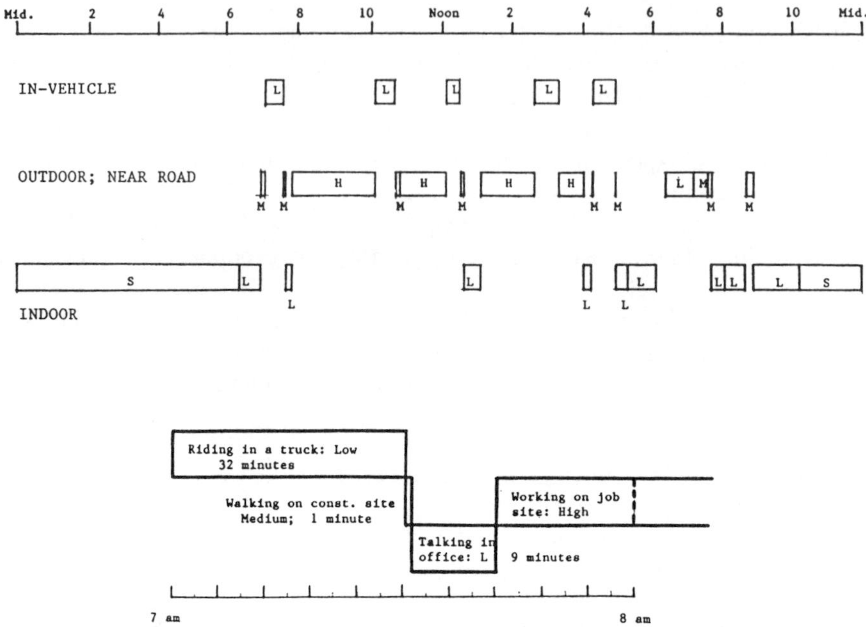

Figure 6. Example of activity data available for an outdoor worker in Cincinnati, OH. (From Johnston, T. "A Study of Human Activity in Cincinnati, Ohio," (Palo Alto, CA: Electric Power Research Institute, 1987). With permission.)

and completely ignore the *pattern* of activities associated with individuals. Ozone exposure analysts should probably focus on an active outdoor worker spending eight or so hours outdoors during the peak O_3 period instead of a typical (i.e., indoor) worker who spends less than 1 h outdoors per day. The same holds true for the unusually active child who plays outside most of the summer day rather than the more typical "couch potato" child who stays indoors watching TV. Aggregated data, unhappily, often emphasize the average person rather than those people of most interest from an exposure viewpoint.

For these reasons, human activity summary data are not presented here. For a review of the human activity database, with aggregate statistics, see Reference 78. For a review of the difficulties associated with self-administered recall diaries, see Reference 79. For an example of the type of activity data available in the EPRI/ Cincinnati database, see Figure 6. This figure depicts activities undertaken by an outdoor worker who lives in an air-conditioned house. The pattern depicted is for a workday. The breathing rate categories conform to those mentioned earlier, where:

- S = sleeping
- L = slow (low)
- M = medium
- H = fast (high)

The data are aggregated for presentation purposes into three major µes:

- In-vehicle
- Outdoors and near roadways
- Indoors

The detailed data at the bottom of the figure depict the information provided by the respondent for the 7 a.m.–8 a.m. hour. It is rich in detail. The quantity c_m * t_m * ventilation rate (v_e), associated with each activity, can easily be calculated for each microenvironment or activity for the individual shown. This calculation will be carried out in the next section, after data on v_e are presented.

Ventilation Rate Information

Over 30 different measurement techniques have been used to assess physical activity in the population,[80] but maximal oxygen consumption is probably the indicator best suited for inhalation exposure analyses. This indicator, however, is difficult to measure outside the laboratory, so surrogate indicators are used. Heart rate (HR) is one such surrogate. The relationship between HR and another more directly useful surrogate — minute ventilation, or ventilation rate — is described in the following quote from Terblanche et al.[81]

> ... an increase in activity, as reflected by heart rate (HR) measurements, could modify the estimated airway/lung dose of a pollutant such as ozone by 70 percent. In addition, biological parameters such as HR and minute ventilation (V_e) can provide the necessary data to calculate the effective dose of air pollutants to the lower respiratory system more accurately. Pollutant dose delivered to the respiratory tract varies mainly with anatomical structure, respiratory patterns, and the volume of air circulated (minute ventilation). V_e increases linearly with exercise intensity (increased activity levels) as reflected by HR until about 60 percent of maximum O_2 uptake; at this point a ventilatory "breaking point" is reached beyond which V_e rises in a more accelerated manner. This point corresponds to HR in the range of 130–150 beats per minute (bpm). Since HR and V_e are highly correlated, HR can also be used to estimate V_e for a given activity level, if the relationship between these variables for an individual has been assessed by standardized exercise testing (Reference 81, p. 942; citations removed.)

Terblanche et al. provide selected HR data for teenagers. HR data exist for numerous age/occupation groups that could be used for exposure assessment; see References 3 and 81 to 83. Normally, the HR information has to be "translated" into v_e data (in units of liters per minute, or $l\ min^{-1}$) for direct O_3 dose modeling work. Most of the just-cited references do just that, the translation usually being an empirical one based on regression analyses. [83–86] An example of such a relationship appears as Figure 7.

Ventilation rate data have been compiled for various age and sex categories.[87,88] A sample of selected v_e data appear as Table 11.

Many researchers use another measure of ventilation to better account for interpersonal differences in body size as it affects lung function. This measure is known as equivalent ventilation rate, or EVR. EVR = v_e divided by body surface area, in units of liters per minute per square meter ($l\ min^{-1}\ m^{-2}$).

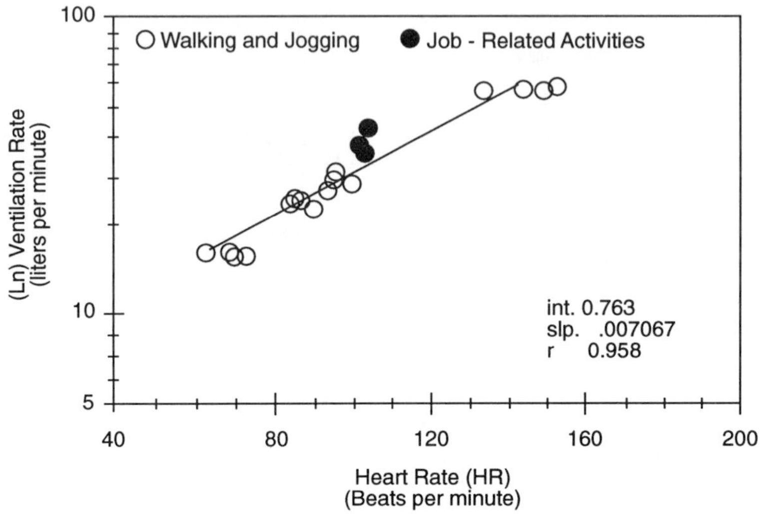

Figure 7. Ventilation rate vs. heart rate: an example for a construction worker: (open circles (j) = walking and jogging; closed circles (v) = job-related activities. (From Linn, W. S., D. A. Shamoo, and J. D. Hackney. *Tropospheric Ozone and the Environment II*, R. L. Berglund, Ed. (Pittsburgh, PA: Air & Waste Manage. Assoc., 1992), pp. 701–712. With permission.)

Body surface area (BSA) is itself a function of body height, weight, and general body shape. BSA data are generally obtained from empirically based equations of the following form:

$$BSA = a * H^b * w^c$$

where H = height (cm)
W = weight (kg)
a, b, c = constants

BSA formulae are reviewed in the USEPA's *Exposure Factors Handbook*.[88] One version of the BSA model was developed by Gehan and George in 1970; its form is:

$$BSA = 0.24 * H^{0.42} * W^{0.51}$$

The USEPA exposure group recommends its use for exposure analyses. The *Handbook* provides percentile distributions of BSA values for children and adults of both sexes that are very useful for calculating normalized ventilation rate (i.e., EVR). In addition, some EVR data are being published directly.[83,85]

The importance of directly including v_e (or EVR) in estimating O_3 exposure can be illustrated by an example. As discussed in the section on "Direct and Indirect Exposure Assessments", most indirect exposure assessments use $c_m * t_m$ as the exposure metric. This product is shown below in Table 12 for the activities emphasized in Figure 6. The table indicates that the person receives 2.52 ppm min

Table 11. Selected Ventilation Rate (v_e) Estimates for Example Activity Levels and Age/Sex Categories

Sex	Age	Activity Descriptor	Representative Activity	Estimated Mean v_e ($l\ min^{-1}$)	Range of Estimated v_e ($l\ min^{-1}$)	Ref.
M/F	<2	Resting	—	0.8	0.3–2.1	88
M	6	Light	—	13.9	5.0–32.0	88
F	12	Moderate	—	26.5	19.6–46.3	88
M	14	Heavy	—	86.9	42.2–121.0	88
F	16	Resting	—	15.2	—	88
F	>18	Resting	—	5.7	4.2–11.7	88
		Light	—	8.1	4.2–29.4	88
		Moderate	—	26.5	20.7–34.2	88
		Heavy		47.9	23.4–114.8	88
M	>18	Resting	—	12.2	2.3–18.8	88
		Light	—	13.8	2.3–27.6	88
		Light	Level walk @ 2 mph; washing	13.0	—	87
		Light	Construction; stack firewood	25.0	—	87
		Moderate	—	40.9	14.4–78.0	87
		Moderate	Climb stairs; spade digging	35.0	—	87
		Moderate	Cycle @ 13 mph	40.0	—	87
		Heavy	—	80.0	34.6–183.4	88
		Heavy	Climbing with load; handball:	63.0	—	87
			Very heavy running @ 10 mph	85.0	—	87

O_3 "exposure" for the hour. (More consistent units would result if concentration were expressed in mass concentration (mg m^{-3}) units instead of volumetric concentration (ppm) units, but the concept is the same in either case.) If the "standard" v_e rate recommended by the USEPA were applied to this exposure estimate, the subsequent "effective exposure" obtained would be about 35 ppm liter for the hour (2.52 ppm min * 13.9 l min^{-1}, which is the per minute breakdown of the 20 $m^3\ day^{-1}$ v_e value recommended by USEPA in its *Handbook*[88]). This value is about 27% of the 130.5 ppm liter "effective exposure" calculation shown in Table 12. Thus, large differences can result in estimating O_3 exposure if v_e is not explicitly — and correctly — considered in an O_3 exposure analysis. This observation is reinforced in the next section.

Estimating O_3 Dose

There are only a few citations in the exposure literature focused on calculating O_3 dose associated with ambient or μe exposure.[89,90] The most recent one, by Kleinman,[1] will be used here to illustrate what factors must be considered in order to rigorously estimate the O_3, dose received by exposed individuals. A quote from his article concisely introduces the topic.

Table 12. Estimating "Effective O_3 Exposure" for an Example 1-h Period for a Working Male

Activity Level	Duration (t_m) (min)	Assigned O^3 Concentration (c_m) (ppm)	"Exposure" (c_m * t_m) (ppm min)	Assigned Vent. Rate (v_e) (l min^{-1})	Effective Exposure (c_m * t_m * v_e) (ppm l)
Low	32	0.02	0.64	13.0	8.3
Medium	1	0.08	0.08	20.0	1.6
Low	9	0.04	0.36	15.0	5.4
High	18	0.08	1.44	80.0	115.2
Totals	60	—	2.52	—	130.5

In addition to being dependent upon dose, the biological effect of ozone is dose-rate (or possibly, concentration) dependent. Thus, all other factors being equal, an exposure to 0.2 ppm O_3 for 4 h produces a lesser effect than does an exposure to 0.4 ppm O_3 for 2 h. Therefore, in order to permit comparisons of biological responses for studies in which subjects were exposed to different concentrations of O_3, the internal thoracic dose was defined as the total mass of O_3 delivered and retained in the thoracic airways (i.e., tracheobronchial and deep lung) per unit body mass, corrected for the increased biological response observed when individuals are exposed to concentrations greater than 0.1 ppm.

Internal thoracic dose may vary significantly among individuals and may yield large variations in biological responses among people exposed to the same concentrations of a pollutant, even though each individual's response may be reproducible. Other factors also play a role; for example, health status, sensitization or adaptation, and person-to-person differences in genetic makeup (Reference 1, p. 310; citations removed).

Kleinman develops an equation to compute internal thoracic dose (ITD) of O_3, in units of micrograms per kilogram ($\mu g\ kg^{-1}$) of body mass.[1] The equation (in modified nomenclature) is:

$$ITD = (c_m * t_m * d) * (v_e / w) * (R) * f(P) * N$$

where d = density of O_3 at STP conditions ($\mu g\ ml^{-1}$)
w = body mass (kg)
R = O_3 retention in the deep lung at v_e
$f(P)$ = fractional(f) penetration(P) of O_3 through the upper respiratory tract at v_e

$$f(P) = (f_o * P_o) + (f_n * P_n)$$

where o = oral breathing
n = nasal breathing

N = non-linearity term used to "correct for" the O_3 dose rate relative to 0.1 ppm O_3. (Note that this variable is dependent upon the biological response being modeled.)

$$N = (\ln 0.1)/(\ln c_m)$$

Kleinman obtained data for all of the variables and exercised the model to calculate fractional changes in group means of FEV_1 (see Chapter 3) for 13 clinical O_3 studies conducted by other pulmonary scientists. Reported and estimated changes in FEV_1 group means showed reasonable agreement ($R^2 \sim 0.5$, $p \leq 0.01$).[1] He then applied the FEV_1 dose model to outputs of the REHEX exposure model (see section on "Specific O_3 Exposure Models" below) to address potential community-wide health effects associated with O_3 exposures. We turn to O_3 human exposure models next.

MODELING HUMAN EXPOSURE TO O_3

This section will present material concerning the indirect O_3 exposure assessment approaches found in the literature. There are only four approaches currently being pursued. They are described in the section on "Specific O_3 Exposure Models". Their outputs are used in various health risk assessments that are discussed in the next chapter.

Before describing particular models, however, we provide a little background on exposure modeling in general. The field of exposure modeling is growing rapidly and it is an interesting intellectual endeavor.

Human Exposure Modeling

The field of human exposure modeling is relatively young. The first rigorous exposure modeling analyses appeared in the mid-1970s, and it was not until the early 1980s that theoretical constructs regarding human exposure to environmental pollution began to be published. This chapter is not the appropriate place to review the intellectual tradition underlying human exposure modeling; the interested reader should consult the material cited in References 8, 54, and 91 to 93 for a discussion of concepts, definitions, and types of exposure models.

Ryan[93] places exposure modeling in the proper perspective:

> The assessment and evaluation of human exposure to pollutants is an important area of investigation from both research and regulatory points of view. The principal rationale for personal exposure assessment is the evaluation of health effects associated with exposure to pollutants. Measurement of personal exposures may be needed for epidemiological investigation and for risk analysis; it may also be needed for the purpose of control and mitigation. Modeling of such exposures often presents

> a cost-effective strategy for assessing both potential exposures experienced by individuals and the effect of control and mitigation strategies. The model is best viewed as a tool for the investigator. It is an aid in understanding exposures and the factors that influence them. (Reference 93, p. 453.)

Most of the modeling efforts focused on O_3 exposure assessment were undertaken for regulatory purposes. All of them suffer from a lack of validation using "real-world" data. This shortcoming probably constrains their usefulness in making generalizations to the entire population regarding O_3 exposure impacts. This point aside, we turn to a review of these models.

Specific O_3 Exposure Models

There are two distinct types of O_3 exposure models: those that focus narrowly on predicting indoor O_3 levels and those that focus on predicting O_3 exposures on a communitywide basis. Often, the second type of model relies on the first type to estimate indoor O_3 exposures.

In general, the first model type uses some version of a mass-balance chemical reactor equation that includes variables to account for:

- Infiltration
- Outdoor make-up air
- Recirculation
- Surface reactivity losses
- Indoor sources

Models of this type focused on O_3 have been around since the early 1970s.[94,95] Since then, there have been numerous versions and improvements.[43,96,97] A recent version of the single-compartment mass-balance equation used by Hayes[97] follows.

$$\frac{dC_i}{dt} = k * a_m(C_o - C_i) \qquad \text{(Infiltration)}$$

$$+k * a_f(E_f C_o - C_i) \qquad \text{(Outdoor make-up air)}$$

$$+\mathrm{k} * \mathrm{a_r}(E_r - 1) * C_i \qquad \text{(Recirculation)}$$

$$-k * K(A / V) * C_i \qquad \text{(Surface losses)}$$

$$+S_{(t)} / V \qquad \text{(Indoor sources)}$$

where C_i = indoor concentration
C_o = outdoor concentration
k = mixing factor (fraction of pollutant mass that is well mixed)
a_m = infiltration flow rate (Ach)

a_f = outdoor make-up flow rate (Ach)
a_r = recirculated air flow rate (Ach)
E_f = efficiency of make-up air filter
(E_f = 1 - fraction of pollutant removed by the filter)
E_r = efficiency of recirculation air filter
K = pollutant reactivity factor (rate constant associated with heterogenous surface decomposition)
A = interior surface area
V = interior volume
$S_{(t)}$ = indoor source generation rate

Usually, a number of assumptions are made to solve this equation, such as perfect mixing indoors and that both indoor and outdoor O_3 concentration profiles are linear between hourly observations, that allow a closed-form exact mathematical solution. Hayes, however, advocates using a recursive form of the equation that provides more flexibility in analyzing time-varying outdoor pollution levels. His recursive form appears in Reference 97 and it has been validated by Weschler et al.[47]

The second exposure model type mentioned above is used for community-level exposure analyses. There are four distinct O_3 human exposure models in this model type. They are generally known as:

1. pNEM/O_3 (based on the NEM series of models)[98-100]
2. SAI/NEM[101-105]
3. REHEX[59,106-108,111]
4. Event probability exposure model (EPEM)[109]

All four models are related to the NEM (NAAQS exposure model) approach in one fashion or another, although significant variations now exist among the models.

NEM is a USEPA exposure model first developed in the early 1980s to complement a probabilistic risk assessment program.[110] There are at least seven versions of the model, although their underlying principles are similar. The major components of the NEM exposure model (and associated risk model) appear in Figure 8. Four major activities are depicted. Outdoor air quality data are obtained from monitoring or modeling data. In most applications of NEM, fixed-site monitoring data are used, although the SO_2 exposure analyses used modeled data. Even if monitoring data are used, hourly values are "transformed" by a suitable relationship so that they better represent air quality outside the various microenvironments (mes) of interest. The "seasonal considerations" box shown in Figure 8 represents those factors that affect both air quality modeling inputs and such me considerations as the probability of having an open window (which affects Ach), internal sources emission rates (especially for space-heating combustion sources), and human activity patterns.

There is a spatial dimension to the air quality component that is not apparent from the diagram. The air quality estimates are spatially disaggregated, no matter

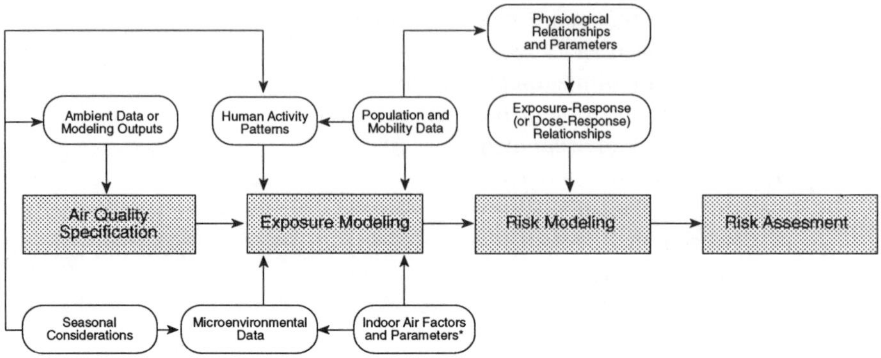

Figure 8. The major components of the NEM series of exposure models and associated health risk assessment procedures. (*Often, this is a mass-balance model in the newer versions of NEM.)

how they are obtained. The spatial dimension may be a grid cell from a dispersion model, a "district" surrounding a monitor, or a "neighborhood type" aggregation of land uses. NEM applications generally use the district approach, where census tracts surrounding a fixed-site monitor are assigned to a particular monitor based on geographic and/or monitor characteristics. The assignment procedures are complex and too detailed to be discussed here; the interested reader should review Reference 98 for more information.

The important point of the NEM spatial dimension is that people can be assigned to a monitor using U.S. Census data. In addition, commuting trips can be assigned among the districts, grid cells, or neighborhood types using census data. (These data are not available for all areas of the U.S., however.) Thus, the NEM model — as well as others — simulate the movement of people through space for work-trip purposes. This allows the analyst to assess exposure in different parts of an urban area. When coupled with human activity data, this allows the analyst to simulate movement in time *and* space.

The "human activity patterns" database contains data for people of all ages, occupations, and health status. Of course, coverage for certain characteristics is better than for others. Thus, individuals in this database can be assumed to represent a "cohort" of people of interest to the analyst for exposure assessment purposes. For O_3, for instance, one of the main groups of interest is exercising children. The database can be disaggregated to obtain a pool of diary days for children who exercise. The disaggregation also can be done for many specific age and sex groups or for specific occupational groups — say, outdoor workers. It can also be done for *some* health status endpoints, such as asthma, but the amount of diary days of data is small. Usually, the specified cohorts used have some relationship to common U.S. Census Bureau categories so that census data can be applied to assign cohorts to exposure districts.

The "exposure modeling" component is where human exposure estimates are calculated. Inputs to this component are numerous and diverse. The population and mobility data input is the Census data mentioned above. The human activity input, in NEM, is a computerized set of approximately 6000 person-days of activity diary data. The data include an estimate of v_e for each activity that appears in the diary. (See section on "Time Budget Information" for more information on this subject.

The "indoor data factors and parameters" input is a set of indoor/outdoor factors in some versions of NEM, but a mass-balance equation in others. The mass-balance approach is used in the latest version of pNEM/O_3. This computation routine "input", along with seasonal considerations, affects how much outdoor air pollution is estimated to come inside and how much indoor sources affect indoor exposures. In pNEM/O_3, values for most variables that affect indoor exposure are obtained by Monte Carlo sampling from empirical distributions of measured data. These variables include:

- Probability of having open windows given the type of air conditioning system found in the home (if any) and the outdoor temperature
- The Ach rate, which itself is affected by the "window status" probability
- Residential surface-to-volume ratio
- Emission rates from any indoor sources and the probability that said sources will be operating or not

All of this results in an estimate of the pollutant concentration experienced by an individual in each me that she or he inhabits. This is the c_m datum mentioned previously. The t_m and v_e data are part of the human activity database. When combined, the exposure modeling component of pNEM/O_3 provides a string of hourly averaged estimates of $c_m * t_m * v_e$ for individuals representing specific cohorts. These hourly estimates can be used for exposure-dose-response modeling purposes or can be used directly for estimating exposures disaggregated by breathing rate category.

It should be noted that pNEM/O_3 could provide $c_m * t_m * v_e$ estimates for time periods as brief as 1 min, but this short time resolution is beyond the health modeling capability at present. In pNEM/CO, time periods as short as 1 min were used to estimate carboxyhemoglobin (COHb) levels in exposed individual's blood. Thus, the pNEM series of exposure models is not constrained to hourly time periods.

Regardless of the time resolution used, outputs of the exposure modeling component are either used directly to estimate exposure or become an input to a risk model. As is discussed in Chapter 5, modeling risk involves its own set of modeling inputs besides exposure estimates. These will not be discussed here. What will be discussed is an output from an early PNEM/O_3 modeling exercise focused on Houston, TX; see Figure 9.

The figure presents decreasing cumulative distributions for the Houston population (2,267,000 people) that experience one or more O_3 exposures to three air quality indicators at all exercise levels. The indicators are:

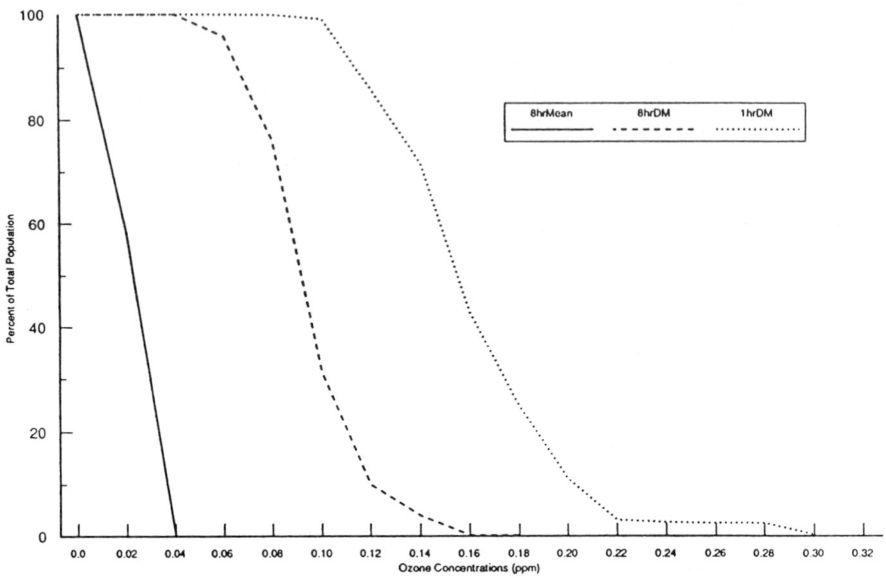

Figure 9. Percent of Houston population that experiences one or more exposures to three O_3 air quality indicators. (From McCurdy, T., et al. *83rd Annual Meeting of the Air and Waste Management Association*, Pittsburgh, June 1990).

- 1-h daily maximum (1hDM)
- 8-h daily maximum (8hDM)
- April-September mean of the 8hDM

These indicators are discussed in more detail in Chapters 2 and 3.

Conforming to intuition, the percent population exposed to various O_3 concentrations decreases as concentration increases for all three indicators. Approximately 86% of the population is estimated to experience a 1-h O_3 exposure of 0.12 ppm, the O_3 NAAQS level, or higher during Houston's O_3 season. The corresponding estimate for the 8hDM indicator of 0.08 ppm is 76%. On the other hand, no one in Houston is estimated to experience a seasonal mean 8hDM > 0.04 ppm. Other outputs from pNEM/O_3 are available, including estimates of the number and relative percentage of total population exposure to O_3 when the O_3 NAAQS is just attained and at specified exercise levels.[100] In brief, these other analyses indicate that O_3 exposures decrease dramatically when the NAAQS is attained and for increasing exercise rate — because fewer people undertake heavy exercise. Space constraints do not allow us to further pursue these other outputs. Instead, we turn to a discussion of SAI/NEM.

SAI/NEM has been used extensively by the American Petroleum Institute,[101-103] the Southern California Edison Company[104] and the Western Oil and Gas Association[105] to evaluate O_3 exposures associated with various precursor emission strategies proposed by the South Coast Air Quality Management District. SAI stands for Science Applications, Inc., a technical consulting firm with extensive experience in developing and applying photochemical reaction models to predict ambient O_3 concentrations. In addition, SAI has also assisted the USEPA in developing its O_3 risk assessment modeling effort (see Chapter 5). SAI's work for the industry groups mentioned above also includes an O_3 health risk assessment component, but that is not of interest here.

SAI/NEM is based upon an early version of NEM. While both models have the same structure (and computer code, to a certain extent), SAI modified NEM significantly. Their version has more districts, more mes, and a more detailed mass-balance model to predict indoor O_3 concentrations than does pNEM/O_3 or earlier versions. On the other hand, SAI's human activity data file is out of date and inflexible. Thus, the two models are quite different and probably would produce different estimates if they were applied to the same urban area.

To date, SAI/NEM has been used only in California's South Coast Air Basin (SOCAB). This area is large and includes all or part of the Los Angeles, Anaheim, Riverside, and San Bernardino metropolitan areas. About 10 million people were included in SAI's South Coast area exposure modeling effort.[104] Both 1hDM and 8hDM indicators were evaluated for three historical time periods (1975, 1979, and 1983) and three future emission control scenarios. Outputs from the control scenarios are ignored here.

Excerpts from SAI's 1988 report[104] succinctly state some of their modeling results.

> [1-hour exposures.] Of the three years studied, risk and exposure levels were highest in 1979, with the levels in 1983 returning to those prevailing in 1975 or below...Exposure and risk levels decreased substantially from 1979 to 1983 in interior portions of the basin, where ozone risk levels are typically the highest. Although ozone risk levels are lower than in the interior, they have increased in coastal portions of the basin during the same period, with levels in the mid-basin remaining largely unchanged...Nearly everyone in the SOCAB (96 and 99 percent) experienced 1-hour average ozone levels above the Federal standard (0.12 ppm) during 1979 and 1983, up from less than 75 percent in 1975. Per capita annual hourly exposures (total occurrences divided by population) to concentrations above the Federal standard peaked at 78.2 in 1979, up from about 59 in 1975, but returned to 55.5 in 1983. About half of basin residents experienced 1-hour ozone concentrations above first-stage alert levels (0.20 ppm), from less than 47 percent in 1975, up to 56 percent in 1979, and back down to 48 percent in 1983.
>
> Basinwide and in every region, there has been a steady increase in the number of people exposed to concentration levels over the NAAQS (0.12 ppm), possibly due more to population growth than to increased ozone. In apparent confirmation of this, a significant decrease is calculated between 1979 and 1983 in the basinwide number of occurrences (people-hours) of such exposures, back to near 1975 levels. Regional differences in this finding are noted, with coastal occurrences increasing steadily over the period, mid-basin occurrences remaining largely unchanged from 1979 to 1983 (but above 1975 levels), and interior occurrences peaking in 1979, with 1983 levels returning to 1975 levels or below.
>
> [8-hour exposures.] Even accounting for periods spent indoors where ozone levels are significantly lower than outdoors, a sizeable percentage of basin residents were exposed to 8-hour average ozone levels above the concentration of the 1-hour

> Federal standard (0.12 ppm), from less than 34 percent in 1975, up to 47 percent in 1979, and down to 25 percent in 1983. Per capita annual 8-hour exposures (total occurrences divided by population) to concentrations above the 1-hour Federal standard peaked at 21.9 in 1979, up from about 10 in 1975, but dropped to 5.5 in 1983. Further, there were exposures to 8-hour averages even above 0.20 ppm for residents of at least one exposure district during each of these three years, with residents of 8 of 10 districts so exposed in 1979...Basinwide 8-hour exposures to concentrations over the level of the 1-hour Federal standard peaked in 1979, up from 1975, with the number of people exposed and occurrences back to below those in 1975. Generally, a similar pattern is seen in coastal, mid-basin, and interior regions. (Reference 104; pp. 6-4 to 6-7, in part.)

A map showing the various basins modeled in SOCAB for the three time periods appears as Figure 10. Reference 104 contains numerous maps depicting the spatial variability of various exposure and risk measures modeled in the Basin.

Both the Houston and South Coast exposure results indicate that a large percentage of the population experience O_3 exposures at or above the USEPA's NAAQS standard level of 0.12 ppm. This finding is reinforced by the third modeling effort that will be discussed: another SOCAB analysis using the REHEX model.

The REHEX model adopted some of the NEM computer code but uses a more detailed geographic resolution.[108] In the REHEX application to the SOCAB area, geographic space was subdivided into 32 exposure districts. This, of course, provides for a much finer resolution of ambient O_3 air quality than was accomplished in the two modeling exercises discussed previously.

Like the NEM models and SAI/NEM, REHEX explicitly uses home/work trip data to "move" people through the region during their day. Human activity data come from the California Air Resources Board's recent activity telephone survey of approximately 3000 persons in the state. Thus, REHEX uses California-specific activity data in its modeling exercise.

The in-transit and outdoor mes are emphasized in the REHEX model. Analysis of the activity data led the REHEX modelers to separate these data by season of the year and day of the week, but to aggregate labor categories as time outdoors varied less with labor class than with season or day of the week. The final database includes about 1600 activity profiles used for modeling purposes.

The REHEX modelers found that California residents spend a lot of time outdoors and in-transit. Their findings are reproduced in Table 13. Note the very large amount of time spent outdoors by children! These *average* times are more than double those usually cited as being the mean time spent outdoors by children. The larger summer estimate is especially important with respect to O_3 exposure, especially in the Los Angeles area with its very high daily O_3 peaks occurring in the June-through-September time period.

Table 13 also indicates that time-in-transit varies little by season or time of week for adolescents and adults. The time spent by children in this me does not vary by season and is much less: about 26 min/day on average.

REHEX, like one version of NEM,[98] calculates O_3 exposure — and dose —using discrete distributions of hourly averaged air quality. While the attendant calculation

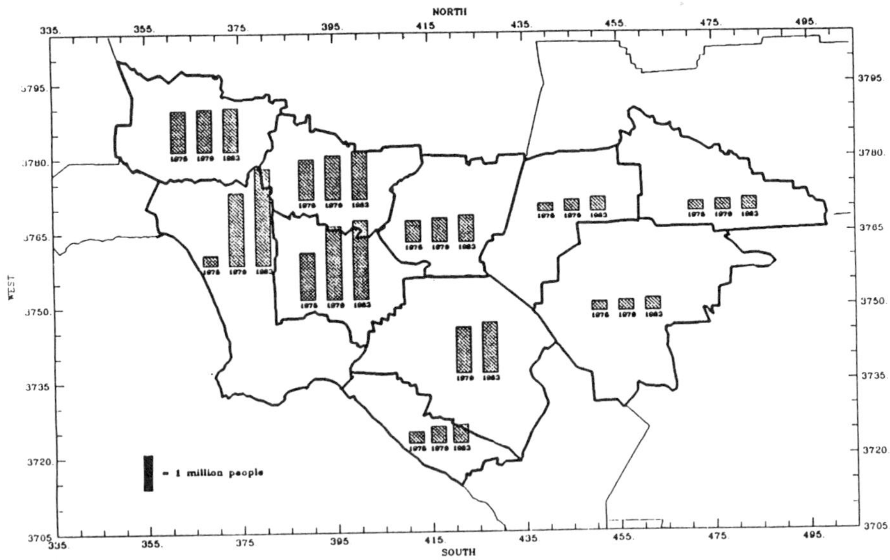

Figure 10. Number of people exposed to 1-h O_3 concentrations above 0.12 ppm (NAAQS), any exercise level: 1975, 1979, and 1983. (From Hayes, S. R., B. S. Austin, and A. S. Rosenbaum. *A Technique for Assessing the Effects of ROG and NO_x Reductions on Acute Ozone Exposure and Health Risk in the South Coast Air Basin* (San Rafael, CA: Systems Applications, Inc., 1988). With permission.)

procedure is fast using this technique, sequential O_3 exposures cannot be estimated or analyzed. Thus, this approach can only be used to evaluate hourly O_3 exposures. The procedure would have to be extensively modified to analyze, for instance, 8-h O_3 exposures. Another shortcoming of the REHEX approach is the use of indoor-outdoor ratios to estimate indoor O_3 concentrations. Many exposure models now use a mass-balance approach to estimate indoor pollutant concentrations.

REHEX contains an exposure-dose relationship that allows analysts to directly estimate discrete, hourly averaged O_3 dose levels in exposed individuals. Dose is defined to be the amount of O_3 that reaches the lung. It is calculated as the product of a dose factor and μe O_3 concentration. The dose factor varies by ventilation rate of the exposed individual, which in turn is defined by the individual's exercise state (activity level).[108]

REHEX has been used extensively by the South Coast Air Quality Management District to analyze O_3 (and other pollutant) exposures in the SOCAB area. The analyses include various control measures for various time periods, but results for only the current situation are of interest here. Figure 11 presents REHEX data on the number of 1-h people-occurrences at selected O_3 concentration levels. The total population analyzed is about 12 million people. REHEX analysts estimate that in the mid-1980s, 98% of SOCAB residents were exposed to O_3 levels above the federal NAAQS level of 0.12 ppm at least one time per year.[106] This percentage is between the 96 and 99% estimated by SAI to exceed this concentration in the same area in 1979 and 1983, respectively.[104]

Table 13. Average Time Spent in Two Microenvironments Employed in Rehex Modeling of O_3 Exposures in the South Coast Air Basin (SDCAB)

Demographic groups	Season span	Weekly considerations	Time outdoors (h day^{-1})	Time in-transit (h day^{-1})
Children	June–Sept.	Everyday	4.6	0.4
	Oct.–May	Everyday	3.5	0.4
Adolescents	All Seasons	Everyday	1.4	1.8
Adults	June–Sept.	Weekdays	1.8	1.9
		Weekdays	2.1	2.0
	Oct.–May	Weekdays	1.0	1.8
		Weekdays	1.2	1.9

From Lurmann, F., and S. Colome. "An Assessment of Current and Future Human Exposure to Ozone in the South Coast Air Basin," (Santa Rosa, CA: Somoma Technology, 1991). With permission.

The REHEX analysis also estimates that more than 50% of the population will experience O_3 levels greater than 0.20 ppm, which is California's "Stage I alert level". (SAI's estimates for this were 47% in 1975, 56% in 1979, and 48% in 1983; see above.)

REHEX analysts present exposure estimates for individual districts, cities, and demographic groups. A quote from one of their reports is of interest.

> The exposure estimates clearly show that some demographic groups are exposed disproportionately to elevated ozone concentrations. For example, while there is little difference in fractions of demographic groups that receive exposures to concentrations above 12 pphm [0.12 ppm] one or more times per year (because more than 90% of all groups are exposed to these levels), there is considerable difference for concentrations above 20 and 35 pphm [0.20 and 0.35 ppm]. The data° show that for concentrations above 20 pphm infants and the elderly experience the lowest ozone exposures per capita (because they spend less time outdoors than other demographic groups), and school-age children, college students, and adults working outdoors experience the highest ozone exposure per capita. At the highest concentration levels, children and college students receive the highest per capita exposure. They receive higher per capita exposure than adults working outdoors at these levels because most of the highest concentrations occurred on weekends and our time-activity model has children and college students spending more time outdoors than adults on weekends (Reference 106, p. 5-7).

The REHEX analysts also found significant O_3 exposure differences among people living in the various municipalities included in the SOCAB area. This finding corresponds to SAI's results.

The last O_3 exposure model to be discussed is International Technology (IT) Corporation's Event Probability Model (EPEM). This model cannot provide *distributions* of O_3 exposure for any specified population group — even for the 1-h time period. Instead, it estimates the probability that a person selected at random (from a specified cohort or the population as a whole) will experience a particular exposure situation.[109] It is an efficient approach to estimate, for instance, the

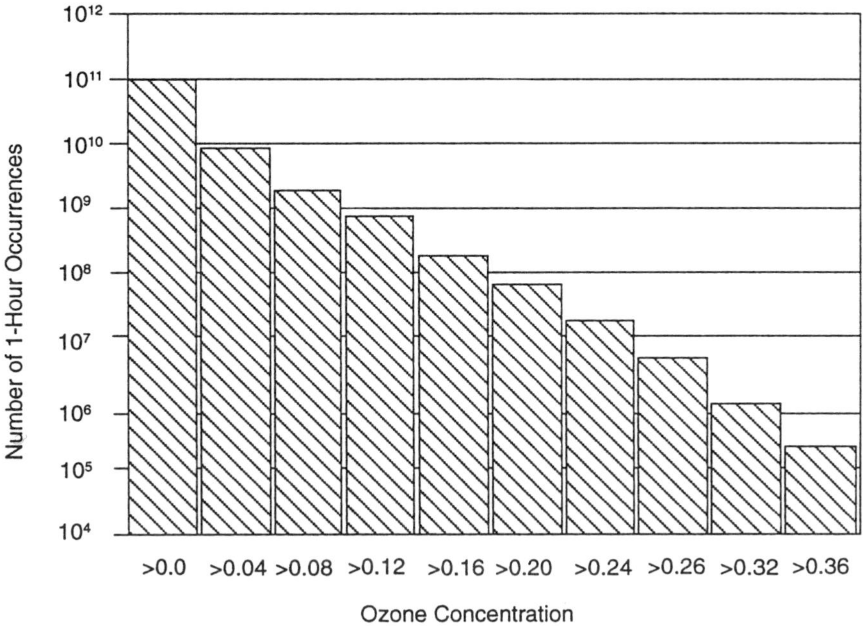

Figure 11. Estimate of the annual distribution of O_3 exposure in the SOCAB region from 1984 to 1986. (From Winer, A. M., et al. *Characterization of air Pollution Exposures in the California South Coast Air Basin: Application of a New Regional Human Exposure (REHEX) Model* (Riverside, CA: University of California, 1989). With permission.

probability that someone would experience the "maximum possible exposure" (MPE) in a year. For O_3, the MPE might be the highest hourly averaged ambient concentration monitored at a fixed site in an area. Other definitions of the event of interest are possible. One might be the 95th percentile O_3 concentration seen in an area, for instance. As currently structured, however, EPEM could not estimate the probability that a person selected at random would experience five or more O_3 exposures at or above some selected concentration value. In other words, there is no multi-day continuity in the model. EPEM could estimate the probability that a person selected at random would experience a maximum 8hDM at or above some specified level, but it has not been so exercised.

The approach used in EPEM is straightforward. The probability that a randomly selected individual experiences a particular event is the summation of the product of two separate variables relating to an "inclusion probability" and an "event probability". In both cases, probability is defined to be a relative frequency observed in an appropriate database. In symbols, modified from IT's nomenclature:

$$pEx_e = \sum_i pI_i * pEv_i$$

where pEx_e = probability of exposure to a defined event *e* (see below) by a randomly selected individual in an area

pI_i = probability that a randomly selected person is included in a particular cohort *i*

pEv_i = probability that a specified event will be experienced by a particular cohort *i*

Definition of the event, *Ev*, is multidimensional. *Ev* is a specified hourly averaged O_3 concentration level *c* from a monitor located in a particular district *d* for a specified time period T seen by people undertaking a specified activity *a*. For example, it might be the 4:00–5:00 p.m. O_3 concentration seen on July 31, 1991 at the monitor located in district 1 seen by people engaging in outdoor exercise. The analyst must then estimate what proportion of each cohort group_i experiences that event. To obtain pEv_i, the EPEM analyst samples from cohort-specific activity data to estimate the number of person-hours undertaking activity *a* that would be in district 1 at time T, and divides this by the total number of cohort_i person-hours of activity *a* in all districts at time T.

It is easier to obtain pI_i. It simply is:

$$pI_i = \frac{\text{population in cohort}_i}{\text{total population}}$$

It should be noted that a cohort is defined by age/sex/occupation status and by a home-and-work district combination. For example, a cohort could be males, 21–45 years of age that work outdoors, and who live in district 1 and work in district 3. The human activity database described above under pNEM/O_3 is used with local census-provided transportation data to develop the cohort-specific activity file.

To date, EPEM has been utilized only in Houston, TX. The exercise resulted in an estimated probability of 0.015 (1.5%) that a randomly selected Houston resident would experience the MPE monitored in 1988 in that urban area. Multiplying this fraction by the total population of 2.27 million people in Houston gives a total "person-event" MPE O_3 exposure estimate of approximately 34,000 people. This estimate, by the way, is for people outdoors for the *entire* hour at *any* exercise level. If a person was outdoors in the particular district experiencing the MPE for 59 min, he or she would not be counted. Obviously, the entire-hour criterion is stringent; it leads to a systematic underestimation of health risks associated with O_3 exposure.

SUMMARY

This chapter has reviewed the various factors that must be considered when undertaking an O_3 exposure assessment. The only extensive direct O_3 exposure assessment using personal (actually portable) monitors undertaken to date is the 1981 Houston study conducted by the University of Texas School of Public Health. The study showed that personal O_3 exposures often are higher than indoor or outdoor monitors would indicate. Personal exposures greater than the NAAQS level of 0.12 ppm were infrequently observed in the Houston study.

A number of "summer camp" studies were reviewed in this chapter. Fixed monitors in these locations provide a reasonable approximation of personal O_3 exposure to the campers and staff as they go about their daily activities. The 0.12-ppm NAAQS O_3 level was frequently exceeded in various camps both in California and New Jersey. High 8-h exposures also were recorded in these studies.

Indoor microenvironmental monitoring studies were reviewed. Indoor O_3 concentrations greater than 0.12 ppm are infrequently monitored in air conditioned spaces even at high outdoor air exchange rates (>2 Ach). That O_3 level is not experienced often, even in mes having open windows. A major reason for indoor O_3 levels being lower than outdoor levels is that O_3 reacts with many indoor surfaces, including air handling ductwork, and is reduced.

The various factors that go into developing an indirect O_3 exposure assessment were described. These factors include indoor-outdoor ratios, outdoor air exchange rates, and human activity information—including breathing rate data. Typical values for most of these factors were provided. The mass balance equation used to estimate indoor O_3 concentrations was also described.

The final section reviewed four O_3 exposure modeling efforts. Large-scale O_3 exposure modeling has been undertaken in only two U.S. cities to date: Houston and Los Angeles. The modeling efforts indicate that almost everyone living in these two urban areas experience O_3 levels in excess of the U.S. ambient standard level of 0.12 ppm. The residents are also frequently exposed to 8-h concentration levels thought by many health scientists to cause adverse health effects in selected subgroups of the population. Thus, the modeling exercises indicate that people may be adversely affected by high O_3 exposures as they go about their daily lives.

The next chapter describes how exposure results are used to estimate health risks associated with everyday O_3 exposure. The pivotal role that exposure assessments play in estimating "head count" risk becomes quite obvious as the risk analysis work unfolds.

REFERENCES

1. Kleinman, M. T. "Effects of Ozone on Pulmonary Function: The Relationship of Response to Dose," *J. Exp. Anal. Environ. Epi.* 1:309–325 (1991).
2. Lippmann, M. "Health Effects of Ozone: A Critical Review," *J. Air Waste Manage. Assoc.* 39:672–695 (1989).

3. Raizenne, M. E., and J. D. Spengler. "Dosimetric Model of Acute Health Effects of Ozone and Acid Aerosols in Children," in *Atmospheric Ozone Research and Its Policy Implications,* T. Schneider, et al., Eds., (Amsterdam: Elsevier, 1989), pp. 319–329.
4. Ambient Standards Branch. "Review of the National Ambient Air Quality Standards for Ozone: Assessment of Scientific and Technical Information," U.S. EPA Report (1989).
5. McKee, D. "Health Effects of Ozone: Critical Review Discussion Papers," *J. Air Waste Manage. Assoc.* 39:1185–1186 (1989).
6. Lippmann, M., and G. D. Thurston. "Exposure Assessment: Input into Risk Assessment," *Arch. Environ. Health* 43:113–123 (1988).
7. Duan, N. *Application of the Microenvironmental Monitoring Approach to Assess Human Exposure to Carbon Monoxide* (Santa Monica: Rand Corporation, 1985).
8. Duan, N., A. Dobbs, and W. Ott. "Comprehensive Definitions of Exposure and Dose to Environmental Pollution," in *Total Exposure Assessment Methodology* (Pittsburgh: Air & Waste Management Assoc., 1990), pp. 166–195.
9. Penrose, W. R., et al. "A Personal Exposure Monitor for CO and O_3," paper presented at the *Air and Waste Management Assoc. Symposium on the Measurement of Toxic and Related Pollutants,* Durham, NC, May 1991.
10. Avol, E., et al. "Progress in Personal and Area Ozone Sampling Using Solic Reagents," in *Total Exposure Assessment Methodology* (Pittsburgh: Air & Waste Management Assoc., 1990), pp. 408–417.
11. Koutrakis, P., et al. "Measurements of Personal Exposure to Ozone, Acid Aerosols and Gases, and Fine Particulates," in *Total Exposure Assessment Methodology* (Pittsburgh: Air & Waste Management Assoc., 1990), pp. 360–367.
12. Mulik, J. D., et al. "Using Passive Sampling Devices to Measure Selected Air Volatiles for Assessing Ecological Change," paper presented at the *Air and Waste Management Assoc. Symposium on the Measurement of Toxic and Related Pollutants,* Durham, NC, May 1991.
13. Linn, W. S., E. L. Avol, and J. D. Hackney. "Effects of Ambient Oxidant Pollutants on Humans: A Moveable Environmental Chamber Study," in *The Biomedical Effects of Ozone and Related Photochemical Oxidants,* S. D. Lee, et al., Eds., (Princeton: Princeton Scientific, 1982), pp. 125–137.
14. Avol, E. L., et al. "Ozone Exposure and Lung Function in Children Attending a Southern California Summer Camp," paper presented at the 83rd Annual Meeting Air and Waste Management Association, Pittsburgh, June 1990.
15. Avol, E. L., et al. "Ozone Exposure and Lung Function: A Southern California Summer Camp Study," in *Tropospheric Ozone and the Environment*, R. L. Berglund, et al., Eds., (Pittsburgh: Air & Waste Management Assoc., 1991), pp. 90–99.
16. Lioy, P. J., T. A. Vollmuth, and M. Lippmann. "Persistence of Peak Flow Decrement in Children Following Ozone Exposures Exceeding the National Ambient Air Quality Standard," *J. Air Poll. Control Assoc.* 35:1068–1071 (1985).
17. Spektor, D. M., et al. "Effects of Ambient Ozone on Respiratory Function in Active Normal Children," *Am. Rev. Respir. Dis.* 137:313–320 (1988).
18. Spektor, D. M., and M. Lippmann. "Health Effects of Ambient Ozone on Healthy Children at a Summer Camp," in *Tropospheric Ozone and the Environment,* R. L. Berglund, et al., Eds., (Pittsburgh: Air & Waste Management Assoc., 1991), pp. 83–89.
19. Berry, M., et al. "Accumulated Exposure to Ozone and Measurement of Health Effects in Children and Counselors at Two Summer Camps," *Environ. Res.* 54:135–150 (1991).

20. Spektor, D. M., et al. "Effects of Single- and Multiday Ozone Exposure on Respiratory Function in Active Normal Children," *Environ. Res.* 55:107–122 (1991).
21. Stock, T. H., et al. "Exposure Estimates for the Houston Area Asthma and Runners Studies," in *The Biomedical Effects of Ozone and Related Photochemical Oxidants*, S. D. Lee, et al., Eds.,(Princeton: Princeton Scientific, 1982), pp. 539–548.
22. Kotchmar, D. J., et al. "Exposure Estimates for the Houston Area Asthma Study," *Environ. Monit. Assess.* 2:129–138 (1982).
23. Stock, T. H., et al. "The Estimation of Personnel Exposures to Air Pollutants for a Community-Based Study of Health Effects in Asthmatics — Design and Results of Air Monitoring," *J. Air Poll. Control Assoc.* 35:1266–1273 (1985).
24. Contant, C. F., Jr., et al. "Estimation of Individual Ozone Exposures Using Microenvironment Measurements," in *Evaluation of the Scientific Basis for Ozone/Oxidant Standards,* S. D. Lee, Ed. (Pittsburgh: Air Pollution Control Association, 1985), pp. 250–257.
25. Johnson, T. R. "Analysis of Second Data File from Houston Asthmatic Study," (Durham, NC: PEI Associates, July 1989).
26. Weschler, C. J., H. C. Shields, and D. V. Naik. "Indoor Ozone Exposures," *J. Air Poll. Control Assoc.* 39:1562–1568 (1989).
27. Allen, R. J., R. A. Wadden, and E. D. Ross. "Characterization of Potential Indoor Sources of Ozone," *Am. Ind. Hyg. Assoc.* 39:466–471 (1978).
28. Mueller, F. X., L. Loeb, and W. H. Mapes. "Decomposition Rates of Ozone in Living Areas," *Environ. Sci. Technol.* 7:342–346 (1973).
29. Cass, G. R., et al. "Protection of Works of Art from Damage Due to Atmospheric Ozone," *Atmos. Environ.* 25A:441–451 (1991).
30. Druzik, J. R., et al. "The Measurement and Model Predictions of Indoor Ozone Concentrations in Museums," *Atmos. Environ.* 24A:1813–1823 (1990).
31. Allen, R. J., and R. A. Wadden. "Analyses of Indoor Concentrations of Carbon Monoxide and Ozone in an Urban Hospital," *Environ. Res.* 27:136–149 (1982).
32. Sabersky, R. H., D. A. Sinema, and F. H. Shair. "Concentrations, Decay Rates, and Removal of Ozone and Their Relation to Establishing Clean Indoor Air," *Environ. Sci. Technol.* 7:347–353 (1973).
33. Shair, F. H., and K. L. Heitner. "Theoretical Model for Relating Indoor Pollutant Concentrations to Those Outside," *Environ. Sci. Technol.* 8:444–451 (1974).
34. Hales, C. H., A. M. Rollinson, and F. H. Shair. "Experimental Verification of Linear Combination Model for Relating Indoor-Outdoor Pollutant Concentrations," *Environ. Sci. Technol.* 8:452–453 (1974).
35. Thompson, C. R. "Measurement of Total Oxidant Levels at Riverside Community Hospital," *Arch. Environ. Health* 22:514–517 (1971).
36. Thompson, C. R., E. G. Hensel, and G. Kats. "Outdoor-Indoor Levels of Six Air Pollutants," *J. Air Poll. Control Assoc.* 23:881–886 (1973).
37. Davies, T. D., et al. "Indoor/Outdoor Concentrations at a Contemporary Art Gallery," *J. Air Poll. Control Assoc.* 31:135–137 (1984).
38. Peterson, G. A., and R. H. Sabersky. "Measurements of Pollutants Inside an Automobile," *J. Air Poll. Control Assoc.* 25:1028–1032 (1975).
39. Moschandreas, D. J., J. Zabransky, and D. J. Pelton. "Comparison of Indoor and Outdoor Air Quality," (Boston: Geomet, Inc. Report EA-1733, 1981).
40. Lebowitz, M. D., et al. "Indoor-Outdoor Air Pollution, Allergen and Meteorological Monitoring in an Arid Southwest Area," *J. Air Poll. Control Assoc.* 34:1035–1038 (1984).

41. Chan, C.-C., et al. "Commuter's Exposure to Volatile Organic Compounds, Ozone, Carbon Monoxide, and Nitrogen Dioxide," paper presented at the *82nd Annual Meeting of the Air and Waste Management Assoc.,* Anaheim, CA, June 1989.
42. Berk, J. V., et al. "Impact of Energy-Conserving Retrofits on Indoor Air Quality in Residential Housing," paper presented at *74th Annual Meeting of the Air Pollution Control Association,* Philadelphia, June 1981.
43. Nazaroff, W. M., and G. R. Cass. "Mathematical Modeling of Chemically Reactive Pollutants in Indoor Air," *Environ. Sci. Technol.* 20:924–933 (1986).
44. Berk, J. V., et al. "The Effects of Energy-Efficient Ventilation Rates on Indoor Air Quality in an Ohio Elementary School," Lawrence Berkeley Laboratory Report EEB-VENT 80-9 (1980).
45. American Society of Heating, Refrigerating, and Air Conditioning Engineers (ASHRAE). *Ventilation for Acceptable Indoor Air Quality: ASHRAE Standard 62–1989R* (Atlanta: ASHRAE, 1989).
46. Weschler, C. J., H. C. Shields, and D. V. Naik. "Indoor-Outdoor Ozone Measurements:1988–1991," paper presented at the *84th Annual Meeting of the Air and Waste Management Association,* Vancouver, June 1991.
47. Weschler, C. J., H. C. Shields, and D. V. Naik. "Indoor Ozone: Recent Findings," in *Tropospheric Ozone and the Environment II,* R. L. Berglund, Ed. (Pittsburgh: Air and Waste Management Assoc., 1992), pp. 681–700.
48. Committee on Indoor Pollutants, National Research Council, *Indoor Pollutants* (Washington, D.C.: National Academy Press, 1981).
49. Matthews, T. G., et al. "Impact of Heating and Air Conditioning System Operation and Leakage on Ventilation and Intercompartmental Transport: Studies in Unoccupied and Occupied Tennessee Valley Homes," *J. Air Waste Manage. Assoc.* 40:194–198 (1990).
50. Moschandreas, D. J., et al. *Indoor Air Pollution in the Residential Environment,* U.S. EPA Report EPA-600/7-78-229a (1978).
51. Versar, Inc. "Indoor Air Pollution; A Scoping Study," U.S. EPA Report (1987).
52. Yocom, J. E., "Indoor-Outdoor Air Quality Relationships," *J. Air Poll. Control Assoc.* 32:500–520 (1982).
53. Berk, J. V., et al. "The Effects of Energy-Efficient Ventilation Rates on Indoor Air Quality in an Ohio Elementary School," Lawrence Berkeley Lab. Report EEB-VENT 80-9 (1980).
54. Committee on Advances in Assessing Human Exposure to Airborne Pollutants. *Human Exposure Assessment for Airborne Pollutants* (Washington, D.C.: National Academy of Sciences, 1991).
55. Quackenboss, J. J., and M. D. Lebowitz. "The Utility of Time/Activity Data for Exposure Assessment: Summary of Procedures and Research Needs," paper presented at the *82nd Annual Meeting of the Air and Waste Management Association,* Anaheim, June 1989.
56. Robinson, J. P. "Time-Diary Research and Human Exposure Assessment: Some Methodological Considerations," *Amos. Environ.* 22:2085–2092 (1988).
57. Robinson, J. P., et al. "Activity Patterns of California Residents and Their Implications for Potential Exposure to Pollution," California Air Resources Board Report CARB-A6-177-33 (1989).
58. Phillips, T. T., P. L. Jenkins, and E. J. Mulberg. "Children in California: Activity Patterns and Presence of Pollutant Sources," paper presented at the *84th Annual Meeting of the Air and Waste Management Association,* Vancouver, June 1991.

59. Lurmann, F., and S. Colome. "An Assessment of Current and Future Human Exposure to Ozone in the South Coast Air Basin," (Santa Rosa, CA: Somoma Technology, 1991).
60. Koontz, M. D., and J. P. Robinson. "Population Activity Patterns — St. Louis Study," *Environ. Monit. Assess.* 2:197–212 (1982).
61. Schwab, M., A. P. S. Terblanche, and J. D. Spengler. "Self-Reported Exertion Levels on Time/Activity Diaries: Application to Exposure Assessment," *J. Exp. Anal. Environ. Epi.* 1:339–356 (1991).
62. Roth Associates. "A Study of Activity Patterns Among a Group of Los Angeles Asthmatics," (Palo Alto, CA: Electric Power Research Institute Report, 1988).
63. Adair, J. H., and J. D. Spengler. "Time Activity and Exposure Assessment: The Six City Indoor Air Quality Experience," paper presented at the *82nd Annual Meeting of the Air and Waste Management Association,* Anaheim, CA, June 1989.
64. Freeman, N. C. G., J. M. Waldman, and P. J. Lioy. "Design and Evaluation of a Location and Activity Log Used for Assessing Personal Exposure to Air Pollutants," *J. Exp. Anal. Environ. Epi.* 1:327–338 (1991).
65. Schwab, M., et al. "Activity Patterns Applied to Pollutant Exposure Assessment: Data from a Personal Monitoring Study in Los Angeles," *Tox. Indus. Health* 6:517–532 (1990).
66. Schwab, M., et al. "The Time/Activity Component of the Kanawka County Health Study," in *Total Exposure Assessment Methodology* (Pittsburgh: Air and Waste Management Assoc., 1990), pp. 118–129.
67. McMaster, L. R., and D. O. Hinton. "Wearable Personal Activity Logger (PAL)," in *Total Exposure Assessment Methodology* (Pittsburgh: Air and Waste Management Assoc., 1990), pp. 335–344.
68. Moschandres, D. J., and S. Relwani. "The Shadow Sensor: An Electronic Activity Pattern Sensor," *J. Exp. Anal. Environ. Epi.* 1:357–367 (1991).
69. Hartwell, T. D., et al. *A Study of Carbon Monoxide Exposure of Residents of Washington, D.C. and Denver, Colorado*, U.S. EPA Report EPA-600/S4-84-031 (1984).
70. Hartwell, T. D., et al. "A Study of Carbon Monoxide Exposure of the Residents in Washington, DC," paper presented at the *77th Annual Meeting of the Air Pollution Control Association, San Francisco,* June 1984.
71. Akland, C. G., et al. "Measuring Human Exposure to Carbon Monoxide in Washington, D.C. and Denver, Colorado during the Winter of 1982–83," *Environ. Sci. Technol.* 19:911–918 (1985).
72. Johnson, T. *A Study of Personal Exposure to Carbon Monoxide in Denver, Colorado,* U.S. EPA Report 600/S4-84-014 (1984).
73. Johnson, T. "A Study of Personal Exposure to Carbon Monoxide in Denver, Colorado," paper presented at the *77th Annual Meeting of the Air Pollution Control Association, San Francisco,* June 1984.
74. Schwab, M. "The Influence of Daily Activity Patterns on Differential Exposure to Carbon Monoxide Among Social Groups," in *Research Planning Conference on Human Activity Patterns,* T. H. Starks, Ed., EPA Report EPA-600/4-89-004 (1989), pp. 18-1 to 18-21.
75. Thomas, J., and J. V. Behar. "Basic Activity Patterns Structure for Modeling Pollution Exposure," in *Research Planning Conference on Human Activity Patterns,* T. H. Starks, Ed., EPA Report EPA-600/4-89-004 (1989) pp. 4-1 to 4-20.
76. Johnston, T. "A Study of Human Activity in Cincinnati, Ohio," (Palo Alto, CA: Electric Power Research Institute, 1987).

77. Adair, J. H., and J. D. Spengler. "Assessing Activity Patterns for Air Pollution Exposure Research," in *Research Planning Conference on Human Activity Patterns,* T. H. Starks, Ed., EPA Report EPA-600/4-89-004 (1989), pp. 6-1 to 6-19.
78. Ott, W. R. "Human Activity Patterns: A Review of the Literature for Estimating Time Spent Indoors, Outdoors, and in Transit," in *Research Planning Conference on Human Activity Patterns*, T. H. Starks, Ed., EPA Report EPA-600/4-89-004 (1989), pp. 3-1 to 3-38.
79. Stock, T. H., and M. T. Morandi. "A Comparative Evaluation of Self-Reported and Independently-Observed Activity Patterns in an Air Pollution Health Effects Study," in *Research Planning Conference on Human Activity Patterns,* T. H. Starks, Ed., EPA Report EPA-600/4-89-004 (1989), pp. 5-1 to 5-17.
80. Laporte, R. E., H. J. Montoye, and C. J. Caspersen. "Assessment of Physical Activity in Epidemiological Research," *Pub. Health Rep.* 100:131–146 (1985).
81. Terblanche, A. P. S., et al. "Relationship Between Self-Reported Activity Levels and Actual Heart Rates in Teenagers," *J. Air Waste Manage. Assoc.* 41:942–946 (1991).
82. Kizakevich, P. N., W. J. Jochem, and R. L. Beadles. "Ambulatory Monitoring of Cardiac and Respiratory Activity," in *Total Exposure Assessment Methodology* (Pittsburgh: Air and Waste Management Assoc., 1990), pp. 588–596.
83. Johnson, T., L. Wijnberg, and J. Capel. *Review and Evaluation of New Research Relating to Population Exposure to Air Pollution* (Durham, NC: PEI Associates, 1990).
84. Shamoo, D. A., et al. "Improved Quantification of Air Pollution Dose Rates by Improved Estimation of Ventilation Rates," in *Total Exposure Assessment Methodology* (Pittsburgh: Air and Waste Management Assoc., 1990), pp. 553–564.
85. Shamoo, D. A., et al. "Activity Patterns in a Panel of Outdoor Workers Exposed to Oxidant Pollution," *J. Exp. Anal. Environ. Epi.* 1:423–438 (1991).
86. Linn, W. S., D. A. Shamoo, and J. D. Hackney. "Documentation of Activity Patterns in 'High-Risk' Groups Exposed to Ozone in the Los Angeles Area," in *Tropospheric Ozone and the Environment II,* R. L. Berglund, Ed. (Pittsburgh, PA: Air and Waste Management Assoc., 1992), pp. 701–712.
87. Anderson, E., et al. *Development of Statistical Distributions or Ranges of Standard Factors Used in Exposure Assessments* (Chapel Hill, NC: GCA Technology, 1984).
88. Exposure Assessment Group. *Exposure Factors Handbook* (Washington, DC: U.S. Environmental Protection Agency, 1989) EPA-600/8-89-043.
89. Mage, D. T., M. Raizenne, and J. Spengler. "The Assessment of Individual Human Exposures to Ozone in a Health Study," in *Evaluation of the Scientific Basis for Ozone/Oxidant Standards,* S. D. Lee, et al., Ed. (Pittsburgh: Air Pollution Control Association, 1985), pp. 238–249.
90. Lioy, P. J., and R. V. Dyba. "The Dynamics of Human Exposure to Tropospheric Ozone," in *Atmospheric Ozone Research and Its Policy Implications,* T. Schneider, et al., Eds. (Amsterdam: Elsevier, 1989), pp. 711–721.
91. Ott, W. R. "Concepts of Human Exposure to Air Pollution," *Environ. Int.* 7:179–196 (1982).
92. Duan, N. "Stochastic Microenvironmental Models for Air Pollution Exposure," *J. Exp. Anal. Environ. Epi.* 1:235–257 (1991).
93. Ryan, P. B. "An Overview of Human Exposure Modeling," *J. Exp. Anal. Environ. Epi.* 1:453–474 (1991).
94. Sabersky, R. H., D. A. Sinema, and F. H. Shair. "Concentrations, Decay Rates, and Removal of Ozone and Their Relation to Establishing Clean Indoor Air," *Environ. Sci. Technol.* 7:347–353 (1973).

95. Shair, F. H., and K. L. Heitner. "Theoretical Model for Relating Indoor Pollutant Concentrations to Those Outside," *Environ. Sci. Technol.* 8:445–453 (1974).
96. Hayes, S. R. "Estimating the Effect of Being Indoors on Total Human Exposure to Outdoor Air Pollution," *J. Air Poll. Control Assoc.* 39:1453–1461 (1989).
97. Hayes, S. R. "Use of An Indoor Air Quality Model to Estimate Indoor Ozone Levels," *J. Air Waste Manage. Assoc.* 41:161–170 (1991).
98. Paul R. A. et al. *National Estimates of Exposure to Ozone Under Alternative National Standards* (Research Triangle Park, NC: U.S. Environmental Protection Agency, 1986).
99. Johnson, T. R., et al. "Estimation of Ozone Exposure in Houston Using a Probabilistic Version of NEM," paper presented at the *83rd Annual Meeting of the Air and Waste Management Association, Pittsburgh,* June 1990.
100. McCurdy, T., et al. "Preliminary Analyses of Ozone Exposures in Houston Using pNEM/O_3," paper presented at the *84th Annual Meeting of the Air and Waste Management Association,* Vancouver, June 1991.
101. Hayes, S. R., C. Seigneur, and G. W. Lundberg. *Numerical Modeling of Ozone Population Exposure: Application to a Comparison of Alternative Ozone Standards* (San Rafael, CA: Systems Applications, Inc., 1984).
102. Hayes, S. R., and G. W. Lundberg. *Further Improvement and Sensitivity Analysis of an Ozone Population Exposure Model* (San Rafael, CA: Systems Applications, Inc., 1985).
103. Austin, B. S., et al. *Further Development and Application of an Improved Population Exposure Model* (San Rafael, CA: Systems Applications, Inc., 1986).
104. Hayes, S. R., B. S. Austin, and A. S. Rosenbaum. *A Technique for Assessing the Effects of ROG and NO_x Reductions on Acute Ozone Exposure and Health Risk in the South Coast Air Basin* (San Rafael, CA: Systems Applications, Inc., 1988).
105. Hayes, S. R., and A. S. Rosenbaum. *An Examination of Acute Ozone Health Risk and Exposure Effects of Proposed SCAQMD Rule 1109 Using the SOCAB 6–7 June 1985 Ozone Episode* (San Rafael, CA: Systems Applications, Inc., 1988).
106. Winer, A. M., et al. *Characterization of Air Pollution Exposures in the California South Coast Air Basin: Application of a New Regional Human Exposure (REHEX) Model* (Riverside, CA: University of California, 1989).
107. Lurmann, F. W., et al. "Development of a New Regional Human Exposure Model (REHEX) Model and Its Application to the California South Coast Air Basin," paper presented at the *82nd Annual Meeting of the Air Pollution Control Association,* Anaheim, CA, June 1989.
108. Lurmann, F. W., A. M. Winer, and S. D. Colome. "Development and Application of a New Regional Human Exposure (REHEX) Model," in *Total Exposure Assessment Methodology* (Pittsburgh: Air and Waste Management Assoc., 1990), pp. 478–498.
109. Johnson, T. R., et al. "The Use of Activity Diary Data to Estimate the Probability of Exposure to Air Pollution," in *Tropospheric Ozone and the Environment II,* R. L. Berglund Ed. (Pittsburgh: Air and Waste Management Assoc., 1992), pp. 713–724.
110. Biller, W. F., et al. "A General Model for Estimating Exposures Associated with Alternative NAAQS," paper presented at the *74th Annual Meeting of the Air Pollution Control Association,* Philadelphia, June 1981.
111. Lurmann, F. W., S. D. Colome, and H. Hogo. "Modeling Current and Future Human Exposure to Ozone in Southern California," in *Tropspheric Ozone and the Environment II*, R. L. Burglund, Ed. (Pittsburgh, PA: Air and Waste Management Assoc., 1992), pp. 725–744.

CHAPTER 5

HEALTH RISK ASSESSMENT OF OZONE

Ronald G. Whitfield, Harvey M. Richmond, Stan R. Hayes, Arlene S. Rosenbaum, Thomas S. Wallsten, Robert L. Winkler, Mariska J.G. Absil, and Paula Narducci

INTRODUCTION

As discussed in Chapter 3, there is an extensive health effects database reporting a wide variety of physiological and adverse health effects associated with short-term, prolonged or subchronic, and chronic exposures to ozone (O_3). This chapter will focus on the U.S. Environmental Protection Agency's (USEPA) efforts to assess health risks associated with acute (1- to 2-h) and chronic (1- and 10-seasons) exposure to O_3. The current U.S. national ambient air quality standard (NAAQS) for O_3 of 0.12 ppm, with one expected exceedance allowed per year, is largely based on concern for acute respiratory effects that have been reported in many controlled human exposure studies. The acute risk assessment[1,2] addresses these effects and provides risk estimates for two categories of effects (pulmonary function and respiratory symptoms) for individuals engaged in heavy exercise based on results from several 1- and 2-h controlled human exposure studies.[3-5]

Longer-duration (6- to 8-h) controlled human exposure studies[6,7] of lung function and respiratory symptom effects have shown a greater response in moderately exercising healthy subjects exposed to O_3 than the responses elicited by 1- to 2-h exposures. Researchers have observed an inflammatory response and

0-87371-475-X/94/$0.00+$.50
© 1994 by Lewis Publishers

increased airway reactivity in these prolonged exposure studies; the reader is referred to Chapter 3 for a qualitative assessment of these effects.

During the last decade, there also has been growing concern regarding the possibility that repeated or long-term exposure to elevated O_3 levels may be causing or contributing to irreversible chronic lung injury. Numerous animal toxicology studies discussed in Chapter 10 of the Air Quality Criteria for O_3 and Other Photochemical Oxidants[8] (hereafter referred to as the criteria document, or CD) and Chapter 3 of this book indicate that O_3 exposure at sufficiently high concentrations over a prolonged period of time results in a variety of morphological effects. Researchers have reported a characteristic lesion occurring at the junction of the conducting airways and the gaseous exchange tissues (i.e., centriacinar region) in monkeys, rats, mice, and guinea pigs after exposure to elevated levels of O_3. These lung structural changes include distal airway narrowing, remodeling of the centriacinar airways, and thickening of the alveolar septa. Recently, USEPA completed the first phase of the chronic lung injury risk assessment for O_3;[9-11] both quantitative and qualitative judgments of six scientists involved in animal toxicology and human autopsy research on chronic health effects of O_3 were obtained. The results of this assessment, both qualitative and quantitative judgments, are not being used in the ongoing review of the O_3 NAAQS, but provide important insight about the possible magnitude of chronic health effects associated with long-term O_3 exposure.

Assessing the health risks for individuals exposed to criteria pollutants requires assessing the probabilities of certain fractions of these populations suffering well-defined health effects under alternative exposure levels. The probabilities must be assessed in a formal, defensible manner that is open to public scrutiny. They will generally reflect two sources of uncertainty — one deriving from the characteristics of the data and the other from a lack of knowledge. The former type of uncertainty reflects measurement and sampling error, and probabilities are generally calculated by means of standard procedures. Although this type of uncertainty can be reduced through experimental manipulation, it cannot be eliminated completely in any finite study.

The latter type of uncertainty reflects the paucity, incompleteness, and indirect nature of much of the available data. For example, it is frequently necessary to draw inferences about a health effect in a specified population on the basis of epidemiological or clinical data on different populations under varying or different exposure conditions, or on the basis of laboratory data on other species or on *in vitro* preparations. Statistical techniques cannot quantify this type of uncertainty because it is judgmental. Uncertainty can, however, be quantified using appropriate procedures to encode judgmental probabilities. Different experts will assess the uncertainty differently, depending on their faith in the implicit or explicit theories underlying their extrapolations from the data to the effect in question and on their perception of the distance over which the extrapolation must be made. Therefore, the procedures must accommodate and represent a possible divergence of judgments. The level of this type of uncertainty, as well as the degree of divergence among experts, can be reduced.

Early Research on Quantitative Risk Assessment Techniques Sponsored by the Office of Air Quality Planning and Standards

The Office of Air Quality Planning and Standards (OAQPS) first became interested in quantitative risk assessment techniques in the late 1970s. Feagans (OAQPS staff member) and Biller (consultant to OAQPS) described two types of risk models: benchmark risk and headcount risk.[12,13] Soon after this initial work by Feagans and Biller, OAQPS sponsored research outside the USEPA that resulted in a series of reports that described a wide variety of risk assessment approaches.[14] At about the same time, OAQPS also sponsored research involving the stability, coherence, and accuracy of an individual's probabilistic judgments.[15,16] Other research was supported to continue development of ideas presented by individuals at SRI International and Woodward-Clyde Consultants. Carbon monoxide was the subject criteria pollutant. This research resulted in reports by Smith et al.[17] and Keeney et al.[18] The OAQPS decided to pursue the Woodward-Clyde approach for a lead risk assessment that began in 1984.[19,20] O_3 risk assessments that began in 1987, and still continue, also benefited from these efforts.[1,2,9-11]

Motivation

Conducting a risk assessment in which the two types of uncertainty mentioned earlier are accommodated responds to a challenge issued by former USEPA Administrator William Ruckelshaus in a speech at Princeton University:[21]

> ... If I am going to propose controls that may have serious economic and social effects, I need to have some idea of how much confidence should be placed in the estimates of risk that prompted those controls.

In the same talk, he went on to propose some principles for reasonable discussion about risk:

> First, we must insist on risk calculations being expressed as distributions of estimates and not as magic numbers that can be manipulated without regard to what they really mean. We must try to display more realistic estimates of risk to show a range of probabilities. To help do this we need new tools for quantifying and ordering sources of uncertainty and for putting them in perspective.
> Second, we must expose to public scrutiny the assumptions that underlie our analysis in management of risk.

Generally, the information relevant to a particular health effect associated with an environmental agent will be indirect and incomplete, and experts will differ with respect to the associated judgmental uncertainty. The degree of agreement among the experts will be a measure of how firm the probability estimates are and will provide useful information to policy makers and the public.

The probability encoding sessions with each expert can include discussions of the relevant data. These discussions supplement the probability distributions and provide guidance on their use. Some of the risk assessments discussed in this chapter include both probability judgments from the experts and summaries of the associated discussions.

Probability Encoding

It must be emphasized that judgmental probability encoding is unnecessary whenever adequate direct data are available. However, more often than not, such data are unavailable, and the only alternatives to encoding judgmental probabilities are to treat the added uncertainty qualitatively — which currently cannot be done in a rigorous manner, or to ignore it altogether, which is indefensible.

For encoded judgmental probabilities to be useful, they must satisfy two important criteria. The first criterion is that judgments must be obtained from a set of experts who span the range of respected opinion. The problem of establishing the range of opinion and selecting appropriate experts is difficult and judgmental in nature; however, the relevant issues tend to be debated frequently and publicly enough that such decisions are possible. For the chronic O_3 lung injury risk assessment, the USEPA's OAQPS with assistance from the Environmental Criteria and Assessment Office (ECAO) selected the experts whose judgments were to be encoded. Although their names are given in this report, their individual judgments and discussion summaries are identified by code letter only. In this way, users of the risk assessment can decide whether the full range of opinion is represented, and the individual experts can feel free to give their best responses without worrying that they may involve themselves in endless arguments or discussions with those who may disagree.

The second criterion is that an individual's encoded probability judgments must be stable (barring new information), be coherent in a well-defined way, and accurately represent his/her uncertainty. Research relating to these issues has been reviewed by Wallsten and Budescu,[15] and an experiment specifically addressing them in a context similar to the work described here was performed by Wallsten et al.[16]

RISK ASSESSMENT PARADIGM

An exposure model and a health model are essential components of the risk model, as shown in Figure 1. The exposure model accounts for human contact with a specific criteria pollutant.* Pollutants may enter the human body through one or more pathways: inhalation, ingestion, or absorption. The contact may be described either in terms of intensity or a cumulative dose over a specific period of time. The OAQPS strives to combine monitor data (of pollutant concentrations in ambient air) with human activity data to estimate exposures or doses. In the

* The criteria pollutants are ozone, lead, carbon monoxide, sulfur oxides, nitrous oxides, and particulate matter.

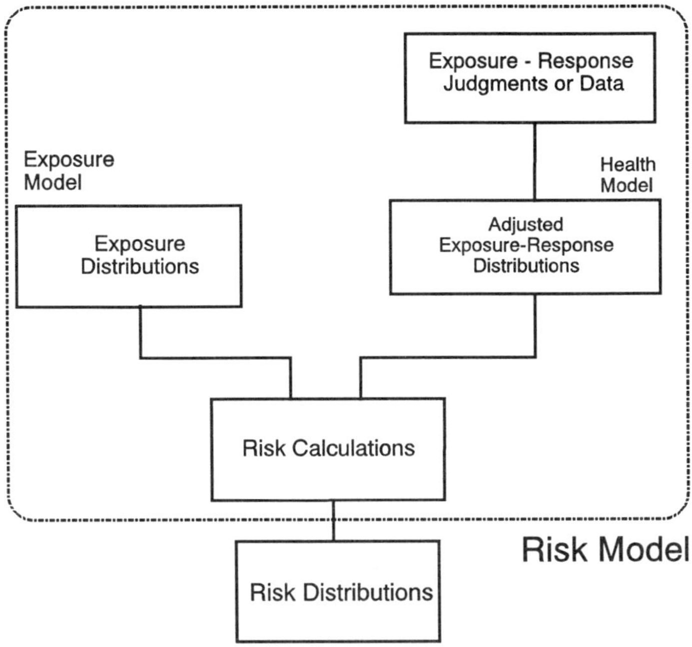

Figure 1. Risk assessment process.

absence of sufficient data to build a model, a hypothetical model can be used to perform risk calculations.

A health model accounts for human reaction to contact with a criteria pollutant. Reactions include symptoms (e.g., coughing or chest pains) or a physiological measure (e.g., change in forced expiratory volume — the volume of air that can be exhaled in 1 s). A health model can be based either on data or judgment. One important aspect of a health model is the "most sensitive group" — people believed to be most at risk either because they are highly reactive (e.g., children whose physiological development may be impaired by exposure to O_3) or more frequently exposed (e.g., outdoor workers). The Clean Air Act requires that NAAQS be set at a level which, in the judgment of the Administrator, will protect the most sensitive population groups with an adequate margin of safety.

Types of Risk Estimates

The OAQPS considers two types of risk models: benchmark risk (a measure of the risk that unhealthful air will occur upon NAAQS attainment) and headcount risk (a measure of the risk posed by the exposures experienced by individuals as they go about their daily lives, moving from place to place both indoors and outdoors). The risk of unhealthful air may be thought of as a hazard or the presence of a source of danger without regard to the number of people who may or do come in contact with it. The risk posed to an individual by personal exposure to O_3 may be thought of as the chance that, having encountered the pollutant, a harmful health event will occur. Here, the pollutant of interest is O_3.

The benchmark risk model calculates the risk (i.e., the probability) that a time-averaged O_3 concentration will exceed a given benchmark concentration k or more times in a given period at some location within a geographic area. A benchmark is the time-averaged O_3 concentration that will cause the occurrence of a specific adverse health effect in up to a given percentage of a sensitive population (e.g., heavy exercisers) under given conditions of exposure.[12] Benchmark risk, which is calculated assuming that all members of the sensitive population are exposed outdoors under identical exposure conditions, is a measure of the hazard posed by elevated ambient O_3 levels. The benchmark risk for a specific standard is the risk associated with exactly attaining the standard.

Headcount risk[12] calculates measures of the risk posed by personal exposure to O_3, taking into account people's lifestyles (e.g., movements among different microenvironments, indoor and outdoor differences, age and occupation, physical activity, and residence-to-work commuting patterns). Headcount risk estimates the chance that, once having encountered O_3 exposures, the health endpoint of concern occurs. Using estimates of population exposure, the headcount risk model estimates the fraction of the sensitive population that may experience the health endpoint of concern. These fractions can be used to obtain expected headcount risk — the expected number (probability weighted) of people experiencing the health endpoint in a given period of time. A variation of the headcount model is to count occurrences of a health endpoint. It is possible that an individual may experience a health endpoint more than once in a given period of time (e.g., chest discomfort or cough).

Full mathematical expressions of these definitions are provided elsewhere.[1,2,12,13] A simplified mathematical expression for the headcount risk model is given later in this chapter.

The next two sections present examples of two types of health models: those that are based on data and those that are based on judgment. Data-based models are preferred over judgment-based models. However, in the absence of adequate data, judgment-based models may be needed to address risk assessment issues.

A HEALTH MODEL THAT IS DATA BASED: ACUTE O_3-INDUCED EFFECTS

Hayes et al.[1] analyzed symptom and lung function data from three controlled human exposure studies.[3–5] Health endpoints addressed in these studies included pulmonary function (decrements in forced expiratory volume or FEV_1) and two symptoms (cough and chest discomfort, each of which were characterized by degree: mild, moderate, and severe). The studies were conducted at O_3 concentrations that varied from 0 to 0.4 ppm and involved varying numbers of subjects. The

Table 1. Fraction of Exposed Individuals Experiencing an FEV_1 Decrement of ≥10%

Avol study		Kulle study		McDonnell study	
Conc.[a] (ppm)	Fraction	Conc. (ppm)	Fraction	Conc. (ppm)	Fraction
0.00	0.00	0.00	0.00	0.00	0.00
0.08	0.02	0.10	0.00	0.12	0.12
0.16	0.30	0.15	0.00	0.18	0.18
0.24	0.67	0.20	0.25	0.24	0.42
0.32	0.80	0.25	0.37	0.30	0.55
		0.30	0.25	0.40	0.58

[a] Conc. denotes ozone concentration.

McDonnell study[5] exposed a total of 135 subjects (divided about equally into six groups), the Kulle study[4] exposed 20 subjects, and 55 subjects participated in the Avol study.[3]

Consider the FEV_1 decrement health endpoint. The studies collected data on FEV_1 (the volume of air, in milliliters, that can be exhaled in 1-s) levels before and after heavy exercise (defined by ventilation rate) in O_3. These effects data were converted into response data by computing the percent change in FEV_1 for each subject (taking exercise into account) and counting the number of people that exceeded a specified FEV_1 change. Table 1 and Figure 2 summarize the observed results (but do not indicate anything about uncertainty). Figure 2 shows that higher response rates were observed in the Avol and McDonnell studies than in the Kulle study. At 0.18 ppm O_3, the McDonnell study observed that 18% of the subjects had FEV_1 decrements ≥10%; that percentage response rate corresponds to a fractional response rate of 0.18.

The "true" response rate is not necessarily 0.18; it could be higher or lower. However, a plausible range around the observed rate can be constructed using the number of subjects involved in each study. A beta distribution (a function of the observed response rate, the number of subjects having FEV_1 ≥10%, and the sample size) gives estimates of this uncertainty and results are shown in Figure 3, which shows central 90% credible intervals around each estimate. Also shown are the median (0.5 fractile) exposure-response relationships.* In a sense, there is a 0.9 probability that the true response rate is within the range indicated by each 90% credible interval. For example, the 90% credible interval around the observed response rate of 0.2 for subjects in the McDonnell study exposed at 0.18 ppm O_3 is 0.03 to 0.42, (i.e., there is only 1 chance in 20 that the true response rate could be higher than 0.42, and a 1 in 20 chance that the true response rate could be lower than 0.03). Similar statements could be made about results of the other studies. Furthermore, a credible interval of any size

* Four-parameter logistic functions were fit to the data to determine curves for each of the fractiles. Only 3 of the fractiles (21 were fit) are indicated.

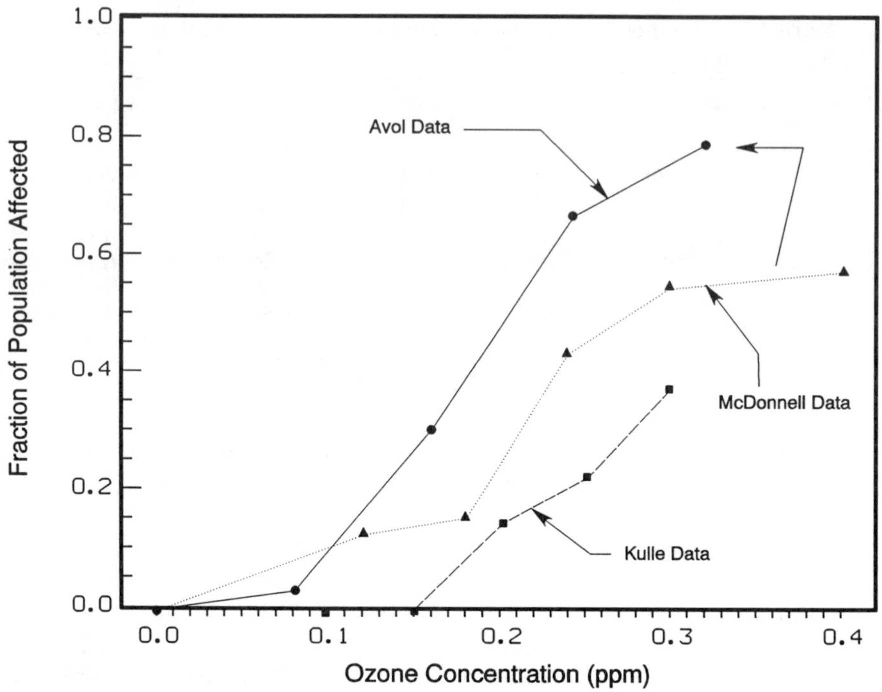

Figure 2. Exposure-response relationships for pulmonary function.

(between 0 and 100%) can be considered. In fact, many credible intervals are calculated and used to estimate risk distributions. Those calculations are discussed later in this chapter.

A HEALTH MODEL THAT IS JUDGMENT BASED: FORMATION OF LESIONS IN THE CENTRIACINAR REGION OF THE HUMAN LUNG

During the recent review of the O_3 NAAQS, the issue was raised as to whether repeated or long-term exposure to O_3 at levels observed in areas such as southern California and the Northeast leads to chronic effects in the form of irreversible lung injury or lung disease. The results of the animal toxicology studies discussed in Chapter 10 of the CD suggest that O_3 exposure at sufficiently high concentrations over a prolonged period results in permanent structural changes to the lung. Among the possible effects of prolonged exposure are narrowing of the distal airways (i.e., thickened respiratory bronchiole walls), remodeling of the centriacinar airways, and thickening of the interalveolar septa resulting in part from increased collagen and epithelial thickening. Inasmuch as there were no direct data available that addressed this issue, the USEPA chose to obtain the judgments of experts.

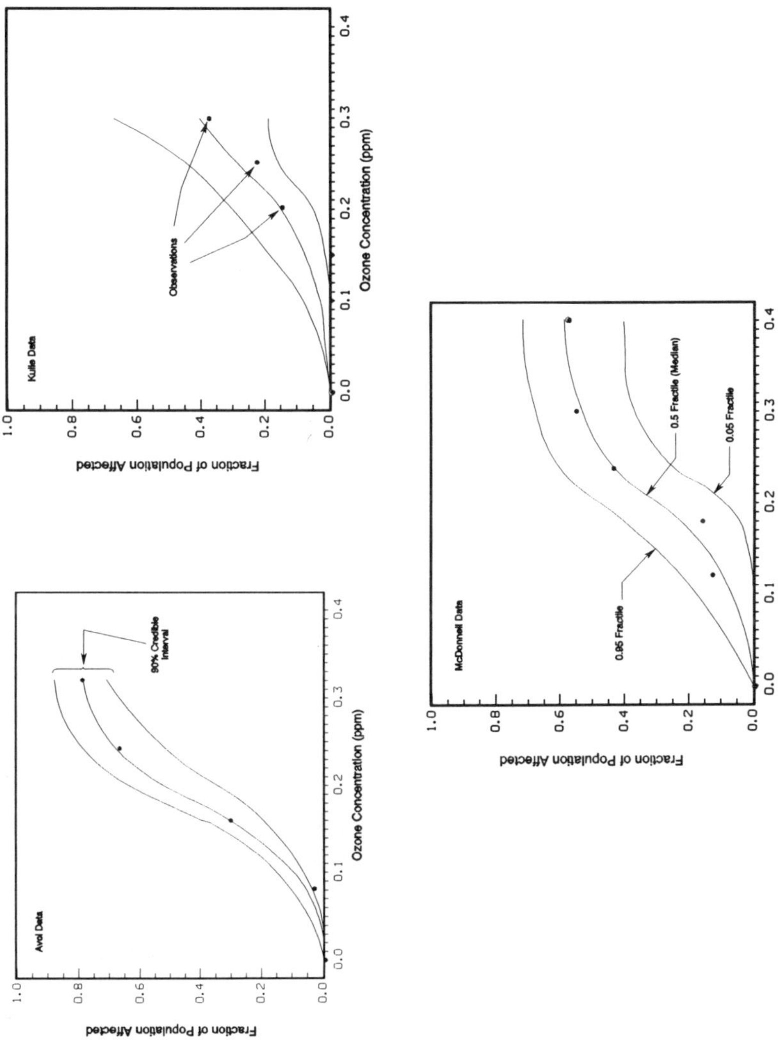

Figure 3. Probabilistic exposure-response relationships for FEV_1 decrement based on three separate controlled human exposure studies.

For the assessment conducted in 1990 and 1991,[9] USEPA chose mild and moderate lesions in the centriacinar region of the lung as the health endpoint for which judgments about probabilistic exposure-response relationships were to be elicited. The centriacinar region of the lung has been shown to be the location of maximum O_3 dose and maximum lung injury in a number of species (i.e., rats, guinea pigs, and monkeys) exposed to elevated O_3 concentrations.[8] The health scientists consulted in the early phases of this assessment agreed that centriacinar lesions are a good indicator of potential chronic lung injury caused by O_3 exposure. For this assessment, the USEPA selected children aged 6 through 18 years and all adult outdoor workers in two urban areas — New York City and Los Angeles — as the populations of interest. These two groups were thought to be at greater risk for chronic effects because their members tend to be outdoors more than most individuals when O_3 concentrations are at their highest. Only results for children in New York City are shown here.

If scientific judgments regarding the risk of lung injury due to long-term O_3 exposure are to be assessed accurately, a carefully designed protocol is necessary. Thus, after appropriate background material and supporting information were compiled, a protocol for encoding the judgments of experts (Appendix A in Reference 9) was developed. The protocol was then pilot tested and modified, and the experts to be interviewed were chosen. Initially, two or three members of the risk assessment team met individually with each expert to encode his or her judgments. Feedback was later provided to the experts. Next, the risk assessment team and the experts met in a workshop format to discuss the quantitative risk assessments and the underlying qualitative factors. After the workshop, members of the risk assessment team again met with each individual to finalize his or her probability judgments. This process is described more fully in the following sections.

Development of the Encoding Protocol

A carefully developed encoding protocol was crucial to the success of the project. It ensured that the questions for the experts were always phrased identically, that specific assumptions and definitions were always the same, and that the encoding process proceeded similarly for each of the participants. Consequently, the differences in probability judgments could be attributed to true differences of opinion and not to differences in assumptions, understandings, or procedures.

In developing the encoding protocol, the assembled background material was used and the assessment team consulted extensively with individuals (named in the acknowledgments in Reference 9) having expertise in different aspects of the problem. Of particular concern were the appropriate health endpoints to consider and the best way to represent exposure conditions. However, many other points also required careful attention. Dealing with chronic rather than acute effects greatly increased the number of factors that had to be considered.

On the basis of the literature and discussions with consultants, a draft of the encoding protocol was prepared. At various points, the draft was circulated to some or all of the participating consultants. Written comments were received in some cases; meetings were held in others. In this manner, several drafts of the protocol were generated and revised.

After several revisions, the protocol was pilot tested with a USEPA staff member and a USEPA contractor. Some weaknesses not previously recognized became apparent during the pilot testing, and the protocol was further revised to correct them. Consultations and discussions among members of the risk assessment team led to some final fine-tuning to create the protocol that was actually used in the encoding process. In the following subsections, each part of the protocol is briefly discussed. The focus of these discussions is on the role played by that particular part of the protocol and on questions that arose during its development.

Brief Summary of the Encoding Protocol

Introduction This section of the protocol presents the motivation for and objectives of the encoding process.

Centriacinar Lesions In this part of the protocol, the health endpoints of interest are defined as mild and moderate centriacinar lesions. The results of animal studies suggest that long-term O_3 exposure can lead to numerous structural changes in the lung, such as narrowing of distal airways, remodeling of the centriacinar airways, and thickening of the interalveolar septa. Each of these health effects is important, and all are correlated. However, it is only when one of them is precisely defined that the concept of a dose-response function can be applied to it. Even then, it is not always clear what level of the endpoint should be considered adverse. It was therefore necessary for the encoding process to select an endpoint for which mild and moderate levels could be precisely defined. The mild level was at the lower end of what might be considered adverse, and the moderate level was somewhat farther along on the continuum.

After much consultation, it was decided to focus on centriacinar lesions, with the two levels defined in terms of the techniques required to detect them. In the protocol, mild centriacinar lesions are defined as "those detectable only by highly sensitive techniques and quantitative analysis" (e.g., morphometric techniques, special stains, and electron microscopy).[9] Moderate centriacinar lesions are defined as "those detectable by qualitative assessment using light microscopy but not necessarily gross techniques".[9] The need for and importance of extrapolation from animal results to humans are emphasized in this section.

Exposure Conditions To deal with chronic effects, it was decided to consider centriacinar damage over a single O_3 season and also over 10 O_3 seasons. Specifying exposure conditions within a season required extensive consultation because of variations in exposure pattern over time. Ambient O_3 concentrations vary diurnally, over days, and throughout the season. Moveover, an individual's personal exposure depends on his/her location and activity level at any point in time. The challenge was to represent typical exposure variations within a season in an easily comprehensible manner while specifying exposure levels in a form suitable for discussing dose-response functions. This task was particularly important because experts differ as to the O_3 exposure characteristics they deem instrumental in causing adverse health effects. It was important not to omit or misrepresent exposure information that might be deemed crucial.

The solution was to specify seasonal mean daily maximum 8-h exposure values for individuals of 0.025, 0.050, 0.075, and 0.100 ppm. For each seasonal mean and for each of two populations in two geographic locations, associated data were provided on the distributions of these personal exposures, on the distributions of personal and ambient daily peak exposures, and on other characteristics of the exposure patterns. As described earlier, these data were derived to be as representative as possible of the actual range of conditions associated with a given seasonal mean value for each population of interest in each geographic location. The histograms illustrating patterns of exposure are explained in this part of the protocol, as are the dosimetry results. The assumptions made regarding exposure history, ambient conditions, and other air pollutants are also stated.

Population at Risk The populations considered for the encoding are all children aged 6 through 18 years and all adult outdoor workers in Los Angeles and New York City. These populations include people with and without respiratory problems and allergies, smokers and nonsmokers, and people exhibiting different levels of activity. It was decided not to stratify the populations along any of these dimensions, although the possibility that one or more experts might wish to do so was not ruled out. (As it turned out, none did.) Therefore, the experts had to assume distributions for these and any other factors relevant to their exposure-response judgments in the populations of interest.

Factors to Consider This part of the protocol serves to remind the experts of factors that may be relevant to their judgments: results reported in the literature, issues concerning extrapolation from animal studies, stages of response, time scales of effects, levels of repair, effects of varying the exposure concentrations over time (e.g., relative importance of cumulative exposure vs. peak exposure), and possible interactions of O_3 with other air pollutants. The experts are asked about their qualitative judgments concerning these factors.

Health Significance This brief part of the protocol simply sets the stage for an exploration of the experts' opinions about the long- and short-term health consequences of lung damage resulting from long-term O_3 exposure. Important questions include how significant the centriacinar lesions are in themselves and what implications they might have for a person's future health.

Factors to Keep in Mind When Making Probability Judgments The purpose of this part of the protocol is to help the experts provide probabilities that accurately represent their judgments. This objective is primarily accomplished through review of some commonly found cognitive biases and simplifying heuristics and through some suggestions on how to try to mitigate their effects. This material is not unique to this protocol, having been used in previous risk assessment protocols. The topics covered include (1) effects related to the order in which evidence is considered or probabilities assessed, (2) the effect of memory on judgment, (3) the tendency to overestimate the reliability of information, (4) the influence of the importance of events on probability judgments, and (5) the difficulty of discriminating between levels of uncertainty.

Session Two: Discussion of Judgments, Health Significance, and Research Needs

This part of the protocol simply notes the plan for a meeting at which the experts are to exchange judgments — both quantitative and qualitative.

EXPERTS

The judgments of the following experts were encoded:

Dr. Daniel Costa — U.S. Environmental Protection Agency
Dr. James Crapo — Duke University
Dr. Charles Plopper — University of California, Davis
Dr. Russell L. Sherwin — University of Southern California
Dr. Walter Tyler — University of California, Davis
Dr. Elaine Wright — General Motors Corporation

The experts were chosen on the basis of their knowledge and familiarity with the health effects of O_3 exposure, as well as their various viewpoints and approaches. The experts were not, and should not have been, chosen randomly; rather, they were selected for their expertise and because, as a group, they represent the range of respected scientific opinion on the questions of interest.

ENCODING PROCESS

Exposure-Response Uncertainties

The probability judgments of the experts regarding 16 exposure-response functions under particular exposure conditions were encoded. The functions represented all possible combinations of severity of lesions (mild or moderate), location (Los Angeles or New York), population (child or adult outdoor worker), and number of seasons of O_3 exposure (1 or 10). The encoded probabilities were to represent each expert's best judgment regarding what the population response rate (i.e., proportion of the population showing the adverse health effect) would be at each of four O_3 levels (i.e., seasonal mean daily maximum 8-h averages of 0.025, 0.050, 0.075, and 0.100 ppm) under the specified conditions.

Specifically, for a given function and exposure level, the expert was to judge the chances of various response rates occurring, taking into consideration the exposure pattern and all other relevant factors. By means of a series of questions to be described later, these judgments were translated into a probability distribution over a response rate. Thus, for each expert and each exposure-response function, four probability distributions over the population response rate were encoded, one at each exposure level.

First Round of Encodings

For the first round of encodings, each of the six experts was visited by two or three members of the risk assessment team. One pair from the team encoded the judgments of three of the experts, and three other team members encoded those of the remaining experts. The time required for these encoding sessions ranged from half a day to a full day. The experts were visited between July 26, 1989 and August 30, 1989.

The basic structure of the six encoding sessions was the same. The experts had received the background material, the dosimetry results, the exposure results, and the protocol prior to the meeting and had been asked to look over these materials. The encoding session began by going through the protocol with the expert and making sure he/she understood the definitions and assumptions. During this review, questions were asked and qualitative judgments obtained about various aspects of the chronic health effects of O_3 exposure, with particular emphasis on the factors to consider and the health consequences. This format encouraged the expert to think about issues highly relevant to the later encoding of probability judgments and also provided valuable qualitative information. Attention was also paid to the dosimetry results and the exposure results. This part of the session lasted from 2 to 4 hours.

Probability judgments were then obtained for 16 exposure–response functions. These functions were considered in pseudo-random order. For each one, the four mean daily maximum 8-h average concentrations were considered, also in pseudo-random order. For each function and each mean exposure level, the expert's probability distribution for response rate (i.e., percentage of the population developing O_3-induced lesions) was obtained.

Typically, the probability encoding process started slowly and then moved along more quickly as the expert became more comfortable with the encoding process and with the notion of expressing judgments in probabilistic form. To help the expert understand the questions, they were phrased in more than one way (e.g., probabilities vs. odds and probability above vs. below a given value). A probability wheel was used as an aid. The wheel is divided into two radial and adjustable segments of different colors. One of the two colors is adjacent to a pointer. The proportion of the wheel's area devoted to one color can be varied continuously from zero to one. For example, the expert was asked, for a given wheel setting, which is more likely: (1) the wheel, if spun, will stop with the pointer on blue, or (2) less than 30% of the population will develop mild lesions at a certain mean exposure level. The relative proportions of the two colors on the wheel are adjusted until the expert can no longer judge either (1) or (2) as more likely. Most of the six experts relied heavily on the wheel; several preferred to simply state their probabilities.

For each function and each mean exposure level, probabilities for several response rates were encoded. After this was done for all four mean exposure levels for a given case, the probabilities were plotted on a graph and shown to the expert. Any inconsistencies were pointed out. For example, there might have been a higher probability for a given response rate at a lower exposure level than at a higher exposure level. Actually, few such inconsistencies were encountered. At this juncture, the implications of the probabilities were pointed out and the expert was asked (regardless of whether there had been any inconsistencies) if the set of probabilities adequately represented his or her judgments. Changes were made, if necessary. Throughout the process, the experts were encouraged to reconsider earlier probability judgments and to change them if they no longer seemed appropriate.

The experts sometimes went through the entire encoding procedure for each new function. However, they sometimes simply used their judgments from a previous case. For example, an expert might judge that the probabilities should be the same for outdoor workers in New York and Los Angeles. In other cases, they simply adjusted previously encoded probabilities from a related function.

When encoding was completed for all cases, the expert was once again offered the opportunity to reconsider his or her judgments. Each expert was then told that further opportunities for modifications would be available when the judgments were sent to the expert and later when the second round of encoding took place.

Chronic Risk Assessment Workshop

On September 16, 1989, the six experts, the risk assessment team, and USEPA staff members met for a workshop. The entire day was devoted to discussions of qualitative and quantitative judgments regarding chronic health effects of long-term exposure to O_3. Workshop participants again went over the issues covered in the protocol and discussed during the first round of encodings. The interactions among the experts added new elements to the picture.

Discussions of the qualitative judgments covered the following topics: definition of health endpoints, nature and significance of health effects, definition of exposure scenarios, dosimetry modeling, sequence of response, time scale of effects, effect of varying concentrations over time, degree and role of repair, interaction with other air pollutants, and discriminatory variables for encoding. The role of the risk assessment team and USEPA staff was to facilitate the discussion. The experts did most of the talking, with all six experts actively participating. Positions were stated and debated, and relevant research too new to be widely available was introduced. There was considerable give-and-take among the experts.

After the qualitative discussions, the experts were given feedback regarding the probability judgments made during the first round of encoding. They were shown the probability judgments of all six experts, but anonymity was maintained. In the presentation, the experts were labeled A, B, C, D, E, and F, and not by name. Each knew only his or her code letter. A general discussion of the probability judgments ensued, with some emphasis on differences among experts. A discussion of important research needs followed.

At the end of the afternoon session, the experts were asked to give their reactions to the workshop itself. They were unanimous in stating that it had been very informative and valuable. One expert claimed that the encoding process and workshop had led to an entirely new way of thinking about research and the adverse effects associated with exposure to O_3.

Second Round of Encodings

The following morning, the second round of encodings took place, with five of the six experts participating. The sixth expert was unavailable that day but subsequently went through the same process. In the second round, members of the risk assessment team met individually with the experts. The probability judgments encoded during the first round were reviewed. The experts were asked to reconsider their judgments in light of what had happened at the workshop. It was emphasized that the intent was to obtain their current judgments and not to force consensus. These sessions proceeded quickly, lasting from 1 to 4 hours.

RESULTS

This section presents both quantitative and qualitative results.* The quantitative results are the judgments of experts about the probability of occurrence of O_3-induced lesions in the centriacinar region of the human lung. These results are presented graphically in two ways: (1) probability distributions over response rate (conditional on exposure level and other factors) and (2) exposure-response relationships. Initially, the judgments of Expert A are presented for illustrative purposes; the judgments of all the experts are then displayed graphically. The qualitative results include explanations for and implications of the judgments made and discussions held at the encoding sessions and workshop.

Quantitative Results

Probability Judgments of Expert A

Judgments were obtained from the six experts about the probability of occurrence of O_3-induced lesions in the centriacinar region of the lung. Probability encoding techniques were used to obtain these judgments in the form of probability distributions over response rate (conditional on exposure level and other factors). Other factors included population (child or adult outdoor worker),

---

* Complete results are reported in Reference 9.

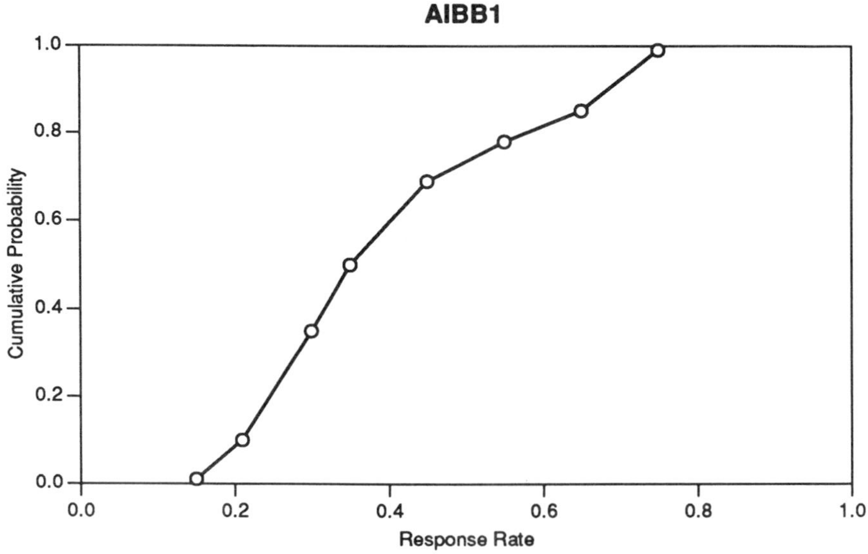

Figure 4. Probability distribution over response rate for the occurrence of lesions in the centriacinar region of the lungs of children at a 0.025-ppm O_3 concentration.

geographic area (Los Angeles or New York), number of exposure seasons (1 or 10), and degree of lesion (mild or moderate).

For example, Figure 4 shows the conditional probability distribution over response rate obtained from Expert A for the occurrence of mild lesions in the centriacinar region of the lung among children living in New York City exposed for one season to O_3 concentrations with a mean daily maximum 8-h average of 0.025 ppm at the mouth. Probabilities were elicited from Expert A for eight possible response rates at the mean concentration of 0.025 ppm under the specified conditions. Adjacent points are connected by straight lines. The set of points and straight lines is referred to as a curve. In Expert A's judgment, the possible response rate varies from 0.15 (with cumulative probability 0.01) to 0.75 (with cumulative probability 0.99). The median response rate (i.e., the rate associated with cumulative probability 0.5) is 0.35. The 0.15 and 0.75 response rates constitute Expert A's 98% credible interval (CI), an interval that (in Expert A's judgment) includes the "true" response rate with probability 0.98. In other words, Expert A believes there is a 50-50 chance that less than 35% of the population will develop O_3-induced lesions for this exposure regimen. Expert A also judges the probability to be 0.98 that between 15 and 75% of the population will experience the effect.

Probability judgments were also elicited from Expert A at mean O_3 concentrations of 0.050, 0.075, and 0.100 ppm for the specified conditions. These judgments and those for 0.025 ppm are compiled in Figure 5. Each of the four curves is a set of points connected by straight lines. To help distinguish between the curves, different symbols are used to represent the encoded points on each curve,

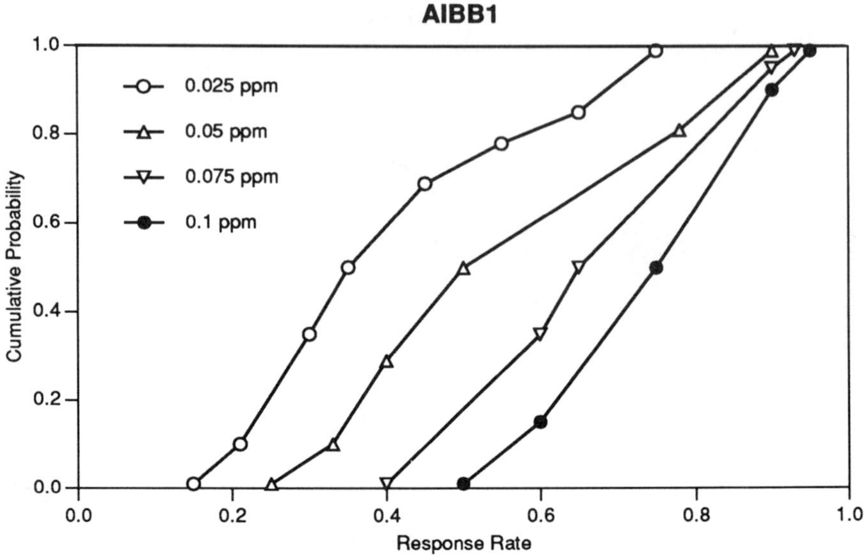

Figure 5. Probability distributions over response rate for the occurrence of lesions in the centriacinar region of the lungs of children at 0.025, 0.050, 0.075, and 0.100-ppm O_3 concentrations.

and different styles are used for the line segments that connect adjacent points. Note that the curves do not intersect and that they are monotonic increasing. This monotonic-increasing property is a necessary condition for a function to be a valid cumulative probability distribution. The nonintersecting property is important from a logical point of view; the intersection of two curves would imply that response rates above the intersection are more likely at the lower than at the higher concentration, and that the converse is true for response rates below the intersection. Also note that the 0.050-ppm curve lies to the right of the 0.025-ppm curve, that the 0.075-curve lies to the right of the 0.050-ppm curve, and so on. This progression is logical because the response rate is expected to increase as the exposure level (i.e., O_3 concentration) increases.

If an expert's assessments violated any of these conditions, he/she was alerted and an attempt was made to resolve the "inconsistency". Thus, in a sense, the expert was "forced" to provide "consistent" probability judgments. The judgments of all the experts are tabulated in Appendix C of Reference 9. Judgments about children and New York City are summarized in the next two subsections.

Exposure-Response Relationships for Expert A

As shown in Figure 5, the median judged response rates at O_3 concentrations of 0.025, 0.050, 0.075, and 0.100 ppm are 0.35, 0.50, 0.65, and 0.75, respectively, for Expert A. These O_3-concentration and response-rate pairs define the median exposure-response relationship shown in Figure 6. An additional point is shown

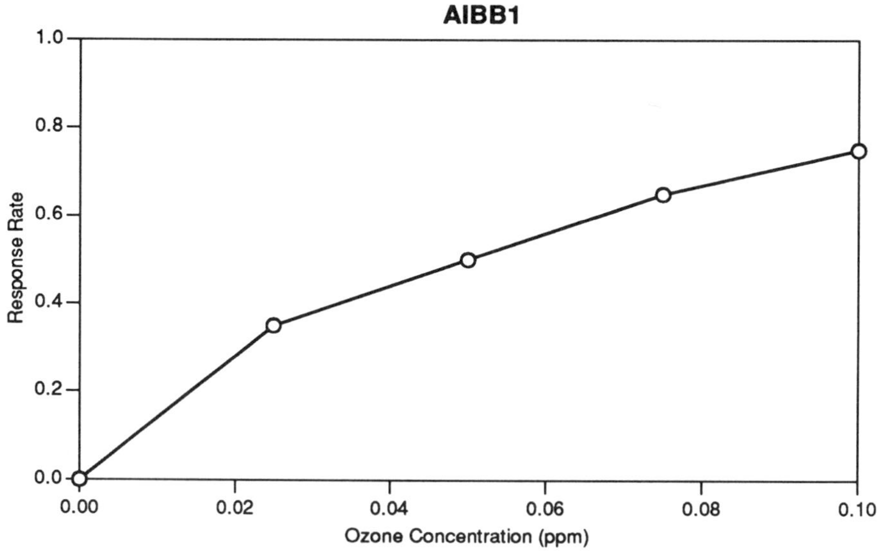

Figure 6. Median exposure-response relationship for Expert A.

at the origin of the graph because no O_3-induced lesions will, by definition, occur at an O_3 concentration of 0 ppm.

Response rates corresponding to cumulative probabilities of 0.05 and 0.95 can be determined from Figure 5 by interpolation. For example, for the 0.025-ppm curve, a response rate of 0.18 corresponds to a cumulative probability of 0.05 (called the 0.05 fractile), and a response rate of 0.71 corresponds to the 0.95 fractile. Thus, the response rates of 0.18 and 0.71 define the central 90% CI at 0.025 ppm for Expert A. In a similar fashion, the 0.05 and 0.95 fractiles at the other O_3 concentrations can be determined and the results drawn as a series of exposure-response relationships like those shown in Figure 7. These curves are monotonic increasing (because the response increases as the exposure increases) and converge at the origin.

Curves like those shown in Figure 7 depict the uncertainty in Expert A's judgments about the relationship between O_3 exposure and the occurrence of O_3-induced lesions in the centriacinar region of the lungs of children living in New York City. If Expert A were certain about this relationship, only one curve would be needed to completely characterize this belief. Since Expert A happens to be uncertain about this relationship, curves at any fractiles of interest could be shown; for simplicity, curves only for the 0.05, 0.5, and 0.95 fractiles are presented.

Recall that the judgments of six experts about the relationship between O_3 exposure and the occurrence of lesions in the centriacinar region of the lung were elicited. Judgments were elicited for conditions determined by all combinations of the occurrence of mild and moderate lesions, the locations of New York and Los

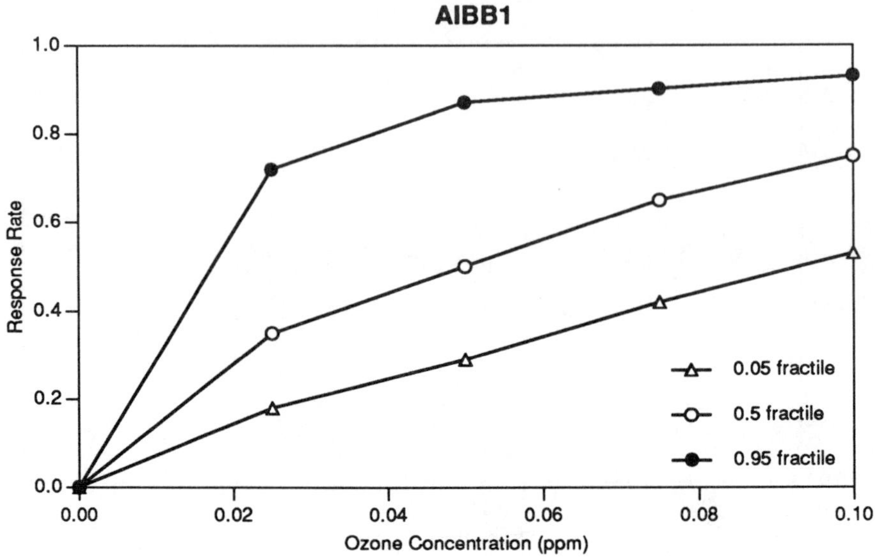

Figure 7. 0.05-Fractile, median, and 0.95 fractile exposure-response relationships for Expert A.

Angeles, the populations of children and outdoor workers, and exposure for 1 and 10 O_3 seasons. The possible combinations of lesion type, city, population of interest, and number of O_3 seasons define 16 cases. To make it easier to refer to a particular case, the following four- or five-character code is used. The format for the code is lcps, where:

- l denotes lesion type (I for mIld or O for mOderate)
- c denotes city (L for Los Angeles, N for New York, or B for Both)
- p denotes population (C for Children, W for Workers, or B for Both)
- s denotes the number of O_3 seasons (1, 10, or B for Both)

For example, INC1 denotes the case for mild lesions, New York City, children, and one O_3 season.

For each case, judgments were sought about four O_3 concentrations: 0.025, 0.050, 0.075, and 0.100 ppm. Thus, there was the possibility of having to encode 64 probability distributions for each expert. However, the experts believed, to varying degrees, that the information was insufficient to distinguish all 64 cases. For example, Expert A did not distinguish between children and outdoor workers, or between Los Angeles and New York City. This pattern held for the formation of mild and moderate lesions and for exposure to O_3 for 1 and 10 seasons. Thus, only four sets of judgments were provided by Expert A, with each set of judgments applying to four cases (i.e., one for each combination of lesion type and number of O_3 seasons). Inasmuch as other experts shared similar beliefs to varying degrees, there is a total of 36 sets of judgments for the group of 6 experts, instead of the largest possible number of 96 sets.

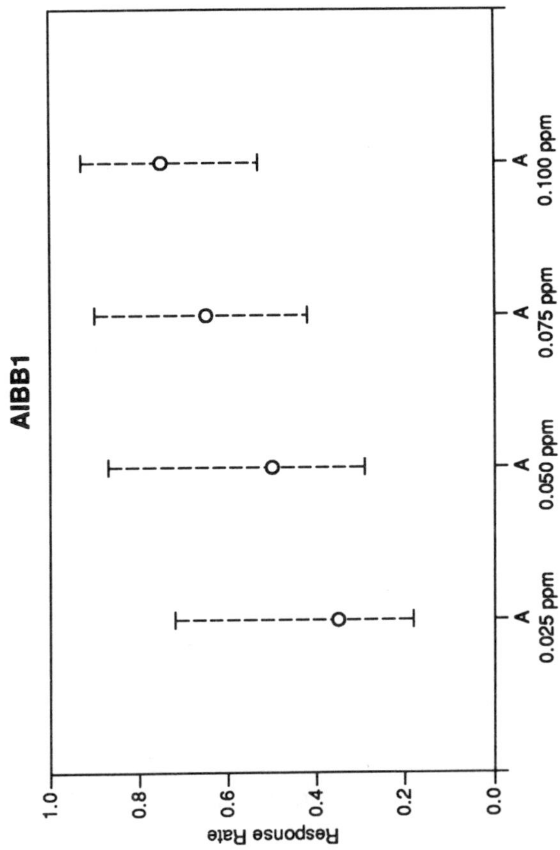

Figure 8. Median and 90% CI judgments of Expert A for mild lesions, Los Angeles and New York, children and outdoor workers, and one O_3 season.

Before presenting the results for each of the experts, readers should view the content of Figure 7 in the format shown in Figure 8. Here, the median and 90% CI are not indicated by "curves". Instead, for each exposure level, circles denote the median judgments, and vertical bars above and below each circle indicate the central 90% CI. Since Expert A does not distinguish between children and outdoor workers, or between Los Angeles and New York, Figure 5 summarizes all of Expert A's judgments about the incidence of mild lesions after one season of O_3 exposure. The notation AIBB1 indicates these facts by containing the letter B (for both) in the positions for population and location.

The set of four median and 90% CI judgments by Expert A is presented in Figure 9. The upper-left panel, which is the case for mild lesions and 1 O_3 season, reproduces Figure 8. Clockwise, the other panels are for mild lesions and 10 O_3 seasons, moderate lesions and 10 O_3 seasons, and moderate lesions and 1 O_3 season, respectively.

Exposure-Response Relationships for All Experts

Results similar to those for Expert A were developed for the other experts. Figures 10 through 13 summarize those results in a way that makes it easier to make comparisons across experts. Figure 10 represents the four combinations of mild vs. moderate lesions and 1 vs. 10 O_3 seasons for children in Los Angeles (cases 1, 2, 9, and 10 in Table 1). The two top panels are for mild lesions, and the two bottom panels are for moderate lesions; the two left-hand panels are for 1 O_3 season, and the two right-hand panels are for 10 O_3 seasons. This figure provides insight into the similarities and differences across experts in their judgments of central tendency (median responses) and in the uncertainty in their judgments (90% CIs). The results for the experts are grouped according to mean O_3 level along the horizontal axis. A symbol (circle, square, or diamond; filled or not filled) denotes the median (0.5 fractile) response elicited from an expert. Vertical bars (of different styles to help distinguish between the experts) above and below the median symbol denote the 0.95- and 0.05-fractile responses elicited. These vertical bars denote the 90% CI about the median. For example, recall from the previous discussion of Expert A's judgments that for case 1 at 0.025 ppm, the median response rate is 0.35 and the 90% CI is 0.18 to 0.71. Those values are indicated in Figure 10 in the area for 0.025 ppm and Expert A. In the figure, the case (ILC1) is indicated above the graph. The legend indicates the set of judgments that applies to the case for each expert. These are AIBB1 for Expert A, BIBC1 for Expert B, and so forth.

The discussion of the quantitative and qualitative judgments of the experts in Reference 9 is organized into the following nine topics: mild vs. moderate lesions, children vs. outdoor workers, Los Angeles vs. New York, 1 O_3 season vs. 10 O_3 seasons, effects of O_3 concentrations, medians, degree of uncertainty, degree of overlap in the 90% CIs, and existence of O_3-induced lesions. The following two subsections provide brief summaries of the quantitative and qualitative results.

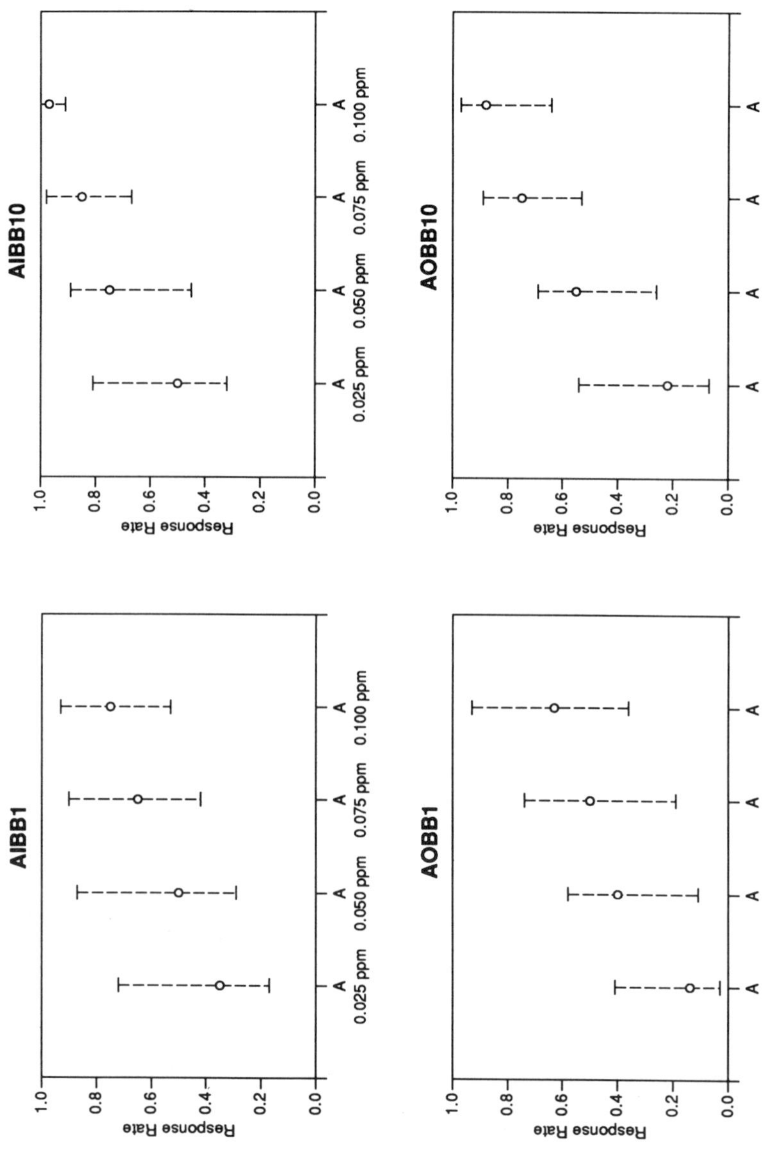

Figure 9. Median and 90% credible interval judgments of Expert A.

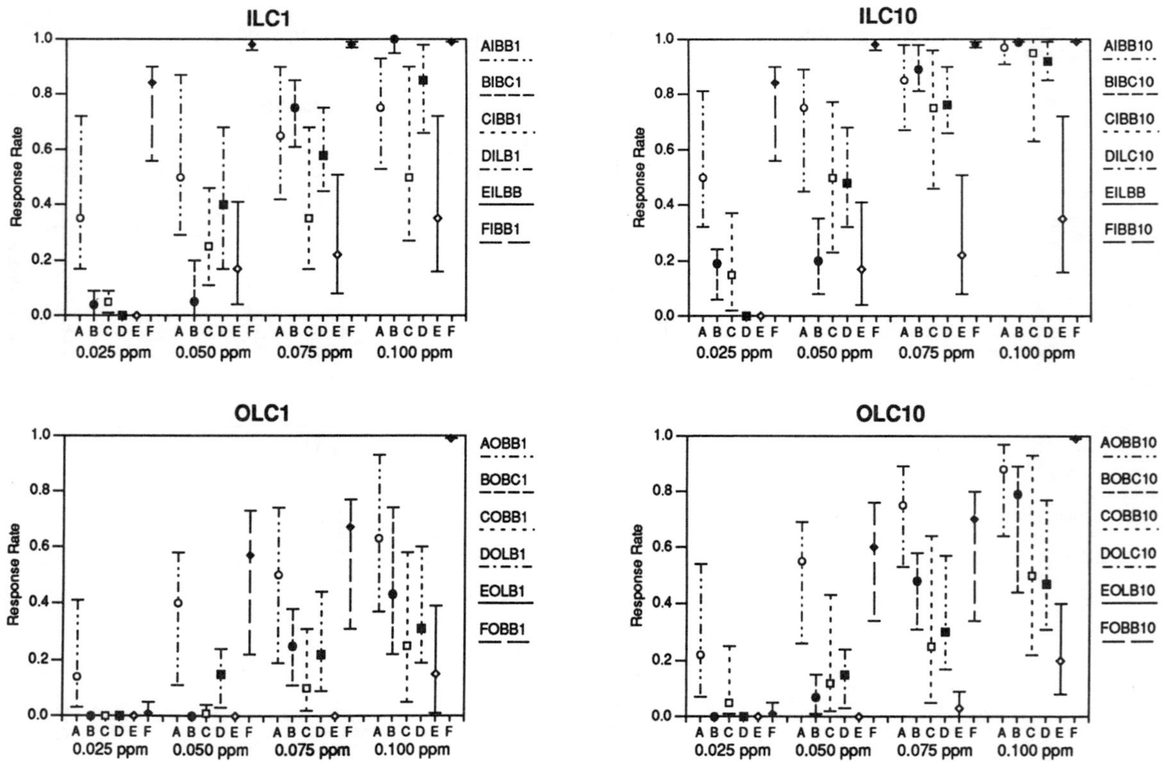

Figure 10. Medians and 90% credible intervals for judgments about mild and moderate lesions, children, Los Angeles, 1 and 10 seasons (summarized are all the judgments for Experts A, C, and F, who did not distinguish populations or locations; differences for Experts B, D, and E as a function of these factors are shown in Figures 11, 12, and 13).

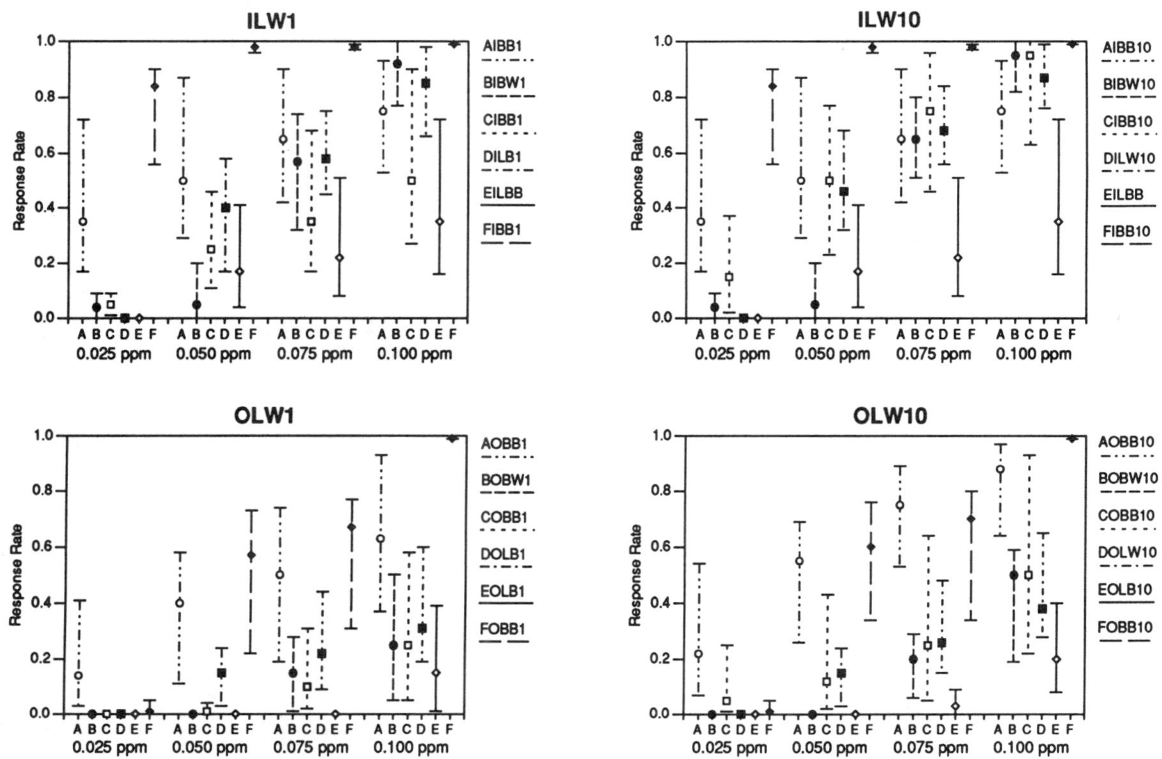

Figure 11. Medians and 90% credible intervals for judgments about mild and moderate lesions, outdoor workers, Los Angeles, 1 and 10 seasons.

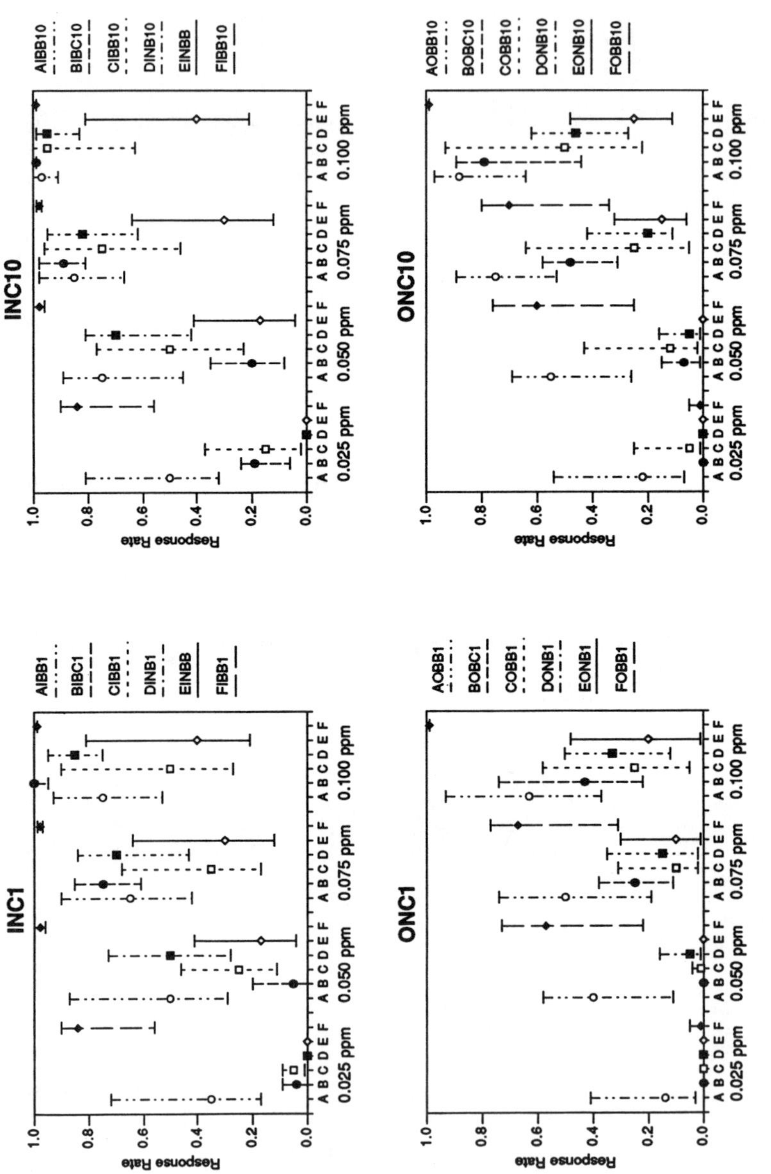

Figure 12. Medians and 90% credible intervals for judgments about mild and moderate lesions, children, New York, 1 and 10 seasons.

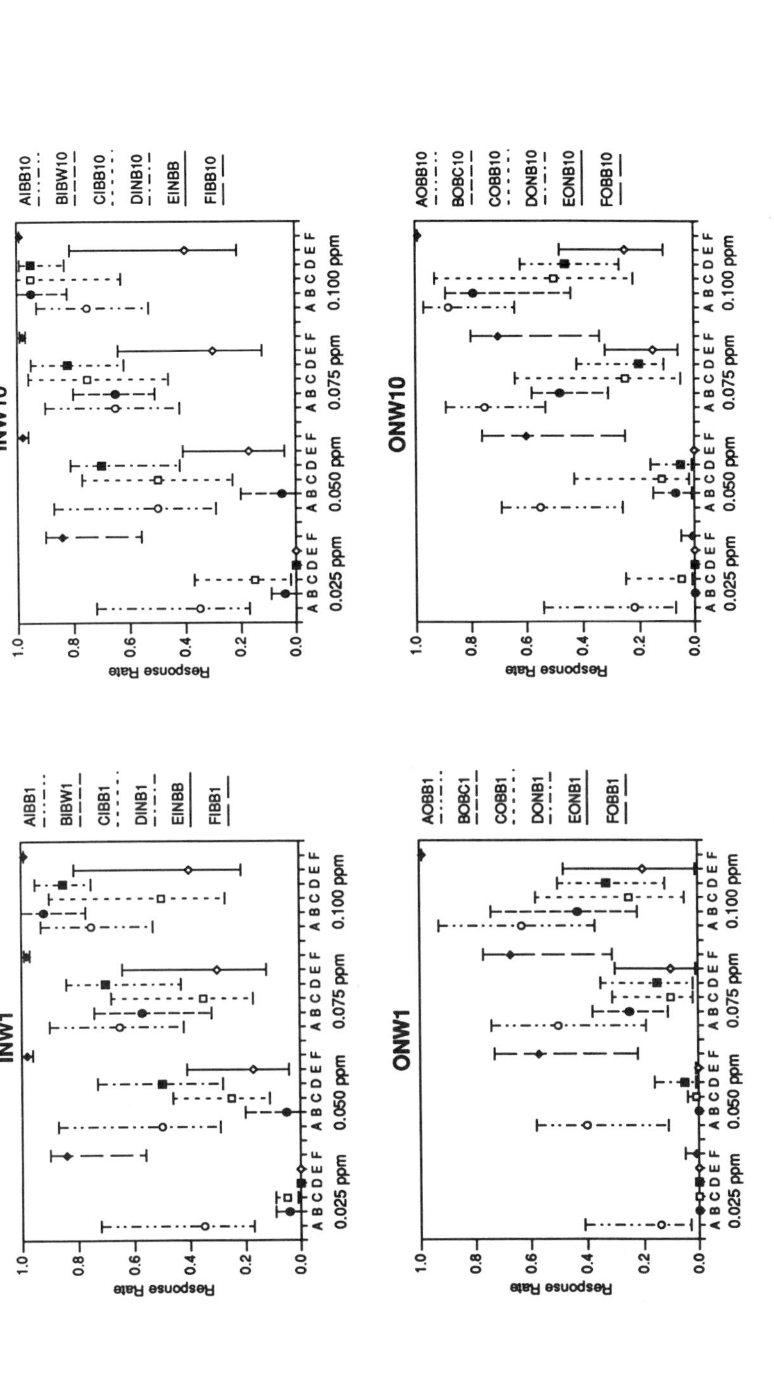

Figure 13. Medians and 90% credible intervals for judgments about mild and moderate lesions, outdoor workers, New York, 1 and 10 seasons.

Summaries of Quantitative and Qualitative Results

Summary of Quantitative Results

The quantitative results consist of numerous sets of probabilistic judgments elicited from six experts about exposure-response relationships under a variety of conditions. Although other probabilities were obtained, the analysis focuses mainly on the median responses for each exposure-response relationship (i.e., on the median of an expert's distribution of the proportion of the population developing centriacinar lesions under the specified conditions) and the 0.05 and 0.95 fractiles of the distribution. The medians provide measures of central tendency, and the 0.05 and 0.95 fractiles form a central 90% CI that indicates an expert's uncertainty about the proportion of the population exhibiting either mild or moderate centriacinar lesions.

The experts differed about the details of specific exposure-response relationships. However, the results indicate that the experts agreed that O_3-induced centriacinar lesions would likely develop under the stated conditions in at least some members of the populations of interest and that, in some cases, they would develop in a substantial proportion of that population. For example, considering exposure for 10 seasons at a mean O_3 concentration of 0.075 ppm, the judgments of five of the six experts implied a probability of at least 0.95 that 45% or more of the population of interest would develop mild centriacinar lesions. At higher O_3 levels, some experts were almost sure that close to 100% would develop lesions. At lower O_3 levels, on the other hand, some experts believed that there would be no response or low response rates. For example, for one season of exposure to O_3 at 0.025 ppm, five out of six experts judged that median response rates would be at or very close to zero for moderate lesions. For mild lesions, the median response rates were judged to be zero for two experts, about 0.05 for two others, and much higher for the remaining two. At or above seasonal O_3 exposure levels of 0.050 ppm, most experts agreed that chronic effects in the form of centriacinar lesions would likely occur in substantial numbers of people.

Certain consistent trends can be observed in the results. For example, response rates were higher for mild lesions than for moderate lesions. The magnitude of this difference varied, ranging from an average of 10% of the population at the low end to 50% at the high end. In a similar fashion, higher response rates were given for exposure over 10 seasons as against 1 season, although one expert did not differentiate between 1 and 10 seasons for mild lesions and a second expert had only very minor differences. At the high end, one expert had median shifts of 0.05–0.45 between 1 and 10 seasons. Finally, and predictably, the expected response rates increased as the mean O_3 level increased, with some experts envisioning large jumps in response at certain points (e.g., moving from 0.050 to 0.075 ppm) and others having a more linear (steady) increase in response rate over the four mean O_3 concentrations used in the study.

The experts showed little inclination to stratify the populations of interest. Only one expert distinguished between children and outdoor workers, and only two experts distinguished between the two locations. Furthermore, even in those few situations, the differences in judgments were relatively small.

The median response rates varied from expert to expert, with one expert tending to have the highest values and another the lowest. The degree of uncertainty, as measured by the width of the 90% CIs, also varied among the experts and for different conditions. The degree of uncertainty was sometimes extremely low, especially when an expert was reasonably sure that the response rate would be near 0 or near 100%. At other times, the uncertainty was quite high, as in the case of a 90% CI from 22 to 93% of the population of interest. Despite such differences, a reasonable degree of overlap was evident among the 90% CIs for the same cases. Three of the experts overlapped in almost all cases and others overlapped frequently.

Summary of Qualitative Results

The experts agreed that the focus on centriacinar lesions was appropriate. Although they sometimes thought about the distinction between mild and moderate lesions in different ways, their reactions to the use of and definitions for mild and moderate lesions were basically favorable.

In considering the role animal data should play in making judgments about the effects of long-term O_3 exposure on humans, certain types of animal studies (e.g., primate studies and studies involving long-term exposures and multiple endpoints) were viewed as more useful than others. The experts agreed that dosimetry modeling is very important, but they expressed concern about some of the values chosen for model parameters and the dosimetry results generated for this particular study. They saw a need for additional dosimetry modeling, animal studies, and epidemiological studies.

The experts noted the potential for differences between children and adults in the formation of O_3-induced lesions, but believed that the evidence available to allow them to differentiate was limited and somewhat mixed. Similarly, four experts judged that the exposure patterns provided for Los Angeles and New York did not allow them to differentiate between the two locations in terms of the effects of long-term O_3 exposure. Given that smokers were included in the populations at risk, some experts expressed concern about differential effects on smokers because of their status with respect to nutrition and chronic obstructive pulmonary disease.

The experts disagreed to some extent about the relative importance of peak vs. cumulative exposure. One expert expressed a concern that not enough is known to discriminate between the effects of different exposure patterns. They could not agree about the role of respites; some experts seemed to view respites as beneficial, whereas others viewed them as making things worse. There were questions about the degree of repair and the possibility of misrepair. More information is needed on the effects of respites.

The evidence regarding the interaction of O_3 with other pollutants and the relationships of such interactions to chronic health effects was judged to be limited. No specific synergies were the object of widespread concern among the experts, who seemed to agree that O_3 plays a very important role in the formation of centriacinar lesions.

There were different points of view about the possible health significance of O_3-induced centriacinar lesions. Most of the experts believed that such lesions could lead to fibrosis or chronic bronchiolitis, but probably not to emphysema. Some experts were concerned about an indirect relationship to cancer (e.g., through reduced ability to clear asbestos or other carcinogens from the lung). Many experts agreed that an important effect of O_3-induced centriacinar lesions is accelerated depletion of lung function reserves with age. This concern is related to how such lesions affect the quality vs. the length of life. Significant decrements in lung function late in life were believed to reduce quality of life by leading to changes in homeostasis and by rendering other pulmonary diseases more serious.

AN EXAMPLE BENCHMARK RISK MODEL AND RESULTS

As mentioned earlier, benchmark risk is calculated by combining exposure-response relationships and probability distributions over ambient O_3 concentrations. This was accomplished[1,2] using estimated O_3 air quality for "as is" conditions and for exact attainment of three alternative NAAQS. The air quality estimates were calculated as part of the USEPA's O_3 NAAQS population exposure project.[23] Benchmark risk for "as is" air quality and three alternative standards (0.08, 0.10, and 0.12 ppm) for O_3 were calculated for 10 urban areas for each of three exposure-response data sets (Avol et al.,[3] Kulle et al.,[4] and McDonnell et al.[5]) referred to earlier in this chapter. (Hereafter, we refer to the studies and the resultant exposure-response relationships as the Avol data set, the Kulle data set, or the McDonnell data set.) A sample of the large amount of results will initially be explained using some results for St. Louis. Then, a few results for the 10 urban areas will be presented.

Results for St. Louis

Benchmark risk results for St. Louis for FEV_1 decrements greater than or equal to 10% are shown in Figure 14. Results are presented in the form of bar charts having three groups of bars corresponding to the three exposure-response data sets considered. Each group comprises four bars. There is one bar each for "as is" conditions and the three NAAQS. Each bar is subdivided into three parts (unfilled, solid filled, or cross-hatched rectangles), each corresponding to a different percentage of the sensitive population (heavy exercisers) responding over and above a very small, hypersensitive percentage that might respond at or below background O_3 levels.

The total height of the bar (the top of the unfilled rectangle) is the risk for the 1%-benchmark case; that is, the probability that at least 1% more of the heavy exercisers would respond (one or more times during the O_3 season) with the

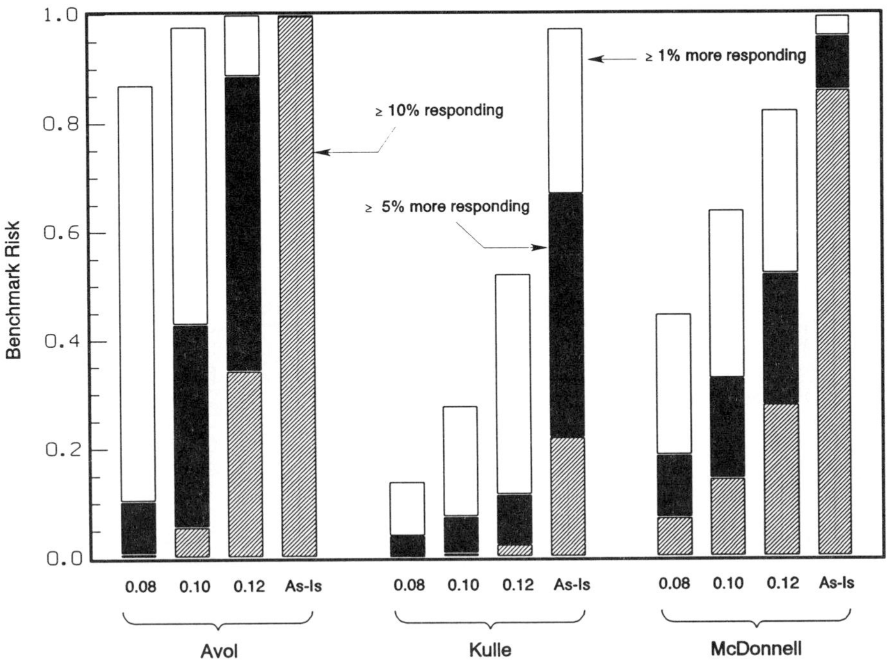

Figure 14. Benchmark risk in St. Louis for FEV_1 decrements of ≥10% under heavy exercise for three exposure-response data sets.

specified health endpoint than under background conditions. The top of the second rectangle corresponds to the risk for the 5%-benchmark case and the top of the third rectangle corresponds to the 10% benchmark case. Several results may be read from it as follows:

- Under "as is" air quality conditions, the probability that at least a 10% FEV_1 decrement would occur in at least 1% more heavy exercisers than at background at least once during the O_3 season is 1 (a certainty) for the Avol and McDonnell data sets and about 0.95 for the Kulle data set.
- Under exact attainment of the current O_3 NAAQS (0.12 ppm), the probability that at least a 10% FEV_1 decrement would occur in at least 5% more heavy exercisers than at background at least once during the O_3 season is nearly 0.9 for the Avol data set, about 0.1 for the Kulle data set, and about 0.5 for the McDonnell data set.

To illustrate insights that may be gained from Figure 14, consider a 20%-benchmark case. Then,

- For the 1%-benchmark case, the benchmark risk cannot be reduced to the specified degree of risk through attainment of any of the alternative standards — except for the 0.08-ppm Kulle case.

Table 2. 5%-Benchmark Risk (FEV_1 Decrement ≥10%) in 10 U.S. Urban Areas

City	Data set	"As is"	Alternative NAAQS (ppm)		
			0.12	0.10	0.08
Chicago	Avol	1.000	0.958	0.791	0.250
	Kulle	0.506	0.152	0.101	0.059
	McDonnell	0.933	0.651	0.456	0.268
Denver	Avol	1.000	0.989	0.773	0.250
	Kulle	0.417	0.176	0.100	0.059
	McDonnell	0.887	0.701	0.450	0.268
Houston	Avol	1.000	0.973	0.796	0.428
	Kulle	0.928	0.211	0.107	0.073
	McDonnell	0.997	0.724	0.489	0.327
Los Angeles	Avol	1.000	0.987	0.796	0.250
	Kulle	0.997	0.152	0.101	0.059
	McDonnell	1.000	0.653	0.458	0.268
Miami	Avol	0.956	0.718	0.698	0.177
	Kulle	0.687	0.145	0.095	0.044
	McDonnell	0.858	0.568	0.441	0.211
New York	Avol	1.000	0.923	0.696	0.201
	Kulle	0.954	0.135	0.094	0.052
	McDonnell	0.999	0.568	0.422	0.241
Philadelphia	Avol	1.000	0.984	0.914	0.493
	Kulle	0.896	0.297	0.136	0.078
	McDonnell	0.990	0.808	0.594	0.348
St. Louis	Avol	1.000	0.886	0.428	0.102
	Kulle	0.669	0.114	0.073	0.038
	McDonnell	0.958	0.521	0.327	0.186
Tacoma	Avol	0.999	0.942	0.537	0.103
	Kulle	0.263	0.121	0.082	0.038
	McDonnell	0.824	0.558	0.365	0.187
Washington, D.C.	Avol	0.999	0.882	0.525	0.100
	Kulle	0.253	0.114	0.081	0.037
	McDonnell	0.811	0.520	0.361	0.184

- For the 5%-benchmark case, reduction to the specified degree of risk can be achieved by attaining either 0.12- or 0.10-ppm standards with the Kulle data set only; however, the specified degree of risk can be achieved by attaining a 0.08-ppm standard for all three data sets.
- For the 10%-benchmark case, the specified degree of risk can be achieved upon attainment of a 0.12-ppm standard with the Kulle data set only; however, the specified degree of risk can be achieved for both the 0.10- and 0.08-ppm standards for all three data sets.

These example results are generally typical of those for the other nine urban areas studied, and illustrate the analysis and interpretation of the individual urban area results. A sample of results for 10 urban areas follows.

A Summary of Benchmark Risk Results for 10 Urban Areas

The interpretation one makes of a given set of benchmark risk results depends critically on the specification of the three primary decision parameters: health endpoint, n%-benchmark definition, and degree of risk. To facilitate assessing the

Table 3. Benchmark Risk Summary for 10 U.S. Urban Areas

		Number of Urban Areas Achieving Degree of Risk								
		1%-Benchmark			5%-Benchmark			10%-Benchmark		
Degree of risk	NAAQS	A	K	M	A	K	M	A	K	M
0.10	As is	—	—	—	—	—	—	—	3	—
	0.12	—	—	—	—	—	—	—	10	—
	0.10	—	—	—	—	6	—	—	10	—
	0.08	—	—	—	—	10	—	—	10	5
0.20	As is	—	—	—	—	—	—	—	4	—
	0.12	—	—	—	—	8	—	—	10	—
	0.10	—	—	—	—	10	—	7	10	3
	0.08	—	5	—	3	10	3	10	10	10
0.30	As is	—	—	—	—	2	—	—	5	—
	0.12	—	—	—	—	10	—	—	10	2
	0.10	—	1	—	—	10	—	8	10	9
	0.08	—	10	—	8	10	8	10	10	10

sensitivity of individual urban area results to the choice of these parameters, a table is presented for different combinations of n%-benchmarks (1, 5, and 10%) and degrees of risk (0.1, 0.2, and 0.3). The table was assembled from benchmark risk tables such as those in Table 2 — which lists benchmark risk in 10 U.S. urban areas for:

- FEV_1 decrement ≥10%
- "As is" ozone air quality and exact attainment of alternative O_3 standards
- A 5% benchmark (i.e., the risk of ≥5% of heavy exercisers responding ≥1 time during the O_3 season

Table 3 is a result. It lists the number of urban areas (≤10) that achieve a specified degree of risk for each combination of the decision parameters. For example, few urban areas are projected to achieve a 0.2 degree of risk upon attaining the current standard of 0.12 ppm for any of the benchmark definitions considered for the Avol and McDonnell data sets. For the Kulle data set, 8 areas are projected to do so for a 5%-benchmark and all 10 for a 10%-benchmark. For a 1%-benchmark, few if any of the urban areas would achieve a 0.2 degree of risk upon attainment for any of the three NAAQS. Results are similar for the 5%-benchmark case. (Only 3 areas are projected to achieve a 0.2 degree of risk for the Avol and McDonnell data sets, even for a 0.08-ppm standard; 8 areas are projected to do so for the Kulle data set for a 0.10-ppm standard, and all 10 for a 0.08-ppm standard.) For a 10%-benchmark, most of the urban areas are projected to achieve the specified degree of risk upon attaining either a 0.10- or 0.08-ppm standard.

AN EXAMPLE HEADCOUNT RISK MODEL AND RESULTS

This section describes the approach and results of an assessment[23] of health risks associated with long-term exposure to O_3. The health endpoint considered is the probability of formation of mild lesions in the centriacinar region of the lung among children living in New York City. The risk model incorporates an exposure model and a health model. The exposure model uses preliminary results of the probabilistic NAAQS exposure model for O_3 ($_pNEM/O_3$), and the health model incorporates the judgments of active researchers about the likelihood of formation of O_3-induced lesions in the human lung. Children and New York City were chosen as the population and city of interest because it is believed that children are more sensitive to O_3 than any other group of people, and New York City is more representative of other urban areas than Los Angeles, the other city for which $_pNEM$ exposure results are available. Risk results are presented for ten exposure distributions generated by $_pNEM/O_3$, two air quality scenarios ("as is" and "attainment"), and two exposure periods (1 and 10 O_3 seasons). The results (1) vary across experts, (2) are not very sensitive to variations in $_pNEM/O_3$ exposure distributions, (3) are lower for attainment conditions than "as is" conditions, and (4) are lower for 1 season of exposure than 10 seasons. Although these results are specific to children living in areas having exposure patterns similar to those found in the Northeast, they are fairly representative of results for outdoor workers in the Northeast and southern California and for children in southern California. The reason for this is that many experts believe that children and outdoor workers respond in a similar fashion to the given exposure patterns, or that exposure patterns in the Northeast and southern California are similar. Some experts held both of these beliefs. These results should help policymakers evaluate the potential for chronic health effects due to O_3 exposure under current conditions and upon attainment of the current 1-h NAAQS.

The Preliminary Exposure Model

An exposure model and a health model are essential components of the risk model, as shown in Figure 1. The exposure model is a preliminary version of $_pNEM/O_3$,[24] the urban scale population exposure model described in Chapter 4. Exposure distributions for seasonal mean daily maximum 8-h average O_3 concentrations of personal exposure were generated using $_pNEM/O_3$ for two scenarios. The first scenario, "as is", is based on actual air quality data from 1983. The second scenario, "attainment", is hypothetical and is based on the assumption that the second highest hourly value at the highest station in a given year is exactly 0.12 ppm. The O_3 season for both scenarios is the months of April through October of 1983. Ten preliminary exposure distributions for the "as is" scenario for children living in New York City under "as is" O_3 conditions generated by $_pNEM/O_3$ are given in Table 4. The distributions are over the fractions of children having personal exposures at each O_3 level. Distributions for the attainment scenario are given in Table 5. (Nearly 1.2 million children are included in these $_pNEM/O_3$ results.)

For example, the column for run 1 in Table 4 shows probabilities of 0.20, 0.55, and 0.24 (which do not sum exactly to 1 because of rounding) for personal exposures of 0.015, 0.025, and 0.035 ppm (midpoints of intervals that are 0.01

Table 4. Exposure Distributions for Children, New York City, and the "As Is" Scenario

	pNEM/O_3 run[a]									
ppm[b]	1	2	3	4	5	6	7	8	9	10
0.015	0.20	0.20	0.20	0.20	0.20	0.20	0.20	0.20	0.20	0.20
0.025	0.55	0.57	0.63	0.60	0.60	0.67	0.55	0.57	0.57	0.55
0.035	0.24	0.23	0.16	0.20	0.20	0.13	0.25	0.23	0.23	0.25

[a] Columns when added may not sum to 1 due to rounding.
[b] O_3 exposure concentration (seasonal mean daily maximum 8-h O_3 concentrations of personal exposure over the 9-month O_3 season) in parts per million; midpoints of intervals that are 0.01 ppm wide.

Table 5. Exposure Distributions for Children, New York City, and the "Attainment" Scenario

	pNEM/O_3 run[a]									
ppm[b]	1	2	3	4	5	6	7	8	9	10
0.015	0.33	0.38	0.67	0.41	0.36	0.33	0.32	0.33	0.34	0.33
0.025	0.67	0.62	0.33	0.59	0.64	0.67	0.68	0.67	0.66	0.67
0.035	0.0	0.0	0.0	0.0	0.0	0.0	0.01	0.01	0.0	0.0

[a] Columns when added may not sum to 1 due to rounding.
[b] O_3 exposure concentration (seasonal mean daily maximum 8-h O_3 concentrations of personal exposure over the 9-month O_3 season) in parts per million; midpoints of intervals that are 0.01 ppm wide.

ppm wide), respectively. The lowest exposures resulted in run 6 and the highest exposures resulted in runs 7 and 10. Interestingly, the fraction of children having personal exposures of 0.015 ppm was exactly the same in all 10 runs. Whether or not variations in the probabilities for exposures of 0.025 and 0.035 ppm affect risk results is a question addressed in this analysis.

Health Model for the Illustration

The health model for the illustration uses a subset of the results presented earlier in this chapter: mild lesions, children, New York City, 1 and 10 O_3 seasons. (The dose-response relationships for these cases are the top two panels of Figure 13.) Inasmuch as the health experts often did not distinguish between children and outdoor workers or between New York City and Los Angeles, these results are, in many cases, representative of results involving children in Los Angeles and outdoor workers in both geographical areas.

These and the other judgments were obtained more than 2 years before preliminary $_pNEM/O_3$ results were available. At that time, it was believed that personal exposures to O_3 would be higher than $_pNEM/O_3$ results indicate, and it was decided to obtain judgments at O_3 concentrations of 0.025, 0.050, 0.075, and 0.100 ppm. Since these levels do not match the levels for preliminary $_pNEM/O_3$ exposure

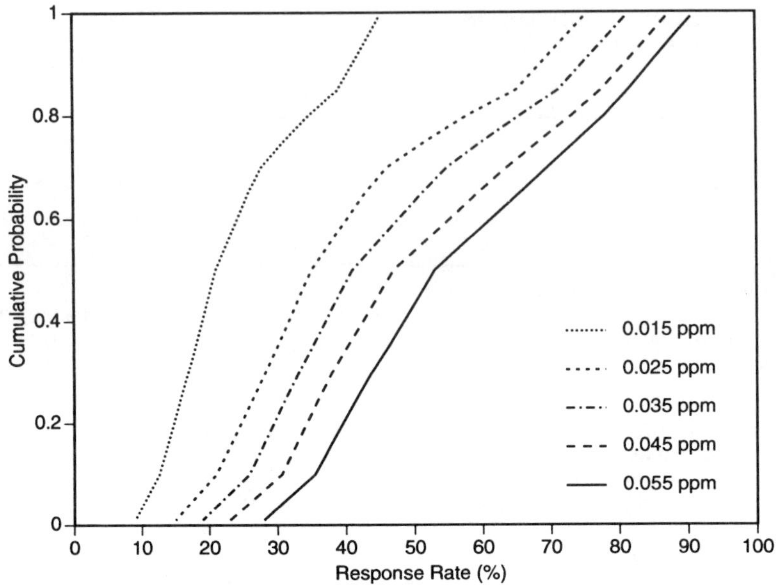

Figure 15. Probability distributions over response rate for O_3 Expert A adjusted to match $_{P}$NEM/O_3 exposure distributions.

results, it was necessary to adjust the encoded judgments. Linear interpolation was used to obtain probability distributions at the required exposure concentrations. The results of adjusting the judgments shown earlier in Figure 5 are given in Figure 15. For example, since 0.015 is 0.6 times 0.025 (i.e., 0.015 ppm is 40% less than 0.025 ppm), the median response at 0.015 ppm is 21% (which is 0.6 times 35%). Note that the distributions appear "smoother" than those in Figure 5 because they were calculated at 21 cumulative probabilities (from 0.05 to 0.95 in steps of 0.05, plus 0.01 and 0.99). Table 5 is a partial listing of the probability distributions over response rate at O_3 concentrations of 0.015, 0.025, 0.035, 0.045, and 0.055 ppm. Results for other experts were obtained in the same manner. (Some $_{P}$NEM/O_3 results not shown here had personal exposures as high as 0.04–0.05 ppm. This is the reason for having a response rate probability distribution at 0.045 ppm. The 0.055 ppm distribution was included to accommodate the possibility that sensitivity analyses might show results in the 0.05-0.06 ppm range.)

Another way to look at the judgments that is more familiar to health scientists is shown in Figure 16. It shows exposure-response relationships for five fractiles (0.05, 0.25, 0.50, 0.75, and 0.95 fractiles). These depict the uncertainty in Expert A's judgments about relationship between O_3 exposure and the occurrence of O_3-induced lesions in the centriacinar region of the lungs of children living in New York City. If the expert had been certain about the relationship between O_3 exposure and the occurrence of O_3-induced lesions in the centriacinar region of the lungs of children living in New York City, only one curve would have been needed to completely characterize this belief.

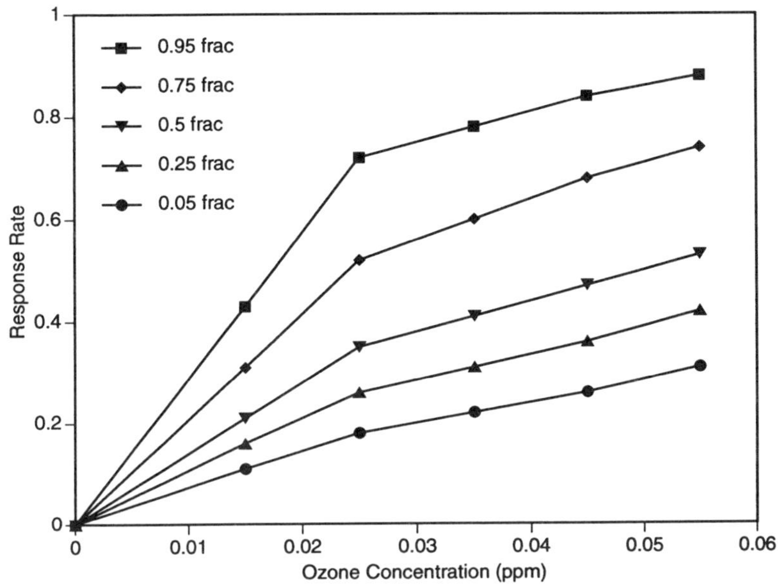

Figure 16. Exposure response relationships for Expert A at five fractiles.

These relationships are constructed from Figure 15. For example, points on the median (0.5 fractile) exposure-response relationship for Expert A come from the median response rates of 21, 35, 41, 47, and 53% at the O_3 concentrations of 0.015, 0.025, 0.035, 0.045, and 0.055 ppm indicated in Figure 15. As would be expected, response rate increases as exposure increases.

Risk Calculations

Risk calculations are conceptually simple. For a particular set of conditions (i.e., lesion type, population, geographical location, number of exposure seasons, air quality scenario, $_{p}$NEM/O_3 run), first calculate the probability-weighted response rate for the 0.5 fractile.

To illustrate the calculations, let P_e denote the fraction of the population having personal exposures of e ppm in a single $_{p}$NEM/O_3 run and let $(RR_k | e)$ denote the exposure-response relationship for the formation of mild lesions among children in New York City exposed for one season at O_3 level e. The formula to calculate risk for the 0.5 fractile is:

$$R_{0.5} = \left[P_{0.015} \times \left(RR_{0.5} | 0.015\right)\right] + \left[P_{0.025} \times \left(RR_{0.5} | 0.025\right)\right]$$
$$+ \left[P_{0.035} \times \left(RR_{0.5} | 0.035\right)\right] + \left[P_{0.045} \times \left(RR_{0.5} | 0.045\right)\right]$$
$$+ \left[P_{0.055} \times \left(RR_{0.5} | 0.055\right)\right]$$

Table 6. Cumulative Probability Distributions over Response Rate (a Fraction) for Expert A, Children, Mild Lesions, New York, and 1 O_3 Season of Exposure at Five Levels of Personal Exposure to O_3

	O_3 concentration (ppm)				
Fractile	**0.015**	**0.025**	**0.035**	**0.045**	**0.055**
0.01	0.09	0.15	0.19	0.23	0.28
0.05	0.11	0.18	0.22	0.26	0.31
.	.	.	.	.	.
.	.	.	.	.	.
.	.	.	.	.	.
0.50	0.21	0.35	0.41	0.47	0.53
.	.	.	.	.	.
.	.	.	.	.	.
.	.	.	.	.	.
0.95	0.43	0.72	0.78	0.84	0.88
0.99	0.45	0.75	0.81	0.87	0.91

Given the exposure probabilities for $_p$NEM/O_3 run 1 in Table 5 and the median exposure-response relationship in Table 6, the median estimate of the probability of formation of mild lesions is:

$$R_{0.5} = (0.20 \times 0.21) + (0.55 \times 0.35) + (0.24 \times 0.41) + (0 \times 0.47)$$
$$+ (0 \times 0.53) = 0.33$$

In general, the risk for the kth fractile is:

$$R_k = \sum_{e=0.015}^{0.055} \left[P_e \times \left(RR_k | e \right) \right]$$

To develop a probability distribution over response rate that accounts for all of the conditions mentioned earlier, repeat the above calculations for any number of fractiles. Results will appear "smoother" if a large number of fractiles are used. Here, 21 fractiles are used to obtain the results presented in the next section.

Illustrative Headcount Risk Results

Figure 17 introduces a convenient format for presenting and discussing results. The figure shows a cumulative probability distribution over population response rate (a fraction) for the formation of mild lesions (among children living in urban areas having "as is" O_3 air quality patterns similar to those in New York City after exposure to O_3 for one season) resulting from the first $_p$NEM/O_3 run and Expert A's judgments. The median of this distribution is a response rate of 33%; the 90% credible interval is 17 to 68%.

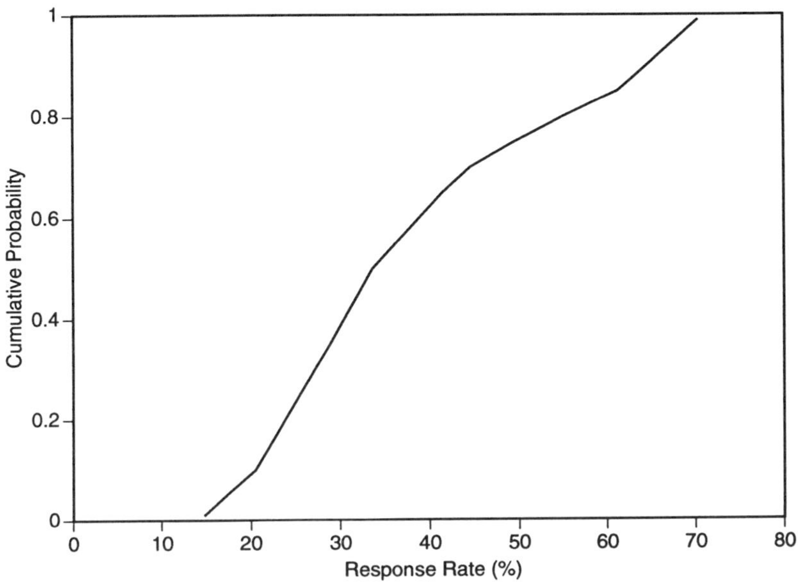

Figure 17. Risk distribution over response rate for Expert A based on the first pNEM/O_3 run for "as is" air quality.

Results for all 10 preliminary $_{p}$NEM/O_3 runs for the "as is" scenario are shown in Figure 18. The distributions over response rate are not distinct at all and look as if they were drawn with a dull rather than a sharp pencil. The 0.05 fractiles vary little (from 16.8 to 17.3%), the 0.95 fractiles vary from 67 to 68%, and the medians vary from 33 to 34%. This means that the variability in exposure distributions from run to run has very little effect on the risk results for this case. As will be seen, this is generally true for other cases involving other experts, attainment air quality, and 10 O_3 seasons.

There is more variability in results for the "attainment" scenario. These results are also shown in Figure 18. The medians are slightly lower than those for the "as is" scenario. They range from 29 to 31%. The 0.05 fractile response rates vary from 15 to 16%, and the 0.95 fractile response rates vary from 60 to 63%.

Results for all experts, mild lesions, New York City, children, 1 and 10 O_3 seasons, and "as is" and "attainment" air quality are shown in Figure 19. There are 4 "cases" (labeled 1 to 4) for each expert, one for each combination of air quality scenario and number of exposure seasons. The figure also indicates the effects, if any, of variability in pNEM/O_3 exposure distributions (from run to run). For each case, the 0.05, 0.5 (i.e., median), and 0.95 fractiles of risk distributions are indicated by rectangles or short horizontal lines. Rectangles indicate the range over which results at a particular fractile vary due to variability in $_{p}$NEM/O_3 runs. A vertical line connects the lower edge of the 0.95 fractile rectangle to the upper edge of the 0.05 fractile rectangle. If little or no variability results from different $_{p}$NEM/O_3 runs, the rectangle becomes a horizontal line.

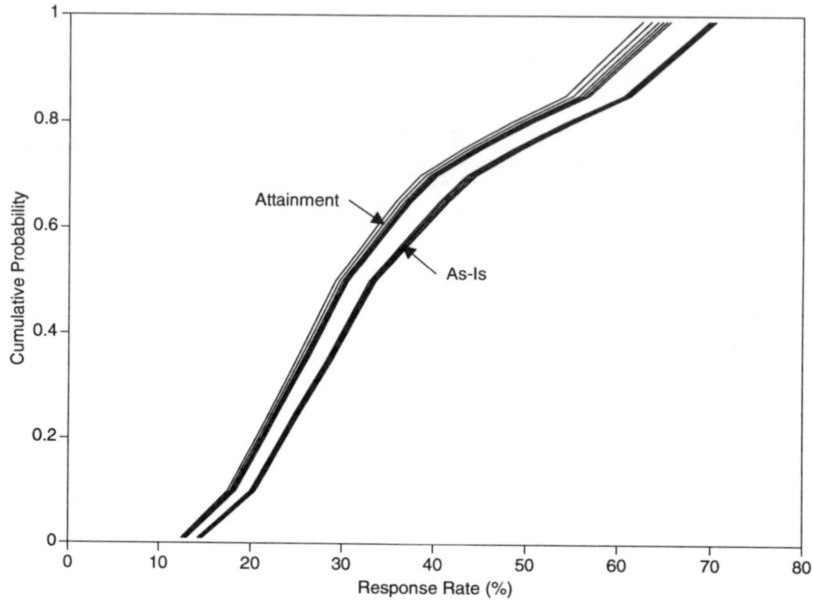

Figure 18. Risk distributions over response rate for Expert A based on all ten pNEM/O_3 runs for "as is" and "attainment " air quality.

For example, there is almost no variability in the 0.05 fractile estimate of risk for Expert C at attainment conditions after 10 exposure seasons. Thus, the 0.05 fractile is indicated by the short horizontal line at a 2% response rate. In contrast, the 0.95 fractile for case 4 (attainment, 10 seasons) and Expert A varies from a minimum of 68% to a maximum of 71%. A rectangle indicates this 0.95 fractile in Figure 19. The top (maximum) and bottom (minimum) of this rectangle are also labeled. One extreme situation is that for Expert D, "as is", and 1 season: the rectangles overlap because variability in the fractiles between runs is large compared to overall variability. This is probably not noteworthy because the risks based on Expert D's judgments are small (the median response rate for this case is between 3 and 4%).

Other observations about Figure 19 include the following:

After one season of exposure:

- Response rates for Experts A and F are substantially higher than those for the other experts.
- At "attainment", there is virtually no response for Experts D and E. For the other

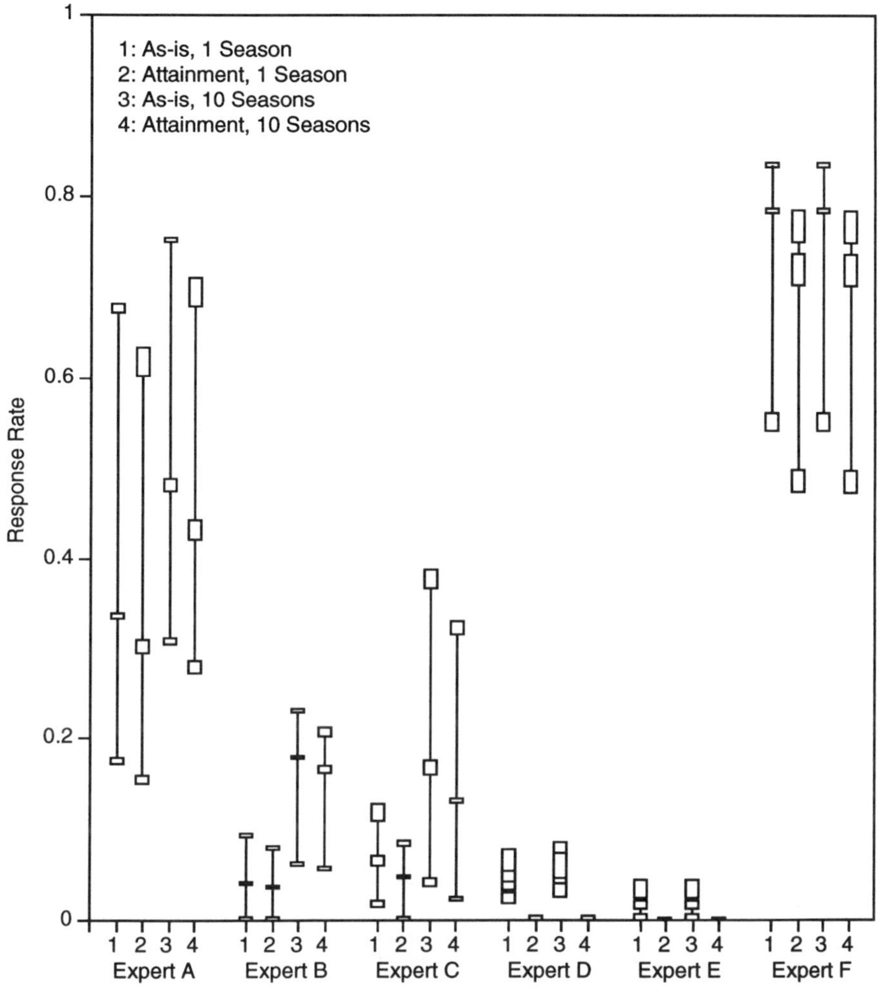

Figure 19. Medians and 90% credible intervals for risk of formation of mild lesions among New York City children after 1 and 10 seasons of exposure to O_3 at "as is" and "attainment " air quality levels.

experts, response rates are only a few percentage points lower (i.e., the distributions are "shifted" toward zero by a few percentage points) than for "as is" conditions.

- For Experts A and F, the 90% credible intervals for response rate range from 15 to 83%. For the other experts, the range is 0 to 12%.

After 10 seasons of exposure:

- The respective risk distributions for Expert A are 12 to 15% higher after 10 seasons than after 1 season.
- Results for Experts E and F are identical to those for 1 season of exposure.
- Median response rates for Experts B and C are more than double compared to 1 season.
- Median response rates for Expert D are slightly higher (for "as is" conditions) compared to 1 season.

"As is" vs. "attainment":

- Except for Experts D and E, "attainment" leads to relatively small reductions in risk compared to "as is" conditions.
- Risks based on the judgments of Experts D and E are virtually zero for "attainment" conditions. In addition, risk estimates at "as is" conditions are small for Experts D and E.
- Although obscured by variability between $_{p}NEM/O_3$ runs, the main effect of "attainment" (for experts other than D and E) is to reduce exposures, and hence risk, for those children who experience the highest levels of exposure.

CONCLUDING REMARKS

The approach described here builds on past risk assessments for criteria pollutants and represents the state of the art as procedures for risk assessments concerning health risks of pollutants continue to be developed and refined. Such procedures have been evolving over a decade of risk assessments conducted by OAQPS. Careful design of an encoding protocol and the ensuing encoding sessions and elicitation of probabilities regarding exposure-response relationships was not new.[19,20] What was new in the chronic risk assessment for O_3 was bringing the experts, the risk assessment team, and the USEPA staff together for a risk assessment workshop that appears to have been a valuable addition for the following reasons.

- The workshop allowed the experts the opportunity to freely exchange viewpoints and perspectives about the scientific evidence and key issues affecting their judgments prior to the final encoding session. In principle and very likely in practice, this format results in more informed judgments in the final encoding session. At the same time, there was no pressure to adopt a consensus viewpoint for the sole sake of consensus. As a result of this exchange and discussion, one expert substantially changed his or her initial judgments.
- The workshop provided an opportunity for the USEPA staff to gain better insight into the reasoning and judgments of the experts regarding various issues (e.g., significance of effect, role of repair, influence of dosimetry modeling, and time sequence of effects) related to chronic O_3-induced lung injury.
- The workshop gave the experts the chance to offer feedback to the USEPA on the various elements of this phase of the risk assessment process, including the protocols, encoding sessions, and supporting analyses that were provided (i.e., exposure and dosimetry modeling).

The risk assessment team and the USEPA staff believe that the exchange of views at the workshop was very illuminating and valuable. Moreover, the experts reacted very favorably to the workshop and were unanimous in stating that it was very informative and valuable to them. Although the experts are very busy, they said that the time at the workshop was well spent. The risk assessment team strongly recommended using such workshops in future risk assessments, whenever feasible.

The most obvious and concrete result of the risk assessment process is the set of probabilistic exposure-response judgments. This set represents the experts' current judgments about the degree of risk for the endpoints and conditions considered in the study. These quantitative results clearly show that chronic O_3 effects in the form of centriacinar lesions are judged by the experts to be likely in at least some members of the populations of interest and, in some cases, in substantial proportions of those populations. Equally important are the qualitative comments made during the encoding sessions and at the workshop. Both types of results serve to summarize the current state of scientific knowledge concerning the chronic effects of O_3 exposure. They also provide new insights for the scientific research community about the future research needed to reduce uncertainties about the chronic health effects associated with exposure to O_3 at elevated levels. The experts participating in the risk assessment have already benefitted in this manner from the risk assessment process. For example, one expert commented that the encoding process and the discussions at the workshop had led to an entirely new way of thinking about research and about the adverse effects associated with O_3 exposure. Much research, by its very nature, focuses on small parts of the overall problem of health risks associated with O_3. The risk assessment process enables researchers and others to step back and think about the bigger picture.

REFERENCES

1. Hayes, S. R., A. S. Rosenbaum, T. S. Wallsten, R. G. Whitfield, and R. L. Winkler. "Assessment of Lung Function and Symptom Health Risks Associated with Attainment of Alternative Ozone NAAQS," prepared by Systems Applications, Inc. for U. S. Environmental Protection Agency, Report No. SYSAPP-87-171 (1987). (For copies, contact Harvey M. Richmond, U. S. Environmental Protection Agency, OAQPS, MD-12, Research Triangle Park, NC 27711, (919) 541-5271.)
2. Hayes, S. R., A. S. Rosenbaum, T. S. Wallsten, R. G. Whitfield, R. L. Winkler, and H. Richmond. "A Health Risk Assessment for Use in Setting the U. S. Primary Ozone Standard," in *Atmospheric Ozone Research and its Policy Implications: Proc. of the 3rd US-Dutch Int. Symp.,* Nijmegen, The Netherlands, May 9–13, 1988, T. Schneider, S. D. Lee, G. J. R. Walters, et al., Eds. (Amsterdam: Elsevier, 1989), pp. 851–867.
3. Avol, E. L., W. S. Linn, T. G. Venet, D. A. Shamoo, and J. D. Hackney. "Comparative Respiratory Effects of Ozone and Ambient Oxidant Pollution Exposure during Heavy Exercise," *J. Air Pollut. Control. Assoc.* 34:804–809 (1984).
4. Kulle, T. J., L. R. Sauder, J. R. Hebel, and M. D. Chatham. "Ozone Response Relationships in Healthy Nonsmokers," *Am. Rev. Respir. Dis.* 132:36-41 (1985).
5. McDonnell, W. F., D. H. Horstman, M. J. Hazucha, E. Seal, E. D. Haak, S. Salaam, and D. E. House. "Pulmonary Effects of Ozone Exposure During Exercise: Dose-Response Characteristics," *J. Appl. Physiol.* 54:1345–1352 (1983).
6. D. H. Horstman, L. J. Folinsbee, P. J. Ives, S. Abdul-Salaam, and W. F. McDonnell. "Ozone Concentration and Pulmonary Response Relationships for 6.6-hour Exposures with Five Hours of Moderate Exercise to 0.08, 0.10, 0.12 ppm," *Am. Rev. Respir. Dis.* 142:1158–1163 (1990).
7. Follinsbee, L. J., W. F. McDonnell, and D. H. Horstman. "Pulmonary Function and Symptom Responses After 6.6-hour Exposure to 0.12 ppm Ozone," *J. Air Pollut. Control Assoc.* 38:28–35 (1988).

8. "Air Quality Criteria for Ozone and Other Photochemical Oxidants," U. S. Environmental Protection Agency, Report No. EPA-600/8-84/020a to 020e (1986). (For copies, contact U.S. Environmental Protection Agency, Center for Environmental Research Information, USEPA/ORD Publications, 26 West St. Clair, Cincinnati, Ohio 45268, (513) 569-7652.)
9. Whitfield, R. G., T. S. Wallsten, R. L. Winkler, H. M. Richmond, S. R. Hayes, and A. S. Rosenbaum. "Assessing the Risk of Chronic Lung Injury Attributable to Ozone Exposure," prepared by Argonne National Laboratory for U.S. Environmental Protection Agency, Argonne National Laboratory Report No. EAIS-2 (1991).
10. Winkler, R. L., R. G. Whitfield, T. S. Wallsten, S. R. Hayes, A. S. Rosenbaum, and H. M. Richmond. "Health Risk Due to Ozone Chronic Lung Injury: An Assessment of Expert Judgments," in *Tropospheric Ozone and the Environment: Papers from an International Conference*, R. L. Berglund, D. R. Lawson, and D. J. McKee, Eds. (Pittsburgh, PA: Air and Waste Management Association, 1991), pp. 161–177.
11. Wallsten, T. S., R. G. Whitfield, R. L. Winkler, H. M. Richmond, S. R. Hayes, and A. S. Rosenbaum. "Chronic Lung Injury Risk Assessment for Ozone," in *Proc. 83rd Annu. Meet. of the Air and Waste Management Assoc.*, (Pittsburgh: Air and Waste Management Association, 1990).
12. Feagans, T. B., and W. F. Biller. "A General Method for Assessing Health Risks Associated with Primary National Ambient Air Quality Standards," U. S. Environmental Protection Agency Draft Report (1981). (For copies, contact Harvey M. Richmond, U.S. Environmental Protection Agency, OAQPS, MD-12, Research Triangle Park, NC 27711, (919) 541-5271.)
13. Feagans, T. B., and W. F. Biller. "Risk Assessment: Describing the Protection Provided by Ambient Air Quality Standards," *Environ. Professional* 3:235–247 (1981).
14. "Four Conceptual Approaches to Health Risk Assessment for Alternative National Ambient Air Quality Standards," U.S. Environmental Protection Agency (1980). (For copies, contact Harvey M. Richmond, U. S. Environmental Protection Agency, OAQPS, MD-12, Research Triangle Park, NC 27711, (919) 541-5271.)
15. Wallsten, T. S., and D. V. Budescu. "Encoding Subjective Probabilities: A Psychological and Psychometric Review," *Manage. Sci.* 29:151–173 (1983).
16. Wallsten, T. S., B. H. Forsyth, and D. V. Budescu. "Stability and Coherence of Health Experts' Upper and Lower Subjective Probabilities about Dose-Response Functions," *Organizational Behavior Human Perform.* 31:277–302 (1983).
17. Smith, A. E., P. C. McNamee, and M. W. Merkhofer. "Development of Decision Analysis Methodology for Health Risk Assessment," final report prepared by SRI International for the U.S. Environmental Protection Agency, Office of Air Quality Planning and Standards (1982). (For copies, contact Thomas R. McCurdy, U.S. Environmental Protection Agency, OAQPS, MD-12, Research Triangle Park, NC 27711, (919) 541-5288.)
18. Keeney, R. L., L. R. Keller, R. K. Sarin, A. Sicherman, and R. L. Winkler. "Development and Application of a Risk Assessment Methodology to Study Alternative National Ambient Carbon Monoxide Standards," final report prepared by Woodward-Clyde Consultants, for the U.S. Environmental Protection Agency, Office of Air Quality Planning and Standards (1982). (For copies, contact Thomas R. McCurdy, U.S. Environmental Protection Agency, OAQPS, MD-12, Research Triangle Park, NC 27711, (919) 541-5288.)

19. Wallsten, T. S., and R. G. Whitfield. "Assessing the Risks to Young Children of Three Effects Associated with Elevated Blood-Lead Levels," prepared by Argonne National Laboratory for U.S. Environmental Protection Agency, Report No. ANL/AA-32 (1986). (Available from NTIS: EE87004219.)
20. Whitfield, R. G., and T. S. Wallsten. "A Risk Assessment for Selected Lead-Induced Health Effects: An Example of a General Methodology," *Risk Analysis* 9:197–207 (1989).
21. Ruckleshaus, W. D. "Risk in a Free Society," *Risk Analysis* 4:157–162 (1984).
22. Paul, R. A., T. Johnson, A. Pope, and A. Ferdo. "National Estimates of Exposure to Ozone Under Alternative National Standards," prepared by PEI Associates, Inc., for the U. S. Environmental Protection Agency, Office of Air Quality Planning and Standards (1986). (For copies, contact Thomas R. McCurdy, U.S. Environmental Protection Agency, OAQPS, MD-12, Research Triangle Park, NC 27711, (919) 541-5288.)
23. Absil, M. J. G., P. Narducci, R. G. Whitfield, and H. M. Richmond. "Chronic Lung Injury Risk Estimates for Urban Areas Having Ozone Patterns Similar to Those in the Northeast," in transactions of the *Specialty Conference on Tropospheric Ozone and the Environment II*, Air & Waste Management Association, Atlanta, GA (1991). (For copies contact Ronald G. Whitfield, Argonne National Laboratory, EID-900, Argonne, IL 60439, (708) 252–8430.)
24. Johnson, T. R., et al. "Estimation of Ozone Exposure in Houston Using a Probabilistic Version of NEM," Paper 90–150.1, in *Proc. 83rd Annu. Meet. Air and Waste Management Assoc.*, (Pittsburgh: Air and Waste Management Association, 1990).

CHAPTER 6

EFFECTS OF OZONE ON CROPS

David T. Tingey, David M. Olszyk, Andrew A. Herstrom, and E. Henry Lee

INTRODUCTION

Research and analysis of the scientific literature indicate that ozone (O_3), alone or in combination with other pollutants, accounts for approximately 90% of the air pollution-induced crop loss in the U.S.[1] Although the majority of O_3 effects research has been conducted in the U.S., O_3 impacts on vegetation have been observed in a several areas (e.g., Great Britain, Scandinavia, central and southern Europe, Japan, Israel, India, and Mexico.).[2,3]

O_3 exerts a phytotoxic effect only if a sufficient amount reaches sensitive cellular sites within the leaf. To do this, it must diffuse from the ambient air into the leaf through the stomata, which exert primary control on O_3 uptake.[4] O_3 injury will not occur if:

- The rate of O_3 uptake is slow enough that the plant can detoxify or metabolize O_3 or its metabolites or
- The plant is able to repair or compensate for the injury[4,5]

Cellular disturbances that are not repaired or compensated for are ultimately expressed as leaf injury, reduced plant growth, decreased yield, changes in crop quality, alterations in susceptibility to abiotic and biotic stresses, and decreased reproduction.

0-87371-475-X/94/$0.00+$.50
© 1994 by Lewis Publishers

Not all O_3 effects on plants are equally important.[6,7] An example of this differential valuation is the common usage, in the vegetation effects literature, of the terms *injury* and *damage*.[2,8] Injury encompasses all measurable plant reactions that do not influence agronomic yield or reproduction. These responses can include changes in metabolism, reduced photosynthesis, leaf necrosis, leaf drop, and altered growth or quality as long as yield is not impaired. Leaf necrosis is one of the earliest and most obvious manifestations of O_3 injury to crops and its occurrence has provided the impetus for abatement or mitigative actions. However, the deleterious effects of O_3 are not limited to visible symptoms.

In contrast, damage includes all effects that reduce the intended human use or value of the plant species, plant communities, or ecosystem. From this perspective, damage is equivalent to an adverse effect and can be segregated into several categories: (1) economic yield, (2) ecological structure and function, (3) genetic resources, and (4) cultural values.[9] In the recent Draft Staff Paper,[10] that the U.S. Environmental Protection Agency (USEPA) prepared as a part of the ongoing review of the National Ambient Air Quality Standards (NAAQS) for O_3, a distinction was made between the relative importance of foliar injury and reduced crop yield. The conclusion reached in the document was that "...greater emphasis will be placed on damage or yield loss than on foliar injury." Owing to difficulties in valuing ecological effects and alterations in genetic resources and cultural values, the focus of this review will be on effects of O_3 on crop yield and its translation into economics effects which follows the same convention as the Draft Staff Paper.[10] Several examples will be used to illustrate the effects of O_3 on crops and how crop loss data can be used in establishing the level and form of NAAQS.

EFFECTS OF O_3 ON CROP YIELD

Diverse experimental procedures — ranging from field exposures without chambers to field exposures in open-top chambers, and to studies done under highly controlled conditions — have been used to study the effects of O_3 on crop plants. In general, the more controlled conditions are most appropriate for investigating specific responses and for providing the scientific basis for interpreting and extrapolating results. These systems are powerful tools for adding to an understanding of the biological effects of air pollutants. However, to assess the impact of O_3 on crop yield and to provide data for economic assessments, deviations from the typical field environment in which the plant is grown should be minimized.

Effects of Ambient O_3

In the 1950s and 1960s, researchers assessed the impact of O_3 on crop yield under field conditions by surveying foliar injury symptoms.[11,12] However, injury symptoms and crop yield are usually not directly proportional due to the importance of various allocation processes and metabolic factors in determining plant yield. For example, significant yield loss can occur with little or no foliar injury.[13-16] Conversely, foliar injury can be much greater than yield loss.[17,18] The injury-yield loss relationship may depend on the O_3 exposure. For example, in field corn, foliar injury occurred at a lower O_3 concentration than yield reductions; but as the O_3 concentration increased, yield was reduced to a greater extent than foliar injury increased.[19]

A number of chamber studies conducted in the U.S. demonstrated that ambient O_3 concentrations were sufficiently high to impair crop yield at many locations of the country. In the late 1960s and early 1970s, research documented that ambient oxidants reduced the yield and quality of citrus, grape, tobacco, cotton, and potato.[20] Subsequent studies substantiated the impacts of ambient O_3 on crop yield (Table 1). Most of these studies have compared yields of plants grown in ambient air with those grown in charcoal-filtered air to determine the impact of ambient oxidants (primarily O_3) on plant yield.

Other studies used chemical protectants to assess the impacts of ambient O_3 on crops while eliminating possible chamber influences on the plant responses (Table 1). In these experiments, some plots were treated with the chemical and others were left untreated. Yield loss was determined by comparing the yield in plots treated with the chemical to the yield in untreated plots. These studies (Table 1) confirmed that ambient O_3 levels, in various locations, were sufficient to decrease crop yields (24 to 41%). Several factors could contribute to the higher yield losses found using chemical protectants than in the chamber studies: (1) higher O_3 concentrations in these studies, (2) different environmental conditions, and/or (3) crops that were more susceptible to O_3 than those used in the chamber studies. Also, when chemical protectants are used, care must be used in interpreting the data because the chemical itself may alter plant growth. The chemical may not be effective against all concentrations of all pollutants in the study area, which could result in an underestimation of yield loss.

Effects of O_3 — Controlled Exposures

In the mid-1970s, the USEPA reviewed the scientific literature to determine the impact of O_3 on vegetation in preparation for promulgation of a NAAQS for O_3. The results of the review and analyses were published in the Air Quality Criteria Document for Ozone and Other Photochemical Oxidants.[20] At that time, several problems precluded a credible comprehensive assessment of the impact of O_3 on agriculture.

1. Relatively few studies had measured the effects of O_3 exposure on crop yield under field conditions. Experimental exposures (concentration and frequency of occurrence) were often unrealistic, rendering inferences to the field conditions inappropriate.
2. O_3 exposure-response data were not available for the economically important crop species. In addition, the few available exposure-response studies were generally not comparable due to important differences in exposure regimes, environmental conditions, response variables, and exposure methods.

Table 1. Examples of Plant Yield Loss Induced by Ambient O_3 Level in Various States

Location	Crop	Yield loss (%)
Chamber studies		
California	Alfalfa	11–28
California	Green pepper	12
California	Lettuce	0[a]
California	Orange (Valencia)	31
California	Strawberry	0[b]
California	Sweet corn (Bonanza)	9
California	Sweet corn (Monarch Advance)	28
California	Tomato	12
California	Wheat	0
Indiana	Muskmelon	21
Maryland	Bean (green)	1
Maryland	Soybean	20
Maryland	Tomato	16
Maryland	Wheat	8
New York	Bean (green)	26
New York	Grape	16
New York	Tomato	33
Pennsylvania	Potato	14[c]
Virginia (1979)	Forbs, sedges, grasses	32
Virginia (1980)	Forbs, sedges, grasses	20
Virginia (1981)	Forbs, sedges, grasses	21
Chemical protectants		
Canada, Ontario	Bean (green)	24
Canada, Ontario	Onion	38
Canada, Ontario	Peanut (PI26861)	20
Canada, Ontario	Peanut (Merah)	0[d]
Canada, Ontario	Potato	36
Canada, Ontario	Tomato	30
Massachusetts	Bean (green)	41
New Jersey	Potato	25

[a] Nonsignificant 20% increase in yield.
[b] Nonsignificant 15% increase in yield.
[c] Total tubers, 11% for Grade One.
[d] Nonsignificant 3% increase in yield.

From References 21–30.

3. Data on the spatial distribution of O_3 in agricultural areas were incomplete and not expressed in a form that allowed estimation of potential O_3 effects on crops.

Consequently, the scientific credibility of the analyses and, hence, the potential to develop a subsequent NAAQS derived from the available data suffered because there were insufficient, high-quality data to reliably determine the effects of O_3 on the yield of major agronomic crops or to determine the resultant economic consequences. Thus, to improve estimates of O_3-induced yield loss in the field, the USEPA initiated the National Crop Loss Assessment Network (NCLAN) in 1980.[1,31]

Table 2. The Major Crops of the U.S. for 1984 Based on Sales Value and Acreage

Crop	Percent of total sales value	Percent of total acreage
Barley	2	3
Beans (dry)	<1	<1
Corn	25	21
Cotton	5	3
Lettuce	<1	<1
Peanuts	2	<1
Potatoes	3	<1
Sorghum	3	5
Soybeans	13	18
Tobacco	4	<1
Turnips	<1	<1
Wheat	11	21

From "Agricultural Statistics 1987," U.S. Department of Agriculture, U .S. Government Printing Office, Washington, D.C. (1987). With permission.

National Crop Loss Assessment Network (NCLAN)

The primary objectives of the NCLAN study were to:

1. Define the exposure-response relationships for the yields of major agricultural crops and O_3 exposure
2. Assess the national economic consequences resulting from the exposure of major agricultural crops to O_3
3. Advance understanding of the cause and effect relationships that determine crop responses to pollutant exposures

Crop Production Data U.S. agricultural statistics were used to identify the major agronomic crops and their production regions (Table 2). Agricultural census data were used to map the principal production areas for wheat, corn, soybean, cotton, sorghum, oat, tomato, potato, tobacco, and hay. The selected crops were grown on approximately 75% of U.S. agricultural cropland and accounted for approximately 70% of U.S. agricultural receipts. For each crop, the major cultivar(s) (as defined by farmer usage) were selected and grown using typical agricultural practices. The crops were studied only within the areas for which they were adapted; for example, cotton was studied only at the southeastern and southwestern research sites.

O_3 Exposure The crops were grown and exposed to O_3 in open-top field exposure chambers using standard experimental protocols at several research sites around the U.S.[1,33] A detailed quality assurance plan was developed to insure data of known quality and to provide measures of data precision.[34] The cultural conditions approximated typical agronomic practices. After the crops were established in the field, open-top field exposure chambers were placed over the plants

to control the O_3 exposure environment without major changes in the microclimatic conditions.[33]

O_3 exposures were programmed (7 or 12 hours daily) to follow the ambient exposure pattern and designed to provide the multiple levels of O_3 exposure needed to develop exposure-response functions. Chambers were supplied with either charcoal-filtered air (control), ambient air, or ambient air supplemented with O_3 to provide three or four exposures greater than ambient. As the ambient O_3 varied from day to day, the base to which additional O_3 was added also varied so that when the concentrations were high in the ambient air, outside the chambers, they were also high inside the chambers and vice versa. This coupling of the O_3 exposures to the ambient environment means that high O_3 concentrations occurred in the chambers when the ambient environmental and air chemistry conditions were conducive for producing elevated ambient O_3 levels. The exposure durations were species dependent, typically from stand establishment until harvest (average of 78 days; range of 21 to 126 days).

Experimental data from 31 NCLAN studies involving 12 crops (Table 2) were used to develop O_3 exposure-response relationships for crop yield.[35,36] The plant response data were analyzed using linear and nonlinear regression approaches. The exposures were typically characterized by a 7-h (9:00 a.m. to 4:00 p.m.) seasonal mean O_3 concentration corresponding to the time period when O_3 was added to the exposure chambers. In summarizing the data, O_3-induced yield loss was usually derived from a comparison of the performance of the plants in charcoal-filtered air. Six studies included multiple cultivars of species and ten studies incorporated a drought stress factor to produce a total of 49 individual response functions.

Selected Results of the NCLAN Study

Examples of relationships between O_3 exposures and plant yield are shown in Figures 1 and 2. These species and cultivars were selected to illustrate the response of several species/cultivars to O_3 as well as the type of year-to-year variation in plant response that may occur. Actual yields (Figure 1) are displayed to show the individual treatment means and the model adequacy of Weibull equations to the actual yields for different species/cultivars and years. The relative yield data (Figure 2) are included to show comparability of O_3 effects on crop yield among replications and/or crops.

Regression equations derived from the NCLAN data can be employed to determine the concentrations predicted to cause a specific yield loss or to estimate the yield loss that would result from a specific O_3 exposure. For example, the O_3 concentrations predicted to cause a 10 or 30% yield loss have been estimated (Table 3). For some species, mean yield reductions of 10% were predicted when the 7-h seasonal mean O_3 concentration was between 0.04 and 0.05 ppm. For Vona wheat and Hodgson soybean, lower concentrations of 0.028 to 0.033 ppm were predicted to cause a 10% yield loss. Grain crops were generally less sensitive to O_3 than the other crops.

A histogram of 7-h seasonal mean O_3 concentrations predicted to cause a 10% yield loss was used to illustrate the differential species/cultivar sensitivity

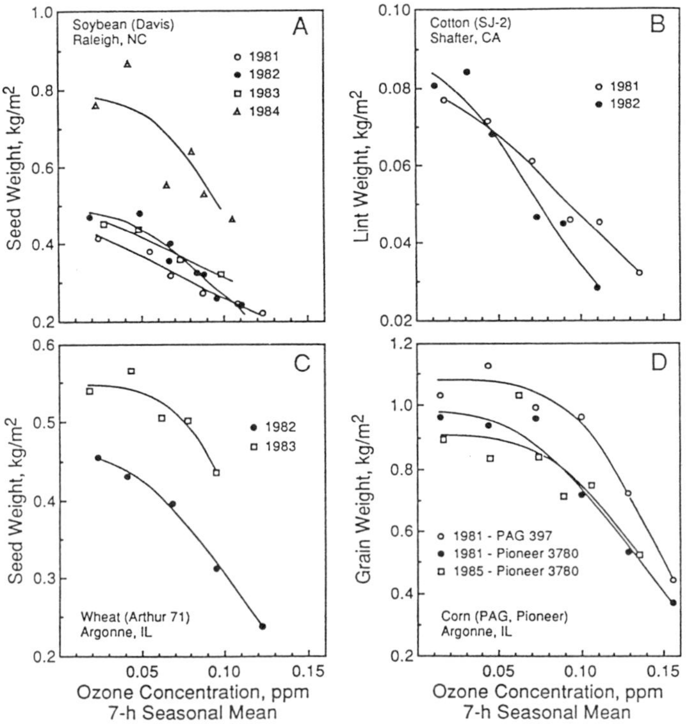

Figure 1. Examples of the effects of O_3 on the yield of several crops/cultivars. The O_3 concentrations are the 7-h seasonal means. Crop yields are expressed as the actual yields. Data are from the NCLAN Data Library.

to O_3 (Figure 3). Approximately 57% of the species or cultivars were predicted to exhibit 10% yield reductions at 7-h seasonal mean concentrations below 0.05 ppm; 35% of plant types were predicted to display a 10% yield loss at 7-h mean concentrations between 0.04 and 0.05 ppm; and 7-h seasonal mean concentrations in excess of 0.08 ppm were required to cause a 10% yield loss in almost 19% of the species or cultivars. The data indicated that approximately 8% of the species or cultivars would display a 10% yield loss at 7-h seasonal mean concentrations below 0.035 ppm, suggesting that these plant types were very sensitive to O_3. The data also demonstrated that sensitivity differences within a species may be as large as differences between species (Table 3). For example, at 0.04 ppm O_3, estimated yield losses ranged from 2 to 15% in soybean and from 0 to 28% in wheat.

APPLICATIONS OF THE NCLAN DATABASE

The NCLAN was the first large-scale field-experimental program to assess the impact of O_3 on crops. Its success led to similar programs being initiated in other countries using the basic NCLAN methodology and experimental design [i.e.,

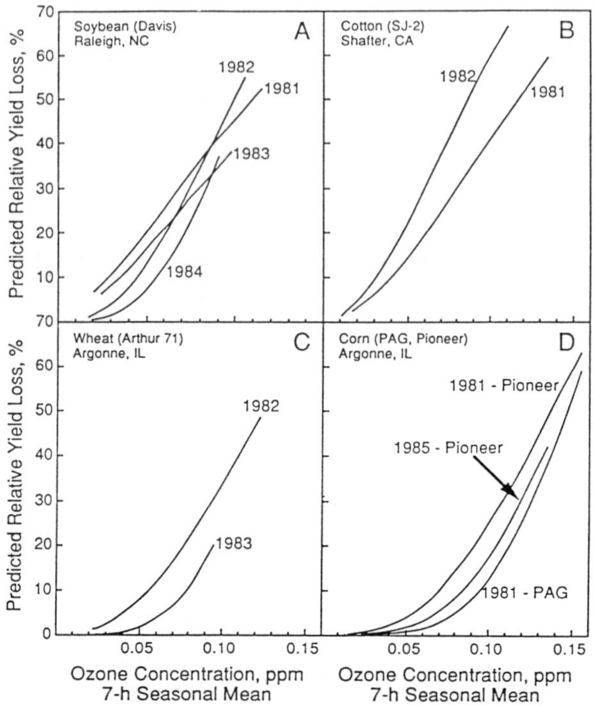

Figure 2. Examples of the effects of O_3 on the yield of several crops/cultivars. The O_3 concentrations are expressed as the 7-h seasonal means. Crop yields are expressed as the predicted relative yield loss (PRYL); that is, yields are relative to maximum yield at zero O_3. Data are from the NCLAN Data Library.

European Crop Loss Assessment Network (EUROCLAN)[38-41]]. The data need of the NCLAN economic and crop loss assessments required spatial estimates of O_3 exposure. Consequently, studies were initiated to spatially estimate O_3 exposures over rural areas.[42-44]

As with all long-term studies, there were experimental and conceptual advances during the course of the study that were not possible to fully incorporate into the study. The most significant accomplishment of the NCLAN program was the development of an electronic database (containing hourly O_3 exposure and individual plot yields) which is available to the scientific community for further analyses and scrutiny. This database has been and continues to be applied in four major areas:

- Economic assessments
- Analyses of exposure indices
- Spatial assessments of O_3 on crop yield
- Analysis and development of ambient air quality standards

Table 3. Summary of O_3 Concentrations Predicted to Cause 10 and 30% Yield Losses and Summary of Yield Losses Predicted to Occur at 7-h Seasonal Mean O_3 Concentrations of 0.04 and 0.06 ppm[a]

	O_3 concentrations, ppm, predicted to cause yield losses of:		Percent yield losses predicted to occur at 7-h seasonal mean O_3 concentration of:	
Species	**10%**	**30%**	**0.04 ppm**	**0.06 ppm**
Legume crops				
Bean, kidney	0.038	0.066	11.2	25.8
Peanut, NC-6	0.046	0.073	6.4	19.4
Soybean, Corsoy	0.048	0.082	6.4	16.6
Soybean, Davis (81)	0.038	0.071	11.5	24.1
Soybean, Davis (CA-82)	0.048	0.081	6.4	16.5
Soybean, Davis (PA-82)	0.059	0.081	2.0	10.4
Soybean, Essex	0.048	0.099	7.2	14.3
Soybean, Forrest	0.076	0.118	1.7	5.3
Soybean, Hodgson	0.032	0.066	15.4	18.4
Soybean, Williams	0.039	0.093	10.4	18.1
Grain crops				
Wheat, Abe	0.059	0.095	3.3	10.4
Wheat, Arthur 71	0.056	0.094	4.1	11.7
Wheat, Vona	0.028	0.041	28.8	51.2
Wheat, Blueboy II	0.088	0.127	0.5	2.8
Wheat, Coker 47-27	0.064	0.107	2.2	8.4
Wheat, Holly	0.099	0.127	0.0	0.9
Wheat, Oasis	0.093	0.135	0.4	2.4
Corn, PAG 397	0.095	0.126	0.3	1.5
Corn, Pioneer 3780	0.075	0.111	1.4	5.1
Corn, Coker 16	0.133	0.175	0.0	0.3
Sorghum, DeKalb-28	0.108	0.186	0.0	2.7
Barley, Poco	0.121	0.161	0.0	0.5
Fiber crops				
Cotton, Acala SJ-2 (81)	0.044	0.096	8.3	16.2
Cotton, Acala SJ-2 (82)	0.032	0.055	16.1	35.1
Cotton, Stoneville	0.047	0.075	4.6	16.2
Horticultural crops				
Lettuce, Empire	0.053	0.075	0.0	16.8
Spinach, America	0.046	0.082	6.8	17.2
Spinach, Hybrid	0.043	0.082	2.6	9.2
Spinach, Viroflay	0.048	0.080	6.0	16.7
Spinach, Winter Bloom	0.049	0.080	5.8	16.5
Turnip, Just Right	0.043	0.064	7.7	24.9
Turnip, Pur Top W. G.	0.040	0.064	10.1	26.5
Turnip, Shogoin	0.036	0.060	13.0	29.7
Turnip, Tokyo Cross	0.053	0.072	3.3	15.6

[a] The yield losses were derived from Weibull equations and are based on the control yields in charcoal-filtered air.

From "Ambient Air Criteria for Ozone and Other Photochemical Oxidcants," Chapter 6, U.S. EPA, Environmental Criteria and Assessment Office, U.S. EPA Report-600/8-84/020cF (1986), pp. 6-1, 6-298. With permission.

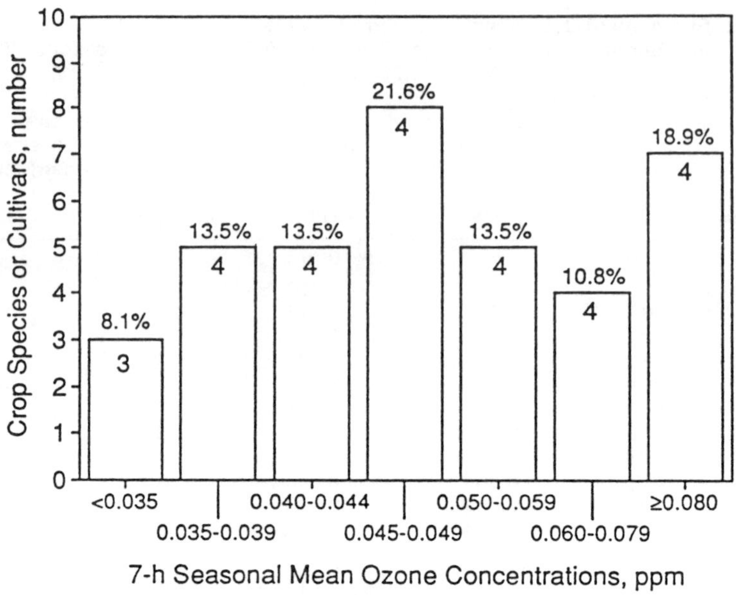

Figure 3. Number and percentage of 37 crop species or cultivars predicted to show a 10% yield loss at various ranges of 7-h seasonal mean O_3 concentrations. Concentration ranges and 10% yield loss data are derived from Table 3. The numbers (in the bars) represent the number of species/cultivars for each concentration range.

Economic Assessments

Owing to the importance of U.S. agriculture on both domestic and world consumption of food and fiber, reductions in crop yields would be expected to adversely affect human welfare. The concern for impacts on human welfare and Executive Order 12292,[45] which mandated that all federal administrative decisions be based on adequate information concerning the need for and the consequences of the proposed governmental actions, has resulted in numerous attempts to assess the economic losses from ambient O_3 to agriculture.

Assessments of O_3-induced damage to agriculture found in the literature range from descriptions of procedures for simple monetary calculations of economic losses to more complex economic assessment methodologies.[37,46-48] The simple procedures calculate monetary effects by multiplying predicted yield or production changes resulting from exposure to O_3 by an assumed constant crop price. However, this approach fails to recognize possible crop price changes arising from yield changes, as well as not accounting for the processes underlying economic response. A more rigorous economic assessment: (1) will capture the economic behavior of producers and consumers (including market adjustments) as they adjust to changes in crop yields and prices that may accompany changes in O_3 air quality; and (2) should accurately reflect institutional considerations, such as regulatory programs, that may result in market distortions.

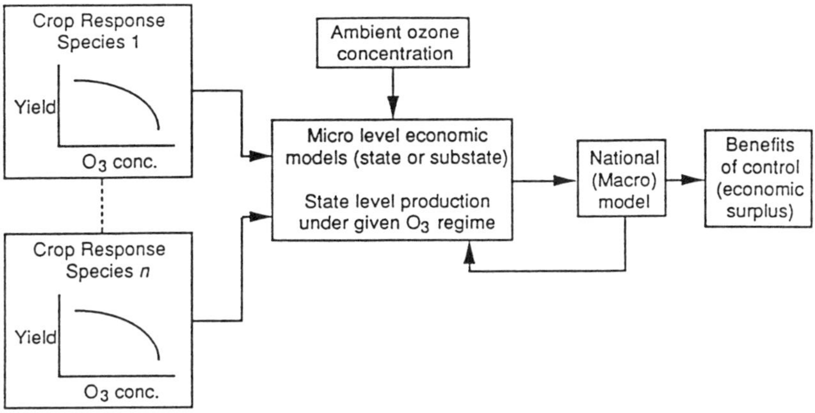

Figure 4. Logic structure and information flow for NCLANs national economic assessments.

The economic consequences of O_3 on agriculture can be assessed only if the reciprocal relations between physical and biological changes and the responses of individuals and institutions are explicitly recognized. Most of the economic assessment studies that use the NCLAN database employ mathematical programming techniques. In this approach, normative optimization models integrating crop exposure-response data are used to estimate the impacts of varying input parameters, such as ambient O_3 concentrations, on the economic response of agricultural systems. The logic flow and data requirements of a typical economic assessment using NCLAN data are illustrated in Figure 4.

Examples of economic assessments based on the NCLAN database are shown in Table 4. It is difficult to directly compare the results of the various studies as the crops, exposure-response information, environmental conditions, and assumed economic conditions differ considerably among studies. Although the conditions under which various studies were conducted may differ, the NCLAN-based studies exhibit common patterns of response and sensitivities to imposed conditions. Results from these assessments illustrate that ambient O_3 levels impose significant costs on society (both producer and consumer). As summarized by Adams et al.,[47] "The continuing utility of the NCLAN data to economists and policy makers is a strong testimonial to the success of this integrated research approach. The NCLAN approach may offer a valuable prototype in designing research efforts to address other bioeconomic issues."

Selection of Exposure Indices Best Related to Plant Yield

The characterization of plant exposures to O_3 has been a major problem in developing a secondary air quality standard to protect vegetation. The difficulty in selecting an appropriate statistic to characterize plant exposure has been

Table 4. Examples of Economic Assessments of the Impact of O_3 on U.S. Agriculture that were Based on the NCLAN Database

Region	Crops	Consumer benefits	Producer benefits	Total benefits
California	18 crops	17	28	45
Corn belt	Wheat, corn, soybeans	2079	−1411	668
Illinois	Corn & soybeans	0	226	226
U.S.	Corn, cotton, soybeans	NR[a]	NR	2220
U.S.	Corn, cotton, soybeans, wheat	NR	NR	2400
U.S.	6 Crops	1160	550	1770
U.S.	5 Crops	NR	NR	1300
U.S.	Soybeans	880	−90	790
U.S.[b]	6 Crops	1047	1590	2637
U.S.[c]	14 crops	683	2097	2780

Note: The benefits are expressed as millions of U.S. dollars (1980). Data derived from Reference 47.

[a] NR = not reported.
[b] The assessment included an adjustment for moisture stress.
[c] The assessment included an adjustment for moisture stress and is expressed in 1982 dollars.

summarized by Jordan et al.[49] and Heagle and Heck.[50] Ambient and experimental O_3 exposures have been presented as seasonal, monthly, weekly, or daily means; peak hourly means; or number of hours above a selected concentration. None of these statistics adequately characterize the relationships among O_3 concentration, exposure duration, interval between exposures, and plant response.

A primary goal of O_3 research is the determination of the functional form relating pollutant exposure and plant response. The quantifying function for this process is frequently termed "exposure-response". An understanding of exposure-response requires three types of information: (1) a measure of plant response, (2) an index describing pollutant exposure, and (3) a mathematical function that relates plant response to exposure. An appropriate exposure index should ultimately be derived from a consideration of the underlying biological basis for the response and a procedure for quantifying the temporal variations in O_3 occurrence.[51] The evaluation and selection of exposure indices is essentially a comparison of the influence of different scalings and weights of the exposure variable on the exposure response (Figure 5). The comparison is based on a set of biological responses that are statistically compared to different mathematical characterizations of exposure. The objective is to select the exposure index that displays the smallest unexplained sum of squared error.[52,53]

In evaluating exposure indices, there are several important factors that need consideration:

- Peak concentrations are more important than low concentrations in determining plant response.[37,54,55]
- Temporal distribution and magnitude of concentrations over the exposure period need to be included. Changes in exposure structure (such as peak shape, time

intervals between exposures, and concentration frequency distributions) alter the magnitude of response.[54,55]

- O_3 effects are cumulative, that is, increasing the duration and/or frequency of the exposure period is expected to cause greater biological response.[37]
- Exposure cannot be characterized as the product of concentration and time because the effect of O_3 on crop yield depends on the cumulative impact of high concentrations during the growing season.[2]

Most of the currently used exposure indices were neither developed from a biological basis, nor were they developed from an experimental approach specifically designed to address all key factors of exposure. The majority of air pollution research experiments (e.g., NCLAN) were designed for quantifying the relationship between pollutant exposure and plant yield rather than for studying the contributing features of exposure. These differences in study objectives may limit, but do not exclude, the use of data sets from existing pollution studies (e.g., NCLAN) in explicitly examining features of exposure and their effect on biological response.

Recently, researchers have conducted retrospective analyses of data sets (mainly from the USEPA's NCLAN program) for comparing alternative O_3 exposure indices for use in relating pollutant exposure and the ensuing biological response.[52,53,56,57] Using mean hourly O_3 data from eight NCLAN data sets, Lee et al.[52] compared 25 exposure indices including: (1) averages over various time periods, (2) sums of hourly concentrations equal to or above selected levels, (3) peak-weighted cumulative indices with various weighting functions, and (4) multicomponent indices. Examples of these various indices (Figure 5) are:

- Averages over time: 7-h seasonal mean (M-7) concentration which is the mean of all concentrations during a 7-h period (9:00 a.m. to 4:00 p.m.) for the specific plant experimental period.
- Sum of hourly concentrations: SUM06 which accumulates all concentrations ≥0.06 ppm for the specific experimental period.
- Peak-weighted concentrations: SIGMOID gives weights of 0.5 or less to concentrations below a specific value (e.g., 0.062 ppm) and weights of 0.5 or greater to concentrations above the specific value. The weighted concentrations are summed over the experimental period.
- Multicomponent indices combine the features of peak-weighted concentrations and includes an additional factor which accounts for changing plant sensitivity (over time) to O_3.

The authors concluded that the optimal indices, based on minimizing the residual sum of squares in exposure-response analyses, were cumulative in nature and gave preferential weights to higher concentrations. With a large NCLAN data set, Lee et al.[53] confirmed these observations that peak-weighted or multicomponent indices were preferred. However, application of the multicomponent indices were not pursued as they were deemed too complex to apply in an air quality standard setting environment.

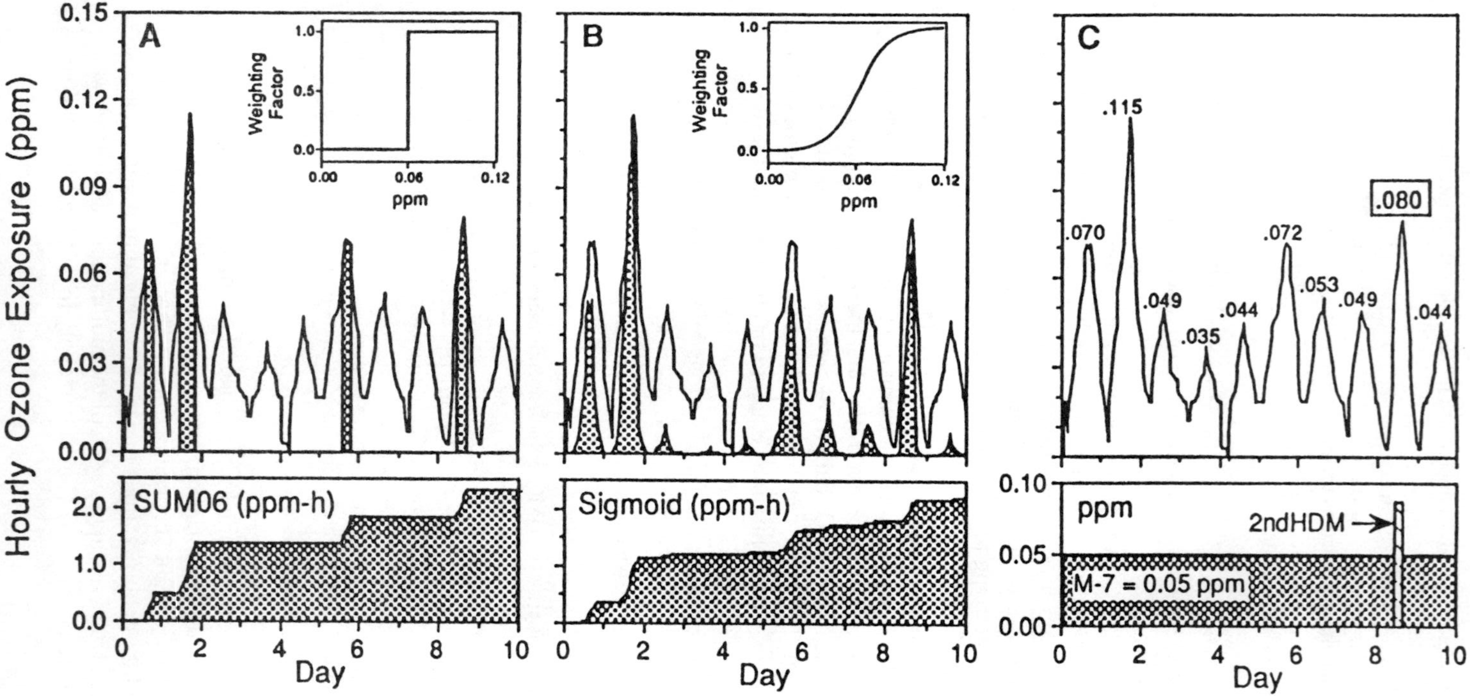

Figure 5. Diagrammatic representation of several exposure indices illustrating how they weight concentration and cumulate exposure.[44] (A) SUM06 — The upper graphic illustrates an episodic exposure profile; the shaded area under some of the peaks illustrates the concentrations ≥0.06 ppm that are cumulated in the index. The insert shows the concentration weighting (0 or 1) function. The lower portion of the graphic illustrates how concentration is cumulated over the exposure period. (B) SIGMOID — The upper graphic illustrates an episodic exposure profile; the variable shaded area under the peaks illustrates the concentration dependent weightings that are cumulated in the index. The insert shows the sigmoid concentration weighting function. The mid-point of the sigmoid weighting scheme was 0.062 ppm. The lower portion of the graphic illustrates how concentration is cumulated over the exposure period. (C) 2ndHDM and M-7 — The upper graphic illustrates an episodic exposure profile. The lower portion of the graphic illustrates that the 2ndHDM considers only a single exposure peak, while the mean applies a constant exposure value over the exposure period.

Lefohn et al.[56] demonstrated the major weakness of the long-term averaging indices — the mean cannot distinguish between exposures to the same concentration but of different durations — using sets of NCLAN data from 2 years of winter wheat and 1 year of soybean. By combining the 2 years of winter wheat into a single set of data, Lefohn et al.[56] evaluated several exposure indices by minimizing the heterogeneity of response to O_3 using a common model that excluded fixed effects to account for year and replicate effects. Hogsett et al.[54] and Tingey et al.[57] have also demonstrated the inability of the experimental mean to describe response.

Using the NCLAN data for cotton, wheat, kidney bean, and potato, Lee et al.[53] corroborated the findings of previous studies[56] that demonstrated the shortcoming of the mean indices in exposure-response modeling when length of exposure is important. The seasons with the longer exposure resulted in greater observed yield losses. The mean index did not differentiate between exposure seasons of differing duration and could not account for year-to-year differences in the magnitude of O_3 effects. Heterogeneity of response can be explained, with varying degrees of success, by differences in temporal exposure patterns and length of exposure. While the cumulative indices (i.e., SUM06 or SIGMOID) did not account for all of the heterogeneity of response among studies of a species, the cumulative indices were better than the mean or peak indices based on less heterogeneity of the Weibull predicted relative yield loss (PRYL) curves.

The evidence favoring the cumulative indices is less clear when studies across multiple sites and cultivars are compared because other confounding factors contribute to the heterogeneity of observed responses.[53] The observed heterogeneity of Weibull PRYL curves for studies of a species indicates that the effects of site, year, environmental, and edaphic variables may be as large or larger than the effects of O_3 and may interact with the multiplicative effects of O_3. The complexity and magnitudes of the effects of O_3 and other factors on biological response are not well understood and limit the use of existing sets of data for evaluating the importance of key features of exposure in determining crop response. The environmental and edaphic effects are least evident in sets of data from replicate studies in consecutive years at the same site because the conditions were more uniform in adjacent years than nonadjacent years.

Since the current NAAQS for O_3 is expressed in terms of the second highest daily maximum (2ndHDM), retrospective studies[52,53,56,57] have compared the 2ndHDM with alternative indices in describing the magnitude of biological response. The 2ndHDM ignores exposure duration and places undue emphasis on a single peak 1-h concentration. Consequently, the current national ambient air quality standard — the second highest daily maximum (2ndHDM) — which measures a peak concentration is poorly related to plant growth. Scientific data strongly suggest that the effects of O_3 are cumulative, and higher concentrations are more important than lower concentrations in determining plant response. These findings further suggest that the appropriate form of an exposure index should be cumulative and emphasize peak concentrations. Also, evidence suggests that exposure patterns with more peaks and higher peak concentrations of O_3 have more severe impacts on plant response.[51,55]

Spatial Assessment of O_3 on Crop Yield

An important application of the NCLAN database is its use in crop loss assessments. To conduct such an assessment, three separate databases are required: (1) ozone exposure, (2) crop distribution, and (3) ozone exposure-response functions. The following example with soybean and wheat illustrate how these databases can be combined to assess the impact of O_3 on crop yield.

Current Rural O_3 Exposure in the U.S.

To maximize geographic coverage and the amount of nonurban monitoring data available, ambient air quality data were obtained from USEPA's Aerometric Information Retrieval System (AIRS) database for all sites that were coded as agricultural, forest, or desert for the years 1982 through 1987. Approximately 90% of the rural sites were agricultural; the remainder were classified as forest or desert. For each site-year, ambient O_3 concentrations were expressed in several forms: the second highest daily maximum (2ndHDM) which is the form of the current NAAQS for O_3, the 7-h seasonal mean (M-7), and the seasonal sum of hourly O_3 concentrations at or above 0.06 ppm (SUM06) (Table 5). The last two indices were calculated for the highest three consecutive months (based on SUM06). The 3-month maximum period captured approximately 70% of the hourly concentrations ≥0.06 ppm[58] and it corresponded reasonably well with the typical experimental durations. The SUM06 and 2ndHDM were based on all hours, while the M-7 used a fixed 7-h window (9:00 a.m. to 4:00 p.m.). Analysis of the ambient monitoring data indicated that less than 15% of mean hourly concentrations at or above 0.06 ppm fell outside of the 7-h window at most monitoring sites.

The mean 2ndHDM (for all site-years) was 0.112 ppm (Table 5) with approximately 30% of the site-years exceeding the current NAAQS for O_3, while the mean M-7 was 0.051 ppm and the mean SUM06 was 23.9 ppm-h. The annual exposure variation among site-years was relatively low (CV of 18 to 27%) for M-7 and 2ndHDM, respectively. In contrast, the among-sites variability for the SUM06 was significantly greater, approximately 52 to 64%. The cumulative index (SUM06) displayed two to three times greater year-to-year variation than the 2ndHDM and M-7. This greater temporal and spatial variability suggests that the cumulative index is more efficient in depicting the actual variation in exposure among sites and years than the 2ndHDM or M-7.

Crop Distribution

The geographic distributions of soybean and wheat production were derived from the 1987 Census of Agriculture.[59] The distributions represent the major production areas for each crop.

Exposure-Response Functions

Composite exposure-response functions were developed for soybean and wheat by estimating the relative yield loss curves for each experiment and regressing the

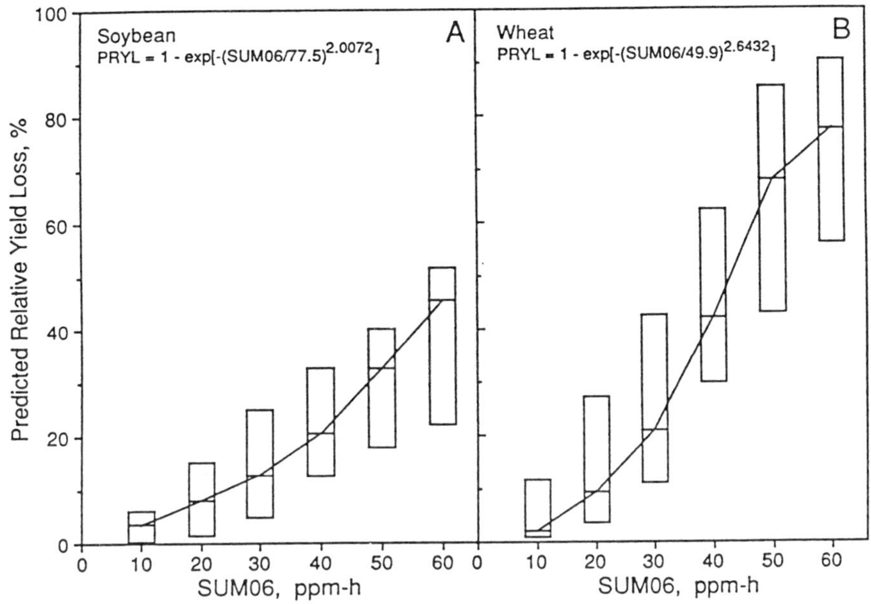

Figure 6. Composite O_3 exposure-response functions for (A) soybean and (B) wheat which illustrate predicted relative yield loss as a function of O_3 exposure. The soybean data were based on 13 data sets and wheat on 6 data sets.

median crop loss estimate against the exposure index (Figure 6). The median yield loss curve provides a valid characterization of the typical O_3 effect for each crop and is not highly influenced by extreme responses across cultivars, sites, and years. The original O_3 monitoring data for each study were reanalyzed to express the exposure in terms of the SUM06 which was used to derive the Weibull and polynomial response models for each experiment and/or cultivar.[53] The SUM06 exposure index was employed in the crop loss assessment study because SUM06 was easy to apply and better correlated with the biological response than other indices, such as the 2ndHM or M-7.[52,53]

The use of the composite exposure-response functions approach differs from Lesser et al.[35] who fit a single exposure-response equation (i.e., Category III model) incorporating all the NCLAN data for each species prior to calculating the predicted relative yield loss. The Category III model for each crop is not a valid characterization of the average O_3 effect on crop production for the crop, but gives a rough summary of the average responses to O_3 in the NCLAN studies.[35] This approach was not used in the analysis of Lee et al.[53] because different functional forms (i.e., Weibull and polynomial models) were required and percentiles other than the median yield loss were of potential interest for economic assessment. Also, heterogeneity of responses across sites, years, and cultivars was evident for some of the indices, particularly the 2ndHDM.

Inasmuch as the various species/cultivars differed in their O_3 sensitivity, the data are shown as a frequency distribution rather than as a single value (Figure 6).

The 50th percentile, the crossbar on each of the vertical bars (connected with a solid line), illustrates the magnitude of predicted relative yield loss that occurred in at least half (50th percentile) of the studies at a given O_3 exposure (SUM06). For example, at a specific O_3 exposure (SUM06) of 30 ppm-h, half of the soybean studies (Figure 6A) showed predicted relative yield losses equal to or greater than 13%. At a SUM06 of 20 ppm-h, the predicted relative yield losses were at least 8% or greater in half of the studies. The lower and upper ends of the bars illustrate the 25th and 75th percentiles, respectively. For example, with an O_3 exposure of 30 ppm-h, three quarters (75th percentile) of the studies showed predicted relative yield losses of 27% or less while a quarter (25th percentile) of the studies showed predicted relative yield losses of 9% or less.

To estimate predicted relative yield loss values in nonmonitored areas, a distance weighting function was used to interpolate the predicted relative yield loss at the 50th percentile at the rural AIRS monitoring sites to all nonmonitored areas to produce a gradient surface over the study region. As with most spatial interpolation techniques, this procedure models change in value solely as a function of Euclidean distance between the geometric placement of known sites. Interpolation does not consider the dynamics of O_3 formation and transport, nor does it consider change in exposure due to time, distance, or elevation from the monitored site. For these reasons, estimates of crop loss over nonmonitored areas can only be considered as a first approximation. Spatial estimates of crop loss were conducted for the years 1982 through 1987.

For example, using 1987 data, the predicted relative yield losses for soybean and wheat associated with ambient air quality for each of the air monitoring sites is shown as coded circles in the top maps of Figures 7A and 8A. The interpolated gradient surface from these data is shown in the bottom map of each figure (7B and 8B). The gradient surface was constrained by the crop distribution data so that predicted losses were shown only for the principle areas where the crop was cultivated.

The Impact of Current O_3 Exposures on Soybean and Wheat Yield

The predicted relative yield losses, using current (1982–1987) air quality, indicated that significant crop losses are expected to occur over the soybean and wheat growing regions (Figures 7 and 8). For both crops, predicted relative yield losses ranged from approximately 0 to greater than 25%. The area-weighted yield losses for soybean and wheat (averaged over the 6 years) are approximately 9.3 and 12.7%, respectively. Annual soybean yield losses for 1982–1987 ranged from approximately 6.9 to 14.8% (Figure 9). The annual yield losses of wheat for 1982–1987 ranged from approximately 9 to 17% (Figure 9). Largest yield losses, for the period 1982 to 1987, were observed in 1983 which was the year with the worst air quality (Table 5). Differential changes in air quality across the crop growing areas are reflected in a trend of increasing soybean losses but decreasing wheat losses since 1983.

For both crops, localized areas of high loss were predicted for areas around Chesapeake Bay and portions of the Ohio and Mississippi river valleys (Figures

A

B

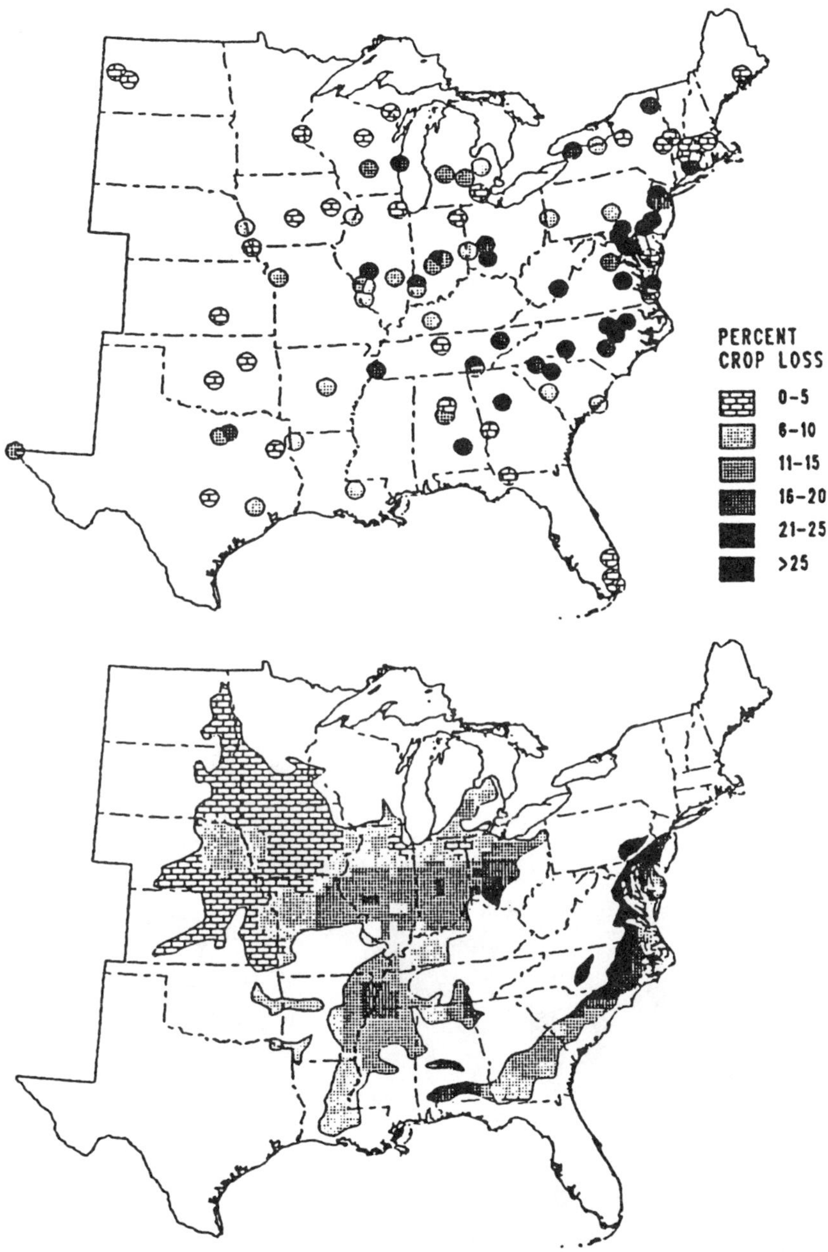

Figure 7. Predicted relative soybean yield loss (%) for 1987. The estimates are based on measured air quality and a composite soybean exposure-response function for O_3 (Figure 6A). The circles on the upper map illustrate the estimated yield loss at each of the rural monitoring sites used to create the gradient surface. The gradient surface on the lower map was interpolated from the monitoring locations using an inverse distance weighting method. The estimated crop losses are likely to be under-estimated as the ambient air quality data were not adjusted for missing values. If the corrections had been made, the measured air quality (SUM06) may have been larger at the various monitoring sites.

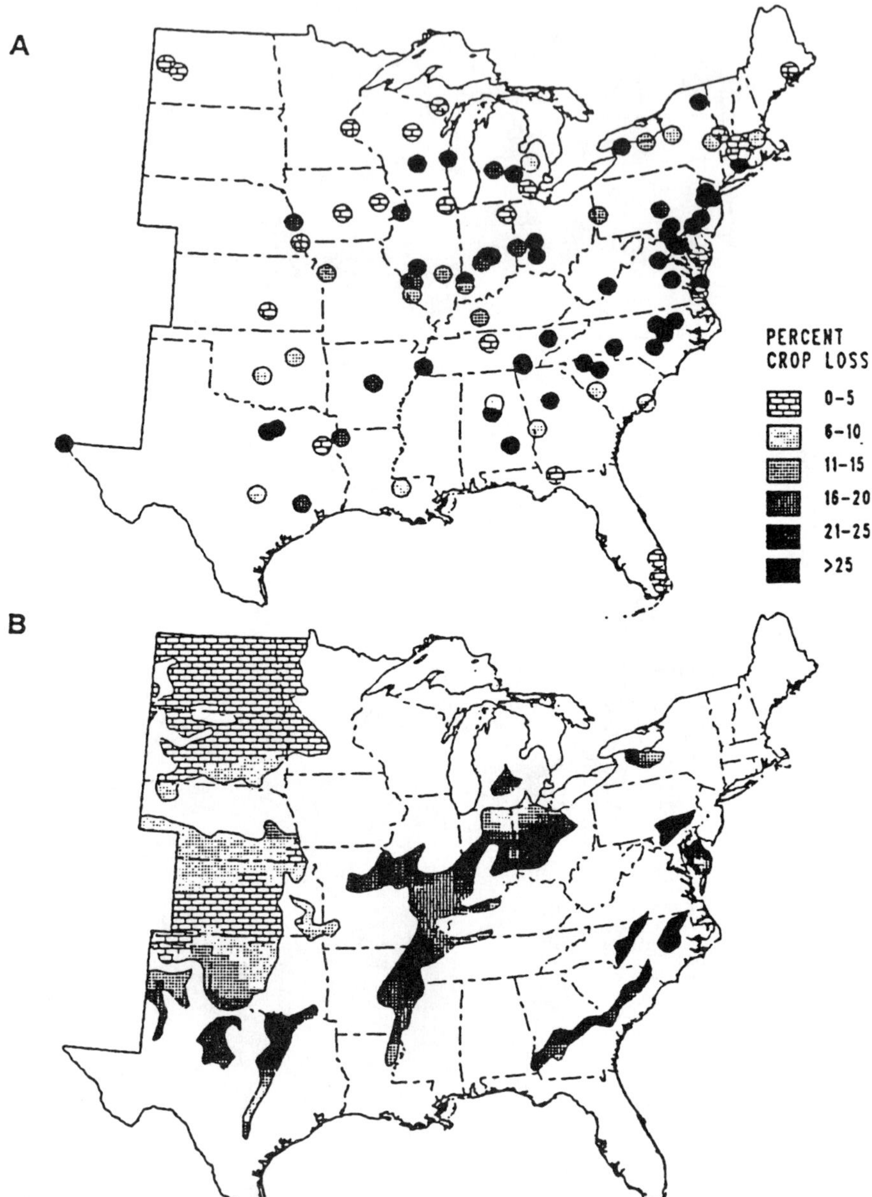

Figure 8. Predicted relative wheat yield loss (%) for 1987. The estimates are based on measured air quality and a composite soybean exposure-response function for O_3 (Figure 6B). The circles on the upper map illustrate the estimated yield loss at each of the rural monitoring sites used to create the gradient surface. The gradient surface on the lower map was interpolated from the monitoring locations using an inverse distance weighting method. Although wheat is grown in other states, they were omitted from the analysis as there were too few air monitoring sites to permit reliable interpolation. The estimated crop losses are likely under estimated as the ambient air quality data were not adjusted for missing values. If the corrections had been made the measured air quality (SUM06) may have been larger at the various monitoring sites.

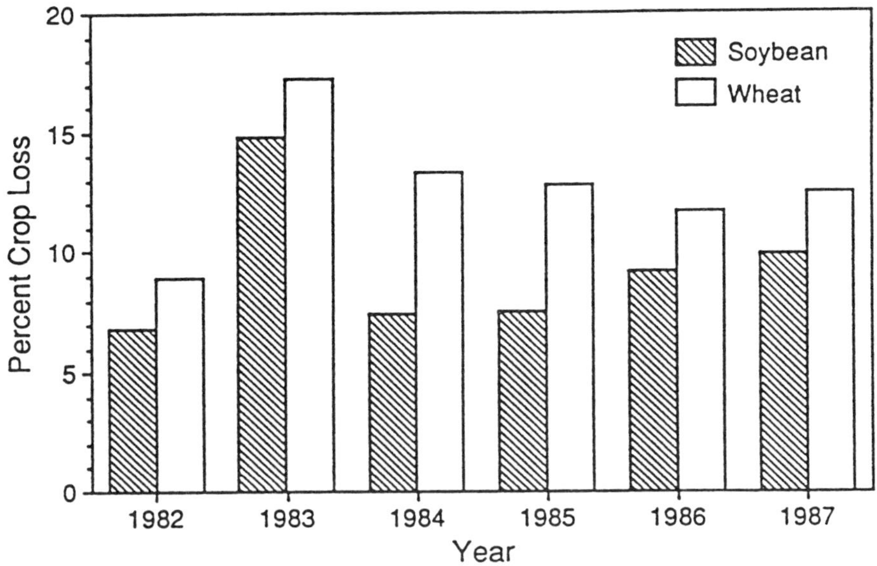

Figure 9. The annual area-weighted average yield loss for soybean and wheat over a 6-year period. The percentages were derived from spatially distributed data as shown in Figures 7 and 8.

7 and 8). For soybean, yield losses of 5 to 10% were forecast for approximately 28% of the area, while yield losses of 20% or greater were anticipated on less than 11% of the area. Similarly, for wheat, yield losses of 5 to 10% were forecast for approximately 18% of the area, while yield losses of 20% or greater were anticipated on approximately 27% of the area.

Development and Analysis of Ambient Air Quality Standards

The promulgation of a NAAQS must consider the scientific evidence for a standard as well as the operational (e.g., deterministic or statistical) form and level. The scientific bases for establishing ambient air quality standards are contained within air quality criteria documents.[37] These documents summarize and evaluate information on "all identifiable effects" ascribed to a pollutant and the concentrations and exposure durations associated with the effects.

The appropriate form (i.e., index) for a standard should have biological relevancy and be based on a response that has importance.[51] The form of the standard should consider the components of exposure that best characterize the biological response of interest. Once the appropriate form is selected, it is necessary to determine the appropriate level of the standard. This involves the determination of the magnitude of effects that occur at different levels of exposure, that is, what is the lowest exposure level at which the magnitude of biological impacts is deemed to be acceptable? For example, the current statistical form of the NAAQS for O_3 is expressed as a daily maximum 1-h concentration which is not to exceed 0.12 ppm on more than 1 day per year on average.[60]

Table 5. The Spatial and Temporal Variation in Ambient O_3 Exposures Expressed in Terms of Several Exposure Indices

Year	Number of sites	2ndHDM mean (ppm)	2ndHDM Cv[a] (%)	M-7 mean (ppm)	M-7 CV (%)	SUM06 mean (ppm-h)	SUM06 Cv (%)	Percent of sites exceeding[b]
1987	117	0.114	22.4	0.053	20.2	26.7	57.1	29.9
1986	117	0.113	21.7	0.051	20.3	23.9	62.8	30.7
1985	124	0.113	22.3	0.051	17.8	24.2	62.0	28.2
1984	111	0.114	24.5	0.051	19.3	24.2	60.0	30.6
1983	115	0.120	26.8	0.054	23.3	29.4	65.3	40.9
1982	115	0.110	27.4	0.049	23.6	22.0	77.0	23.5
Among years[c]	117	0.112	11.2	0.051	10.7	23.9	36.7	30.6

Note: The data are from the EPA's AIRS database for all sites that were coded as agricultural, forest or desert for the years 1982 through 1987. The data are based on a total of 170 sites for a total of 699 site-years. To insure valid data, the following data completeness requirements were used: (1) data for 75% of the hours between 9:00 a.m. and 9:00 p.m. constituted a valid day and (2) 75% valid days within a maximum three-month period constituted a valid site-year.

[a] The coeffient of variation for each of the indices and years provide a measure of the variation among monitoring sites.
[b] Percent of sites that exceeded the current NAAQS for O_3 (0.12 ppm) for a given year.
[c] Among years is a measure of the mean (across years) for each of the indices and the variation in each of the indices among years.

Adequacy of Level of the Current O_3 Standard to Protect Crops

The adequacy of the current standard was evaluated in terms of the degree of protection that it provided to crops. The impact of current (1982 through 1987) O_3 exposures on crop yield was estimated and compared with the crop loss predicted to occur if the current secondary NAAQS for O_3 were attained. Since there is spatial variation in O_3 exposure, the estimates of crop loss were also expressed in a spatial context. For these analyses, similar procedures were used as in the preceding section ("Spatial Assessment of O_3 on Crop Yield") in which databases for O_3 exposure, crop distribution, and O_3 exposure-response functions were combined to provide spatial estimates of crop loss.

- O_3 exposure: air monitoring data from USEPA's AIRS database for 1982 to 1987 were used in the analysis to determine O_3 exposures for rural areas (Table 5). The mean 2ndHDM (for all site-years) was 0.112 ppm with approximately 30% of the site-years exceeding the current NAAQS for O_3.
- Crop distribution: the 12 major annual agricultural crops of the U.S. (Table 2) were used in the evaluation of O_3 effects. Major crop production areas were combined in a Geographic Information System (GIS) to map U.S. agricultural land. The crop production areas were assumed to be constant for the period 1982 to 1987.
- Exposure-response functions: exposure-response functions were developed by regressing crop yield against the O_3 exposure (2ndHDM). Separate exposure-response functions were developed for each of the 49 separate experiments. The individual functions were then combined to provide a composite exposure-response function

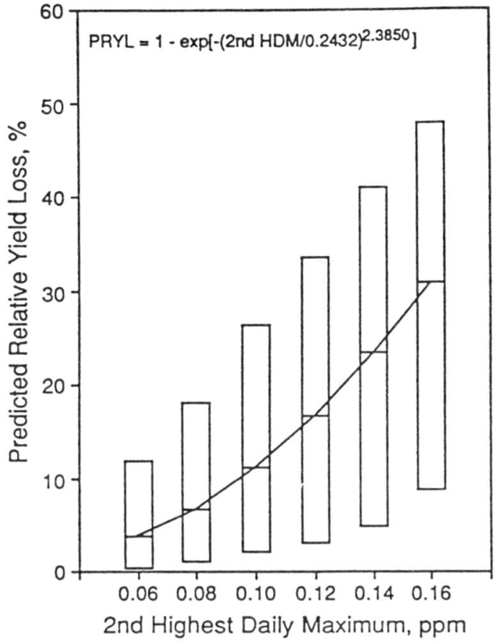

Figure 10. Composite O_3 exposure-response function based on 49 experiments. The function depicts the increase in predicted relative yield loss with increasing O_3 exposure (2ndHDM).

for all crops (Figure 10). Inasmuch as the function was based on a number of separate experiments replicated over time and space with a number of crop species growing under a range of environmental conditions, it incorporated a wide range of variability existing under field conditions.

Predicted Crop Loss— With Current O_3 Levels To assess crop loss, monitored ambient O_3 concentrations (2ndHDM) were combined with the composite predicted relative yield loss function (Figure 10) to estimate crop loss at specific geographic sites. The predicted crop loss associated with the ambient air quality for each of the air monitoring sites is shown as coded circles (Figure 11A). By interpolating these point estimates of crop loss to all areas, a gradient surface is produced (Figure 11B). The gradient surface was constrained by the crop distribution data so that predicted losses are shown only for major agricultural areas.

The assessment of predicted relative yield loss, using current air quality, indicates that significant crop losses may be occurring over large areas of the U.S. (Figure 11). The area-weighted mean predicted relative yield loss was approximately 14.0, with a range of 11.4 to 16% over 1982 to 1987 (Figure 12). The predicted relative yield losses exceeded 20% for approximately 15% of the agricultural area. The magnitude of the predicted yield loss must be viewed with caution as the 2ndHDM is not a robust predictor of crop loss.[52,53] Analyses have established that crop yield is reduced by the growing season-long O_3 exposure which is not well represented by an extreme value statistic such as the 2ndHDM. The 2ndHDM will predict higher agricultural yield losses than a more robust O_3 exposure index such as the SUM06, which accumulates concentration and emphasizes concentrations of 0.06 ppm or greater. Despite this limitation, the conclusion that crop yield is reduced by ambient O_3 (expressed as the 2ndHDM) is consistent with the analysis using a more robust O_3 exposure index (SUM06). The analysis also provides the necessary basis for evaluating the likely impacts of attaining various standards (levels or forms).

Predicted Crop Loss — If Current NAAQS Is Attained Although significant crop losses are predicted to occur over large areas under current air quality, the fundamental question remains: *Would significant crop losses still occur in the U.S. if the current NAAQS for O_3 were attained?* To address this question, NAAQS attainment was simulated for all areas of the U.S. by setting O_3 concentrations (2ndHDMs) greater than 0.12 ppm equal to 0.12 ppm. Using this procedure, O_3 concentrations already in attainment were not adjusted. As in the previous analysis, monitored ambient concentrations of O_3 were combined with the composite predicted relative yield loss function (Figure 10) to estimate crop yield loss at specific geographic sites.

The predicted crop loss for each of the air monitoring sites is shown as a color-coded circle (Figure 13A). Extrapolation of the point predictions to the crop growing area (Figure 13B) reveals that significant crop losses would still occur over large areas of the U.S., even if the current NAAQS were attained at all sites. The area-weighted mean predicted relative yield loss was approximately 11% (vs. 14% with current O_3 levels), with a range of 9.8 to 11.3%, over the 6-year period. That significant yield losses are predicted to occur if the current NAAQS for O_3 were attained is not surprising. Neither the form nor the level of the current NAAQS were specifically developed to protect agricultural crop production. Also, as discussed previously, O_3-crop yield reduction is not well correlated with the 2ndHDM. Consequently, a reduction in the air quality indicator (2ndHDM) is not reflected in yield changes.

Form of a Standard to Protect Crops

The selection of an appropriate form for a standard is important if the resources of interest are to be efficiently protected. If the wrong form is selected, adverse effects may not be prevented and a great deal of time and resources would be expended to attain an inappropriate goal. Considerable effort has already been expended to reduce the concentration of the 2ndHDM; however, this measure of air quality is poorly related to plant growth.[52,53] A primary consideration in the selection of the appropriate form of a standard is the biological basis of the response. O_3 impacts on crop yield are the consequence of the season-long cumulative exposure rather than a single peak event.[37,57,61] Consequently, a

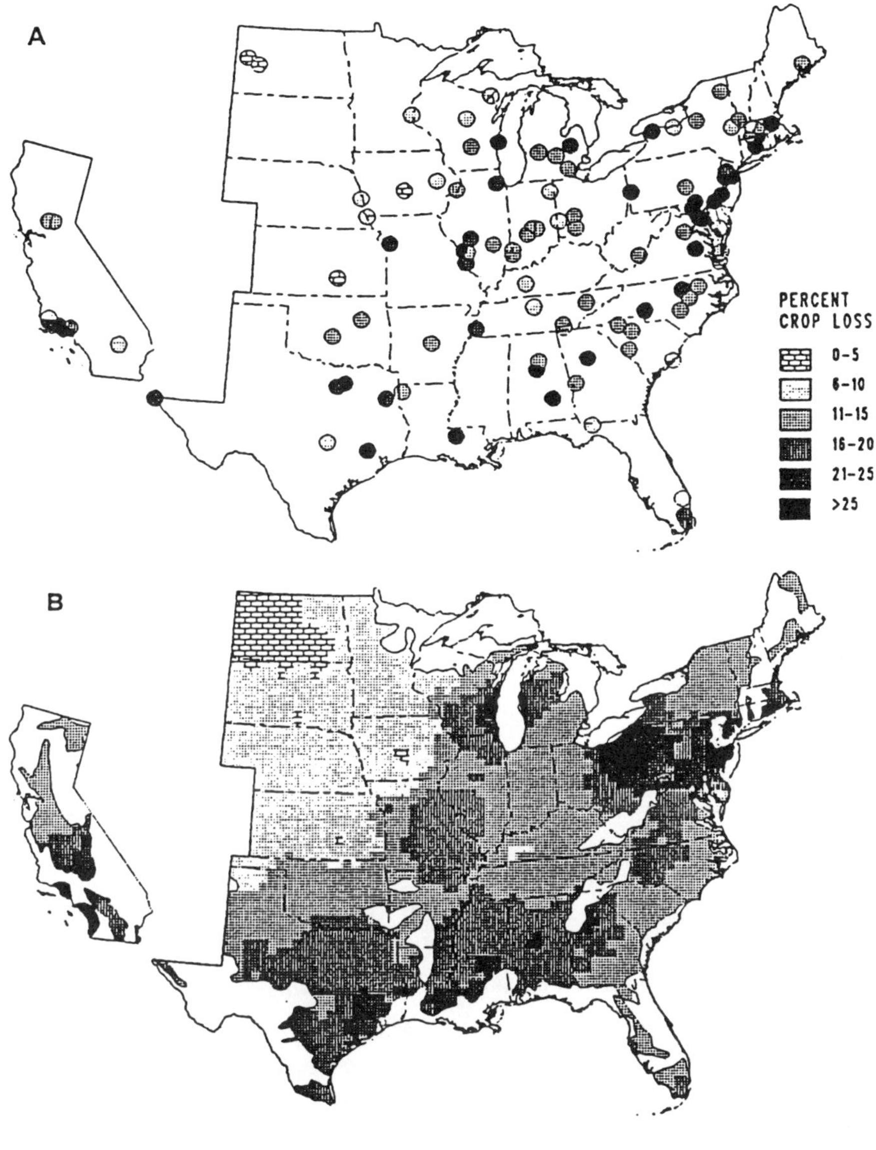

Figure 11. Predicted relative crop loss (%) for 1987. The estimates are based on measured air quality (2ndHDM) and a composite crop exposure-response function for O_3 (Figure 10). The circles on the upper map illustrate the estimated yield loss at each of the rural monitoring sites used to create the gradient surface. The gradient surface on the lower map was interpolated from the monitoring locations using an inverse distance weighting method. Most western states except California were omitted from the analysis as there were too few air monitoring sites to permit reliable interpolation.

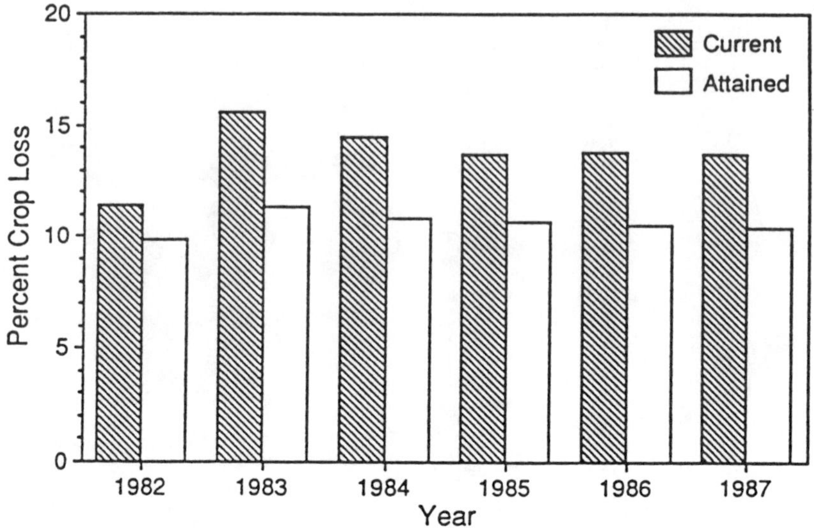

Figure 12. Annual area-weighted average crop loss over a 6-year period. The percentages were derived from spatially distributed data as shown in Figures 11 and 13. The "current" is based on measured ambient O_3 concentrations; for the "attained" category, concentrations greater than 0.12 ppm were set equal to 0.12 ppm, while lower concentrations were not adjusted.

standard should reflect the cumulative nature of the plant growing season exposure.[52-54]

In addition to biological relevancy, the form of the standard should also be capable of capturing year-to-year and site-to-site variability in ambient exposures. An analysis of ambient O_3 exposure (Table 5) showed that the 2ndHDM and M-7 displayed less spatial and temporal variability than the cumulative indices. This greater spatial and temporal variability suggests that the cumulative indices are more efficient in depicting the actual variation in exposure among sites and years than the 2ndHDM or M-7 which showed much less variation among "clean" and "dirty" sites as well as among years.

Possible Forms of the Standard and Ease of Attainment When comparing the proposed levels for 2ndHDM and SUM06 needed to attain the same level of biological protection (e.g., 10% crop loss using the 50th percentile), it is apparent that more sites would meet a SUM06 standard, based on current air quality (Figure 14). For each of the 699 site-years (Table 5), the respective 2ndHDM and SUM06 were calculated and plotted; 21.3% of the site-years (149) would attain a 2ndHDM standard (0.094 ppm), designed to limit crop loss to 10% in half the cases. Of these 149 site-years, 144 would also meet a SUM06 standard (24.4.0 ppm-h) that would provide an equivalent level of protection. In contrast, 53% of the site-years (371) would currently comply with a SUM06 standard that would cause less than 10% crop loss in half the cases, suggesting that it may be easier to attain a cumulative standard.

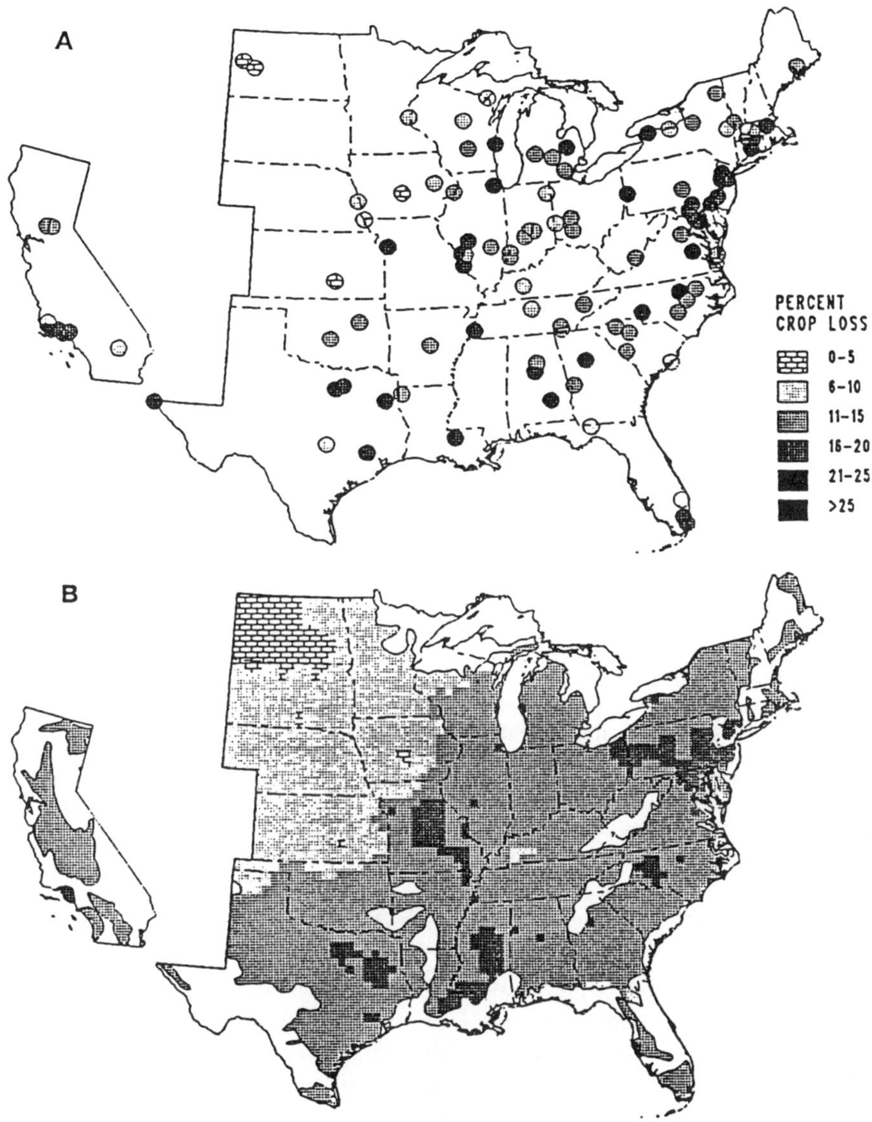

Figure 13. Predicted relative crop loss (%) for 1987, if the current NAAQS for O_3 were attained. To simulate NAAQS attainment, 2ndHDMs greater than 0.12 ppm were set equal to 0.12 ppm; the concentrations already in attainment were not adjusted. As in Figure 11, a composite crop exposure-response function for O_3 (Figure 10) was used. The circles on the upper map illustrate the estimated yield loss at each of the rural monitoring sites used to create the gradient surface. The gradient surface on the lower map was interpolated from the monitoring locations using an inverse distance weighting method. Most western states except California were omitted from the analysis because there were too few air monitoring sites to permit reliable interpolation.

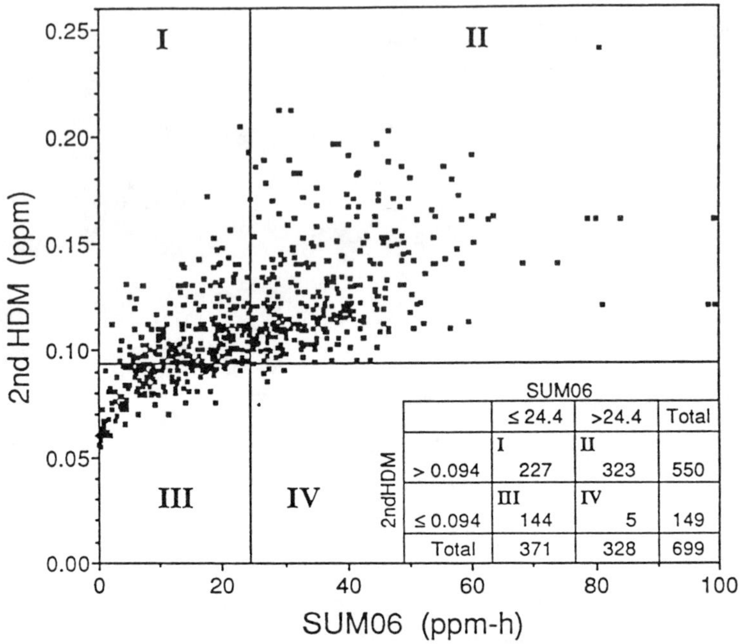

Figure 14. The association between measured O_3 exposures expressed as the 2ndHDM or SUM06. The vertical line is at 24.4 ppm-h (SUM06) and the horizontal line at 0.094 ppm (2ndHDM). These exposure values were selected because they provide an equivalent level of crop protection, that is, limiting crop loss to 10% in half of the cases based on exposure-response functions. The insert shows the number of site-years in each of the four quadrants.

CONCLUSIONS

Ambient O_3 produces significant impacts to crops across the U.S. Field studies have moved from characterization of O_3 foliar injury symptoms to quantification of yield losses using open-top chambers in order to develop O_3 exposure-yield loss functions. These loss functions, in conjunction with ambient O_3 exposure and crop distribution data, have been used to predict a current 14% yield loss (for the years 1982–1987) for major U.S. crops. Yield losses would still occur even if all U.S. sites would attain the current secondary NAAQS. It is apparent, therefore, that the form and level of the current standard is not protective of crop production.

REFERENCES

1. Heck, W. W., O. C. Taylor, R. Adams, G. Bingham, J. Miller, E. Preston, and L. Weinstein. "Assessment of Crop Loss from Ozone," *J. Air Pollut. Control Assoc.* 32:353–361 (1982).

2. Guderian, R., D. T. Tingey, and R. Rabe. "Effects of Photochemical Oxidants on Plants," in *Air Pollution by Photochemical Oxidants*, R. Guderian, Ed. (Berlin: Springer-Verlag, 1985), pp. 127–333.
3. Rey, H. D. L., L. I. de Bauer, J. K. Shibata, and N. M. Mendoza. "Impacto de los Oxidantes Ambientales en el Cultivo de Frijol, en Montecillos, Estado de Mexico," *Agrociencia* 66:83–95 (1986).
4. Tingey, D. T., and G. E. Taylor, Jr. "Variation in Plant Response to Ozone: A Conceptual Model of Physiological Events," in *Effects of Gaseous Air Pollution in Agriculture and Horticulture*, M. H. Unsworth and D. P. Ormrod, Eds. (London: Butterworth Scientific, 1982), pp. 113–138.
5. Tingey, D. T., and C.P. Anderson. "The Physiological Basis of Differential Plant Sensitivity to Changes in Air Quality," in *Ecological Genetics and Air Pollution*, G. E. Taylor, Jr., L. F. Pitelka, and M. T. Clegg, Eds. (New York: Springer-Verlag, 1991), pp. 209–235.
6. Frank, R. "Defining Adverse Health Effects," *Environmental Forum* April pp. 9–11 (1983).
7. Russell, M. "Ozone Pollution: The Hard Choices," *Science* 241:1275–1276 (1988).
8. Guderian, R. *Air Pollution: Phytotoxicity of Acidic Gases and Its Significance in Air Pollution Control*, (Berlin: Springer-Verlag, 1977), pp. 75–97.
9. Tingey, D. T., W. E. Hogsett, and S. Henderson. "Definition of Adverse Effects for the Purpose of Establishing Secondary National Ambient Air Quality Standards," *J. Environ. Qual.* 19:635–639 (1990).
10. "Review of the National Ambient Air Quality Standards for Ozone. Assessment of Scientific and Technical Information," U.S. Environmental Protection Agency, Air Quality Management Division, Office of Air Quality Planning and Standards, Research Triangle Park, NC, U.S. EPA, OAQPS Draft Staff Paper (1988).
11. Millecan, A. A. "A Survey and Assessment of Air Pollution Damage to California Vegetation in 1970," Air Pollution Control Office, Research Triangle Park, NC, U.S. EPA Report APTD-0694 (1971).
12. Millecan, A. A. "A Survey and Assessment of Air Pollution Damage to California Vegetation: 1970 through 1974," California Department of Food and Agriculture, Sacramento, CA (1976).
13. Tingey, D. T., W. W. Heck, and R. A. Reinert. "Effect of Low Concentrations of Ozone and Sulfur Dioxide on Foliage, Growth and Yield of Radish," *J. Am. Soc. Hortic. Sci.* 96:369–371 (1971).
14. Kress, L. W., and J. M. Skelly. "Response of Several Eastern Forest Tree Species to Chronic Doses of Ozone and Nitrogen Dioxide," *Plant Dis.* 66:1149–1152 (1982).
15. Feder, W. A., and F. J. Campbell. "Influence of Low Levels of Ozone on Flowering of Carnations," *Phytopathology* 58:1038–1039 (1968).
16. Adedipe, N. O., R. E., Barrett, and D. P. Ormrod. "Phytotoxicity and Growth Responses of Ornamental Bedding Plants to Ozone and Sulfur Dioxide," *J. Am. Soc. Hortic. Sci.* 97:341–345 (1972).
17. Heagle, A. S., D. E. Body, and G. E. Neely. "Injury and Yield Responses of Soybean to Chronic Doses of Ozone and Sulfur Dioxide in the Field," *Phytopathology* 64:132–136 (1974).
18. Oshima, R. J., O. C. Taylor, P. K. Braegelmann, and D. W. Baldwin. "Effect of Ozone on the Yield and Plant Biomass of a Commercial Variety of Tomato," *J. Environ. Qual.* 4:463–464 (1975).
19. Heagle, A. S., R. B. Philbeck, and W. M. Knott. "Thresholds for Injury Growth, and Yield Loss Caused by Ozone on Field Corn Hybrids," *Phytopathology* 69:21–26 (1979).

20. "Air Quality Criteria for Ozone and Other Photochemical Oxidants," U.S. Environmental Protection Agency, Environmental Criteria and Assessment Office, Research Triangle Park, NC, U.S. EPA Report-600/8-78-004 (1978).
21. Ensign, J., G. Hofstra, and R. C. Roy. "The Impact of Ozone on Peanut *Arachis-hypogaea* Exposed in the Laboratory and Field," *Phytopathology* 75:429–432 (1985).
22. Heggestad, H. E., J. H. Bennett, E. H. Lee, and L. W. Douglas. "Effects of Increasing Doses of Sulfur Dixoide and Ambient Ozone on Tomatoes Plant Growth, Leaf Injury, Elemental Composition, Fruit Yields, and Quality," *Phytopathology* 76:1338–1344 (1986).
23. Musselman, R. C., W. J. Kender, and D. E. Crowe. "Determining Air Pollutant Effects on the Growth and Productivity of Cultivar Concord Grapevines Using Open Top Chambers," *J. Am. Soc. Hortic. Sci.* 103:645–648 (1978).
24. Olszyk, D. M., A. Bytnerowicz, G. Kats, P. J. Dawson, J. Wolf, and C. R. Thompson. "Crop Effects from Air Pollutants in Air Exclusion Systems vs. Field Chambers," *J. Environ. Qual.* 15:417–422 (1986).
25. lszyk, D. M., A. Bytnerowicz, G. Kats, P. J. Dawson, J. Wolf, and C. R. Thompson. "Effects of Sulfur Dixoide and Ambient Ozone on Winter Wheat and Lettuce," *J. Environ. Qual.* 15:363–369 (1986).
26. Olszyk, D. M., G. Kats, C. L. Morrison, P. J. Dawson, I. Gocka, J. Wolf, and C. R. Thompson. "Valencia Orange Fruit Yield with Ambient Oxidant or Sulfur Dixoide Exposures," *J. Am. Soc. Hortic. Sci.* 115:878–883 (1990).
27. Pell, E. J., N. Pearson, and C. Vinten-Johansen. "Qualitative and Quantitative Effects of Ozone and/or Sulfur Dioxide on Field-Grown Potato Plants," *Environ. Pollut.* 53:171–186 (1988).
28. Slaughter, L. H., C. L. Mulchi, E. H. Lee, and K. Tuthill. "Chronic Ozone Stress Effects on Yield and Grain Quality of Soft Red Winter Wheat," *Crop Sci.* 29:1251–1255 (1989).
29. Snyder, R. G., J. E. Simon, R. A. Reinert, M. Simini, and W. W. Heck. "Effects of Air Quality on Foliar Injury, Growth, Yield and Quality of Muskmelon," *Environ. Pollut.* 53:187–196 (1988).
30. Takemoto, B. K., D. M. Olszyk, A. G. Johnson, and C. R. Parada. "Yield Responses of Field-Grown Crops to Acidic Fog and Ambient Ozone," *J. Environ. Qual.* 17:192–197 (1988).
31. Heck, W. W., R. M. Adams, W. W. Cure, A. S. Heagle, H. E. Heggestad, R. J. Kohut, L. W. Kress, I. O. Rawlings, and O. C. Taylor. "A Reassessment of Crop Loss from Ozone," *Environ. Sci. Technol.* 17:573A-581A (1983).
32. "Agricultural Statistics 1987," U.S. Department of Agriculture, U.S. Government Printing Office, Washington, D.C. (1987).
33. Heagle, A. S., L. W. Kress, P. J. Temple, R. J. Kohut, J. E. Miller, and H. E. Heggestad. "Factors Influencing Ozone Dose-Yield Response Relationships in Open-Top Field Chamber Studies," in *Assessment of Crop Loss From Air Pollutants*, W. W. Heck, O. C. Taylor, and D. T. Tingey, Eds. (London: Elsevier Applied Science, 1988), pp. 141–179.
34. Coffey, D. E., J. C. Sprenger, D. T. Tingey, G. E. Neely, and J. C. McCarthy. "National Crop Loss Assessment Network: Quality Assurance Program," *Environ. Pollut.* 53:89–98 (1988).
35. Lesser, V. M., J. O. Rawlings, S. E. Spruill, and M. C. Sommerville. "Ozone Effects on Agricultural Crops: Statistical Methodologies and Estimated Dose-Response Relationships," *Crop Sci.* 30:148–155 (1990).

36. Sommerville, M. C., S. E. Spruill, J. O. Rawlings, and V. M. Lesser. "Impact of Ozone and Sulfur Dioxide on the Yield of Agricultural Crops," in *Tech. Bull. 292* (Raleigh, NC: Dept. of Statistics, North Carolina State University, 1989).
37. "Ambient Air Criteria for Ozone and Other Photochemical Oxidants," U.S. Environmental Protection Agency, Environmental Criteria and Assessment Office, Research Triangle Park, NC, U.S. EPA Report-600/8-84/020cF (1986), pp. 6-1, 6-298.
38. Mathy, P. "The European Open-Top Chambers Programme: Objectives and Implementation," in *Assessment of Crop Loss from Air Pollutants*, W. W. Heck, O. C. Taylor, and D. T. Tingey, Eds. (New York: Elsevier Applied Science, 1988), pp. 505–513.
39. Adaros, G. H., J. Weigel, and H.-J. Jäger. "Impact of Ozone on Growth and Yield Parameters of Two Spring Wheat Cultivars (*Triticum aestivu* L.)," *Zeitschrift Pflanzenkrankheiten Pflanzenschutz* 98:113–124 (1991).
40. Fuhrer, J., A. Egger, B. Lehnherr, A. Grandjean, and W. Tschannen. "Effect of Ozone on the Yield of Spring Wheat (*Triticum aestivum* L., cv. Albis) Grown in Open-Top Field Chambers," *Environ. Pollut.* 60:273–289 (1989).
41. Pleijel, H., L. Skärby, G. Wallin, and G.Selldén. "Yield and Grain Quality of Spring Wheat (*Triticum aestivum* L., cv. Drabant) Exposed to Different Concentrations of Ozone in Open-Top Field Chambers," *Environ. Pollut.* 69:151–168 (1991).
42. Lefohn, A. S., H. P. Knudsen, J. L. Logan, J. Simpson, and C. Bhumralkar. "An Evaluation of the Kriging Method to Predict 7-hr Seasonal Mean Ozone Concentrations for Estimating Crop Losses," *J. Air Pollut. Control Assoc.* 37:595–02 (1987).
43. Knudsen, H. P., and A. S. Lefohn. "The Use of Geostatics to Characterize Regional Ozone Exposures," in *Assessment of Crop Loss from Air Pollutants*, W. W. Heck, O. C. Taylor, and D. T. Tingey, Eds. (London: Elsevier Applied Science, 1988), pp. 91–105.
44. Tingey, D. T., W. E. Hogsett, E. H. Lee, A. A. Herstrom, and S. H. Azevedo. "An Evaluation of Various Alternative Ambient Ozone Standards Based on Crop Yield Loss Data," in *Transactions Tropospheric Ozone and the Environment*, R. L. Berglund, D. R. Lawson, and D. J. McKee, Eds. (Pittsburgh: Air and Waste Management Association, 1991), pp. 272–288.
45. "Federal Regulation (Executive Order 12291 of Feb 17, 1981)," *Federal Register, The President*, 46:33 (19 Feb. 1981) Part 3. pp. 13193–13198.
46. Adams, R. M., S. A. Hamilton, and B. A. McCarl. "An Assessment of the Economic Effects of Ozone on U.S. Agriculture," *J. Air Pollut. Control Assoc.* 35:938–943 (1985).
47. Adams, R. M., J. D. Glyer, and B. A. McCarl. "The NCLAN Economic Assessment: Approach, Findings and Implications," in *Assessment of Crop Loss from Air Pollutants*, W. W. Heck, O. C. Taylor, and D. T. Tingey, Eds. (London: Elsevier Applied Science, 1988), pp. 473–504.
48. Adams, R., J. D. Glyer, S. L. Johnson, and B. A. McCarl. "A Reassessment of the Economic Effects of Ozone on U.S. Agriculture," *J. Air Pollut. Control Assoc.* 39:960–968 (1989).
49. Jordan, B. C., A. C. Basala, P. M. Johnson, M. H. Jones, and B. Madariaga. "Policy Implications from Crop Loss Assessment Research: The U.S. Perspective," in *Assessment of Crop Loss from Air Pollutants*, W. W. Heck, O. C. Taylor, and D. T. Tingey, Eds. (New York: Elsevier Applied Science, 1988), pp. 521–535.
50. Heagle, A. S., and W. W. Heck. "Field Methods to Assess Crop Losses Due to Oxidant Air Pollutants," in *Crop Loss Assessment: Proceedings of E. C. Stakman Commemorative Symposium*, P. S. Teng and S. V. Krupa, Eds. (St. Paul, MN: University of Minnesota, 1980), pp. 296–305.

51. Hogsett, W. E., D. T. Tingey, and E. H. Lee. "Ozone Exposure Indices: Concepts for Development and Evaluation of Their Use," *Assessment of Crop Loss from Air Pollutants*, W. W. Heck, O. C. Taylor, and D. T. Tingey, Eds. (London: Elsevier Applied Science: 1988), pp. 107–138.
52. Lee, E. H., D. T. Tingey, and W. E. Hogsett. "Evaluation of Ozone Exposure Indices in Exposure-Response Modeling," *Environ. Pollut.* 53:43–62 (1988).
53. Lee, E. H., W. E. Hogsett, and D. T. Tingey. "Efficacy of Ozone Exposure Indices in the Standard Setting Process," in *Transactions Tropospheric Ozone and the Environment*, R. L. Berglund, D. R. Lawson, and D. J. McKee, Eds. (Pittsburgh: Air and Waste Management Association, 1991), pp. 225–271.
54. Hogsett, W. E., D. T. Tingey, and S. R. Holman. "A Programmable Exposure Control System for Determination of the Effects of Pollutant Exposure Regimes on Plant Growth," *Atmos. Environ.* 19:1135–1145 (1985).
55. Musselman, R. C., R. J. Oshima, and R. E. Gallavan. "Significance of Pollutant Concentration Distribution in the Response of 'Red Kidney' Beans to Ozone," *J. Am. Soc. Hortic. Sci.* 108:645–648 (1983).
56. Lefohn, A. S., J. A. Laurence, and R. J. Kohut. "A Comparison of Indices that Describe the Relationship Between Exposure to Ozone and Reduction in the Yield of Agricultural Crops," *Atmos. Environ.* 22:1229–1240 (1988).
57. Tingey, D. T., W. E. Hogsett, and E. H. Lee. "Analysis of Crop Loss for Alternative Ozone Exposure Indices," in *Atmospheric Ozone Research and its Policy Implications*, T. Schneider, S. D. Lee, G. J. R. Wolters, and L. D. Grant, Eds. (Amsterdam: Elsevier, 1989), pp. 219–227.
58. Lee, E. H., D. T. Tingey, and W. E. Hogsett. "Exposure Indices Consideration for Rural Ozone Relationships in the U.S.," *American Statistical Association 1989 Proceedings of the Business and Economics*, Section 517–522 (1990).
59. "Census of Agriculture," U.S. Dept. of Commerce, Bureau of the Census, Crops Harvested Maps, Vol. 2, Subject Series, Part 1, Washington, D.C. (1987).
60. "Revisions to the National Primary and Secondary Ambient Air Quality Standards for Photochemical Oxidants," *Federal Register* 44(28):8202–8221 (1979).
61. Cure, W. W., J. S. Sanders, and A. S. Heagle. "Crop Yield Responses Predicted with Different Characterizations of the Same Ozone Treatments," *J. Environ. Qual.* 15:251–254 (1986).

SECTION II

Perspectives on Tropospheric Ozone Pollution

CHAPTER 7

THE CLEAN AIR ACT AMENDMENTS — THE USEPA'S ROLE IN ABATING OZONE AIR POLLUTION

G. T. Helms, Annie Nikbakht, and Willis P. Beal

INTRODUCTION

This chapter traces the evolution of the Federal Clean Air Act and the associated efforts to abate air pollution problems in the U.S., particularly those related to ozone (O_3). Early on, Congress believed that states should have the primary role in ensuring air quality. The focus of the earlier federal legislation was on the development of state and local air pollution control agencies, with the federal government serving in a support role for research and financial assistance. It became clear that more active participation by the federal government was needed; therefore, the U.S. Environmental Protection Agency (USEPA) was formed in 1970 and amendments to the Clean Air Act (CAA) were also passed in 1970 and 1977. Moreover, Congress has recently attempted to protect air quality by approving the most sweeping clean air legislation yet, the Clean Air Act Amendments of 1990, signed by President Bush on November 15, 1990.

In particular, this chapter traces the history of control programs to attain the National Ambient Air Quality Standard (NAAQS) for O_3. A brief history of the development of the CAA and an overview of these most recent 1990 CAA Amendment changes impacting O_3 are also presented. New requirements for states to develop state implementation plans (SIPs) that demonstrate effective reduction

0-87371-475-X/94/$0.00+$.50
© 1994 by Lewis Publishers

of volatile organic compounds (VOCs), precursors to O_3 formation, are noted. New nonattainment (Title I) requirements for O_3 are also discussed in this chapter. Requirements in Title II for mobile source controls are also examined. Likewise, due to the increasing recognition of the role played by oxides of nitrogen (NO_x) in O_3 formation, the most recent requirements for control of NO_x emissions will be examined.

BRIEF HISTORY OF THE ACT

With the passage of initial legislation in 1955, the U.S. Congress first began to recognize the health and welfare impacts associated with the pollution of the nation's ambient or outside air. This first legislative effort was basically restricted to research into the nature and extent of the nation's air pollution problem.

The 1963 CAA authorized grants to state and local agencies to assist them in their control programs. The Act also gave the federal government some limited authority to take action to abate interstate pollution problems.

This authority was strengthened by the 1967 Air Quality Act which gave citizens the statutory right to participate actively in the control process through public abatement hearings. Citizens took advantage of this opportunity to become forceful local advocates for clean air.

While significant control activity was undertaken in the U.S. in the 1950s and 1960s, most notably in southern California, the major national effort to clean up urban smog or O_3 began with the passage of the 1970 CAA Amendments. The 1970 Act established procedures for setting NAAQS, set up a comprehensive system for the development of SIPs to attain the NAAQS, and set up federal programs to control new and hazardous sources. The Act focused on several areas:

1. Section 108 provided for setting NAAQS (the NAAQS for O_3 was initially set at 0.08 ppm)*
2. Section 110 provisions required SIPs to be attained by 1975 and 1977
3. Section 111 established new source performance standards (NSPS) for new sources
4. Section 112 provided for national emissions standards for hazardous air pollutants (NESHAPs) to be set

The 1970 Amendments introduced the concept of air quality management into the vocabulary of air pollution control. This meant, in general, that a state would inventory the emissions put out by its sources, model their impact, and adopt and implement controls that would prevent violations of the NAAQS. For O_3 specifically, the policy direction was toward the control of VOCs because of their importance as a precursor of O_3.

* The earlier 0.08-ppm standard (considers every hour of the year) was an hourly average not to be exceeded more than once per year; however, the current 0.12-ppm standard (based on daily maximum hourly concentration) is a 1-h average for each day.

Table 1. Man-Made Sources of VOCs in 1980

	Millions of metric tons	% of total
Transportation	8.2	38.3
Industrial processes	5.8	27.1
Fuel combustion	2.5	11.7
Misc. solvents	1.9	8.9
Petroleum storage & marketing	1.4	6.5
Other	1.6	7.5
Total	21.4	100.0

From NAPAP Interim Assessment, Vol. II MacKenzie; 1988, 41.

The largest single source of man-made VOCs are transportation sources. Table 1, lists the major VOC sources.

Under the 1970 Amendments, SIPs for the attainment of the NAAQS were due by January 31, 1972. The first attempts to control stationary sources of VOCs were begun under the 1970 Amendments. Similarly, for mobile sources, the Federal Motor Vehicle Control Program (FMVCP) was introduced, which set vehicle emission limits and inspection/maintenance (I/M) programs for motor vehicle emissions of hydrocarbons (for O_3) and carbon monoxide were begun.

The SIP requirement in the 1970 Amendments was reinforced by a federal grant program to assist state and local air pollution control agencies.

After the attainment dates of 1975 (or 1977 if a state had a 2-year extension for O_3 or carbon monoxide), air quality conditions revealed that many cities had not attained the O_3 NAAQS. Additional control measures were needed; thus, the 1977 CAA Amendents were passed. The 1977 Amendments introduced a broad new planning effort for O_3 and called for attainment by 1982, or 1987 if given an extension. One important condition for getting an extension was adding a state I/M program. The 1977 Amendments established a program to designate nonattainment areas under Section 107 as "nonattainment" or "attainment/unclassifiable". This was accomplished in early 1978. The 1977 Amendments also added new source review (NSR) provisions for nonattainment areas, as well as prevention of significant deterioration (PSD) requirements. In addition, the 1977 Amendments called for reasonably available control technology (RACT) on stationary sources. The USEPA subsequently issued 26 control technique guidelines (CTGs) to assist the states in implementing RACT. The focus of the CTGs was to reduce VOC emissions to RACT by control equipment or coating formulations. (Approximately 80% overall emission reductions were called for.) The concept of transportation control measures was also introduced with the 1977 Amendments.

General control requirements were not altered with the 1979 revision of the NAAQS for O_3 to 0.12 ppm. A number of nonattainment areas continued to fall short of attaining the NAAQS for O_3; thus, the USEPA issued calls for SIP revisions under Section 110(a)(2)(H) of the CAA in 1986, 1988, and 1989 for the states to revise plans to provide for attainment as soon as possible.

As 1987 approached, it become apparent to the USEPA that as many as 100 O_3 areas would not attain the NAAQS. On November 24, 1987 (52 FR 45043), the

Table 2. Summary of the Development of the Clean Air Act

Year	Action title	Significant initiatives
1955	Air Pollution Control-Research & Technical Assistance Act	Research on air pollution problem
1963	Clean Air Act	First regulatory legislation
1965	Amendments to CAA	Added Title II, The Motor Vehicle Air Pollution Control Act, emission stds. for new vehicles
1967	Amendments to CAA	Increased stationary source control; gave federal government authority to adopt emission control regulations where needed; provided limited federal enforcement; set up Air Quality Control Regions (AQCRs)
1970	Amendments to CAA	Required EPA to set NAAQS; established action deadlines for development of SIPs to attain by 1975 and 1977; required EPA to set NESHAPs; established NSPS for new sources
1977	Amendments to CAA	Introduced broad new O_3 planning effort; called for attainment by 1982 or 1987 if extension; called for RACT on stationary sources and EPA to issue CTGs; Nonattainment areas to be designated for the first time according to county boundaries instead of AQCRs; added new source review provisions (NSR); provided for sanctions for disapproved SIPs and failure to implement SIPS; concept of transportation control measures introduced
1990	Amendments to CAA	Introduced sweeping changes to the Act; new approach to nonattainment problems (deadlines & penalties for continued nonattainment); establishment of air toxics program; added enforcement powers; established a national permits program; introduced market-based controls

USEPA proposed policy guidance on approval of post-1987 O_3 and carbon monoxide plan revisions for areas not attaining the NAAQS, including Appendix D covering discrepancies and inconsistencies found in the SIPs. Additional clarification of Appendix D, "Issues Relating To VOC Regulation Cutpoints, Deficiencies, and Deviations", known as the "Blue Book", was published on May 25, 1988 and later revised (January 11, 1990). This document provided guidance and clarification so that states could fully implement or adopt their existing required regulations in a nationally consistent way. However, the post-1987 policy was never finalized as Congress began to seriously debate the new 1990 CAA Amendents to address the problem. The latest 1990 Amendments supersede post-1987 policy work and include many elements of the proposed policy.

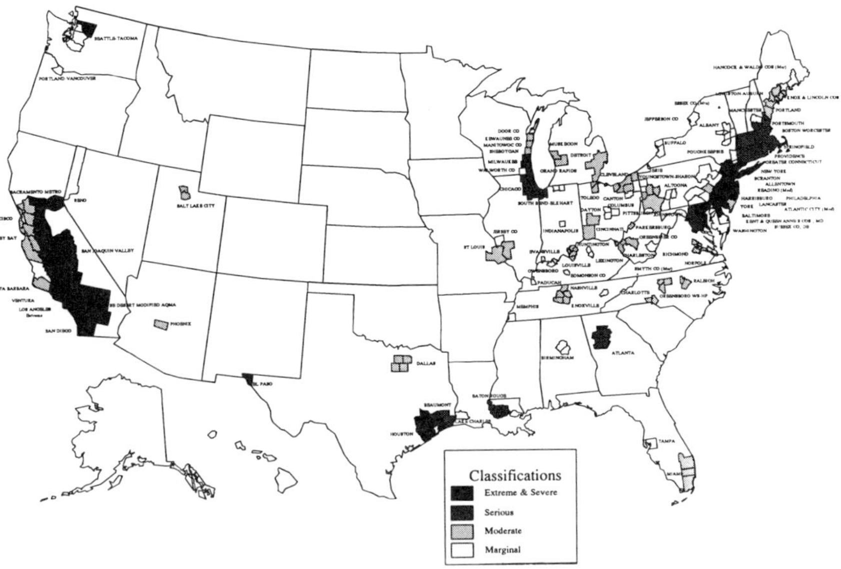

Figure 1. O_3 nonattainment areas.

Table 2 briefly summarizes the development of the CAA over the years.

TITLE I PROVISIONS PROVIDING FOR THE CONTROL OF O_3

Despite the gains made since the first CAA passed in 1970, there were still many areas violating the health-based standards after 1987, the last deadline in the 1977 CAA Amendents. The national map in Figure 1 indicates the scope of this nonattainment area problem for O_3. This continuing nonattainment situation was a clear signal that more needed to be done; but how could anyone be sure that what is done this time will bring better results? The 1990 Amendments contain at least three answers to this question.

First, the previous "one size fits all" approach to addressing all nonattainment problems was discarded and replaced by a system that recognizes the different degrees of the nonattainment problem in the U.S. Controls will be made increasingly more stringent as the severity of the problem increases. Attainment dates are set that will allow the areas with the worst problems to take more time than areas with less serious problems.

Second, a blend of technology-based controls and a long-term planning approach is required. This will insure a quick start with specific required controls placed on air pollution sources to be implemented by everyone, along with more sophisticated planning tools, where appropriate, where more intensive controls are needed to solve more difficult and longer-term air quality problems.

Third, the possibility of not attaining the standards within prescribed dates and failing to meet deadlines and milestones was anticipated this time in the legislation. Earlier legislation assumed that attainment deadlines would be met and provided no guidance on what to do if they were not. The 1990 Amendments anticipated this possible problem and carefully spelled out the consequences, in discussions to follow.

With these changes in mind, the 1990 Amendments establish a classification system for nonattainment areas so that areas with similar air quality problems are grouped together. Areas within each classification are obligated to apply similar control measures and share the same attainment deadline. In this way, the control measures are more closely tailored to the problems than they were in the past.

Areas are first officially designated by the USEPA as either attainment, nonattainment, or unclassifiable (because of insufficient data). Once an area is designated "nonattainment", it is then classified according to its air quality "design value", a numerical value which represents that area's air quality situation. The Congress specified five classifications for O_3 — marginal, moderate, serious, severe, or extreme — all to be based on the area's design value. The entire designation/classification process is quite detailed, but one important date in this process was USEPA's announcement of most designations on November 6, 1991. Areas already designated "nonattainment" under previous legislation at the date of enactment (November 15, 1990) were then designated "nonattainment" by operation of law.

The geographic size of the nonattainment areas is a significant nonattainment issue because one of the reasons for past failures to attain is that the area of control was not large enough to include all of the emissions that caused the problem. To overcome this, the 1990 Amendments required that the boundaries for serious, severe, and extreme O_3 nonattainment areas must be the metropolitan statistical area (MSA) or the consolidated metropolitan statistical area (CMSA) unless the state makes an acceptable showing that the outer areas do not contribute significantly to the problem. This is appropriate because MSAs and CMSAs are metropolitan areas having similar population densities, commuting patterns, commercial and industrial bases, and the like. The Office of Management and Budget has identified these areas. In some cases, MSAs in close proximity are grouped together into CMSAs because of the social and economic interaction within the entire area.

Table 3 shows the design values, the numbers of nonattainment areas across the country, and the time each classification has to attain the O_3 standard. Table 3 shows that areas with the most difficult air quality problems are given the most time to solve their problems. Similarly, the greater the air quality problem, the more stringent the controls that will be applied. Figure 2 shows the cumulative nature of the application of the controls for O_3. O_3 controls begin with *marginal* areas, where one of the main requirements is for states to correct their regulations controlling emissions from stationary sources that had been adopted and implemented in years prior to enactment of the recent Amendments. In general, these are control measures for a wide range of air pollution sources — auto painting, paper coating, dry cleaners, etc — that the USEPA has defined as being "reason-

Table 3. O_3 Classifications

Classification	Design value (ppm)	Number of areas[a]	Time to attain
Marginal	0.121–0.138	39	3 years after enactment
Moderate	0.138–0.160	32	6 years after enactment
Serious	0.160–0.180	16	9 years after enactment
Severe 1	0.180–0.190	5	15 years after enactment
Severe 2	0.190–0.280	3	17 years after enactment
Extreme	>0.280	1	20 years after enactment

[a] Based on 1987–1989 air quality data.

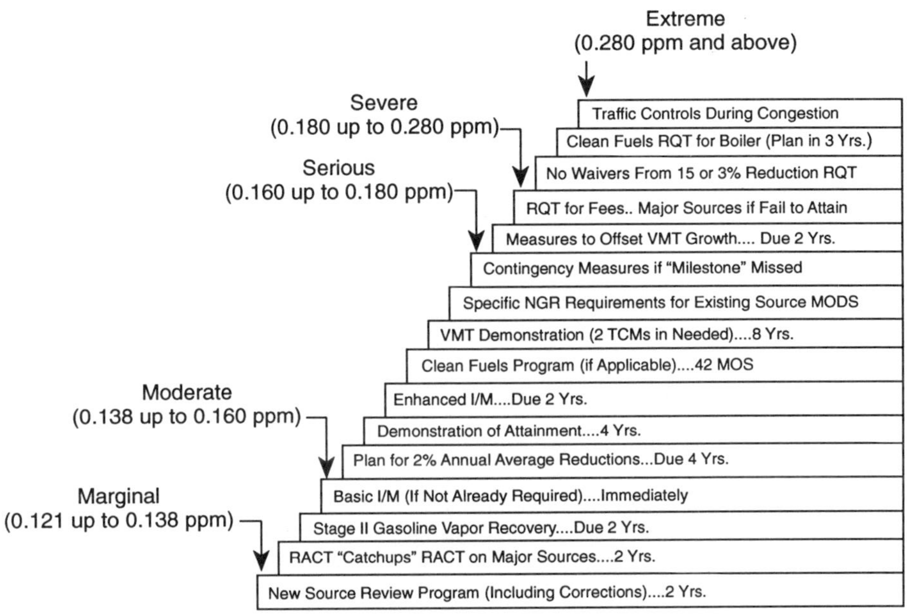

Figure 2. O_3 plan requirements.

ably available control technology" (RACT) for application to these kinds of sources. This means that states must complete requirements and commitments that were previously due. This, plus the other requirements shown, and national measures, such as the controls built into automobiles at the factory (FMVCP) and gasoline vapor pressure regulations, which reduces the amount of gasoline that evaporates into the atmosphere, are expected to be sufficient to bring the marginal areas into attainment by November 1993.

Moderate areas face tougher requirements. To begin with, they must do everything that the marginal areas are required to do. In addition, they must plan for a 15% reduction in VOC emissions within 6 years. The states must adopt and implement rules implementing all existing and future USEPA recommendations on RACT for major sources (CTGs), motor vehicle refueling regulations to control emissions at the gas pump (Stage II gasoline vapor recovery controls), and a basic motor vehicle I/M program if they do not already have one in place.

Serious areas must implement everything required in marginal and moderate areas plus plan for 3% annual VOC reductions after 6 years. Controls on sources of NO_x may be implemented in place of VOC controls if the result is at least equivalent in terms of O_3 reduction. More sophisticated mathematical air quality models must be used to demonstrate attainment by the deadline. Urbanized areas greater than 200,000 population must implement an enhanced vehicle I/M program within 2 years; this program is more demanding than the basic program and includes the use of computerized analyzers and centralized inspection facilities. Most serious areas will also be responsible for implementing a clean fuels program.

Severe areas are responsible for all of the previously mentioned controls, but must also implement measures to offset growth in the number of vehicle miles traveled, since motor vehicles are such an important source of O_3-forming pollutants; adopt requirements for fees on major sources if the area fails to attain; and, in the 9 worst O_3 areas, refiners will be required to manufacture gasoline which produces lower levels of O_3 precursors.

Two categories of areas that do not fit within the basic classification scheme are:

1. *Transitional areas* — areas that were designated nonattainment for O_3 before enactment of the CAA Amendents, but are not now recording violations of the standard. The Administrator had until June 30, 1992 to determine whether the area was still not violating the O_3 standard. At that time, if no violations have occurred, the state must submit a maintenance plan within 12 months of that determination. However, if violations are found, the area becomes subject to the new requirements according to its classification, based on its design value.
2. *Rural transport areas* — emissions from the area do not contribute significantly to the O_3 concentrations in that area or in other areas. Precursor emissions from point, area, and mobile sources in the area must not make a significant contribution to the O_3 concentrations measured in the area or in other areas.

A final major distinction between these Amendments and earlier legislation is the greater specification of the consequences of missing deadlines and milestones. For missing emission reduction milestones, the remedies include a "bump-up" to the next higher classification and having to meet those requirements, additional measures to meet the next milestone, and an economic incentive program.

Besides these requirements which must be met by the state and local agencies, the USEPA is required to develop certain federal measures. Within 3 years, 13 new CTGs were to be issued, including CTGs for aerospace coating and for shipbuilding operations. All existing CTGs were also to be reviewed. Alternative control techniques (ACT) documents for VOC or NO_x sources over 25 tons per year (tpy) are due within 3 years. These will recommend levels of control for those sources. A federal rule on marine vessel loading and unloading is due within 2 years. Finally, a study on the emissions and control options regarding commercial and consumer solvents is due within 3 years. Federal regulation of these emissions is required in subsequent years.

Interstate Transport

The Amendments of 1990 also focused attention on the problem of the transport of O_3 and its precursors. This is important because they can be transported long distances and many areas face an air quality problem that is totally or partially caused by sources that are not in their jurisdiction. The new Amendments establish a transport commission for the Northeast and allow the establishment of others in other parts of the country. Generally, an entire transport region, including the rural areas, is subject to the same requirements as moderate areas.

The 1990 Amendments grant the Administrator the authority to establish interstate transport regions whenever deemed necessary, that is, if there is reason to believe that pollutants from one or more states contribute significantly to violation of the NAAQS in other states. The Amendments include provisions so that states may be added to or removed from the original interstate transport region. At the time of establishing a transport area, the Administrator will establish a transport commission consisting of a minimum: (1) the Governor of each transport state or the Governor's designee; (2) the Administrator or the Administrator's designee; (3) the Regional Administrator (RA) or the RA's designee; and (4) an air pollution control official representing each state in the region, appointed by the Governor.

The transport commission has responsibility for the assessment of the degree of interstate transport of the pollutant or precursors to the pollutant in the transport region, for evaluating strategies for mitigating the interstate pollution, and for making recommendations to the Administrator regarding necessary measures to ensure that plans for the relevant states adhere to Section 110(a)(2)(D) requirements. The commission has the right to request that the Administrator make a finding that SIPs of involved area states do not adequately meet these requirements.

Redesignations/Maintenance Plans

If a state requests to be redesignated to attainment under Section 107(d), then the SIP must also be revised to reflect provisions for maintenance of the NAAQS for at least 10 years after such redesignation. The SIP must include the necessary additional measures to ensure maintenance. At the end of 8 years after the area has been redesignated to attainment, the state must revise the SIP to add an additional 10 years beyond the previous 10-year period required to demonstrate maintenance.

SIPs from states seeking redesignation must also include contingency plans that provide follow-up corrective actions if violations occur in the area once it is redesignated to attainment. The 1990 Amendments allow the Administrator discretion in requiring SIP revisions in the event that the redesignated area does not maintain the NAAQS.

If plans are not approved, not submitted, or not implemented, the USEPA has the choice of restricting federal highway funding or requiring new sources to obtain 2-to-1 offsets to achieve twice as much pollution reductions from existing sources as the new source will put into the atmosphere. At least one of these is to be applied no later than 18 months of a finding if the failures are not corrected within that timeframe. Additionally, the second sanction must be applied at 18 months if a "good faith" effort to correct the problem is not shown or at 24 months if actual failure correction has not occurred. If the corrections are not made within 2 years, the USEPA has the power to promulgate a plan in place of the state plan.

If the attainment date is missed, an area is "bumped up" to the next higher classification, except for severe and extreme O_3 areas, which receive several other penalties, including an excess emission fee program.

This range of tools to handle the entire range of eventualities for missing milestones and deadlines is much more specific than previous legislation and should remove any ambiguity about the consequences of failure to meet these milestones.

Sanctions

The 1977 CAAA provided for two types of sanctions: construction bans and funding restrictions. The construction bans automatically applied when the USEPA disapproved an SIP for failure to meet Section 110(a)(2)(I) as amended in 1977 or when the USEPA found that SIP provisions had not been implemented as specified under Section 173(4). At its discretion under Section 113(a)(5), the USEPA could apply a construction ban upon a state if noncompliance with NSR permitting requirements in a nonattainment area was found. Also, the USEPA could use its discretion in applying the funding restrictions of air grants or highway funding under Section 176(a) and (b) or sewage treatment works funding under Section 316(b).

The 1990 Amendments allow for other sanctions in the event of plan disapprovals, nonsubmittals, and nonimplementation of the SIP. Available sanctions provided for in Section 179(b) include: (1) highway funding restrictions (except if used for safety improvements, transit, ride-sharing, etc.); (2) increased offset emissions for new and modified sources (i.e., up to a 2-to-1 ratio for emission reductions to increased emissions); and (3) air grant funding restrictions (discretion rests with the USEPA). At least one of the first two sanctions mentioned above must be applied if failures are not corrected within 18 months of a finding (both can be imposed if there is not a "good faith" effort) and both after 6 more months.

Emissions Statements

The new Amendments require that owners or operators of each stationary source of NO_x and VOC issue annual statements indicating the actual emissions. SIPs should be revised by November 15, 1992 to reflect this requirement. The first statements are due November 15, 1993 and should cover emissions for the previous calendar year. Emission statements should be certified as to accuracy. Waivers are acceptable for sources emitting less than 25 tons if the state has used USEPA-acceptable emission factors (i.e., those as in USEPA publication AP-42) when making such inventory determinations for these sources.

New Source Review (NSR) Requirements

The 1990 CAA Amendments make numerous changes to NSR requirements of Section 173 of the CAA. The Amendments tighten the requirements on new sources, such as the offset ratio or the amount of reductions needed at existing sources to offset increases in emissions from new sources. For example, permits to construct and operate may only be issued if sufficient offsetting emissions reductions have been obtained. As with the source size threshold, the stringency of these requirements varies with the area's nonattainment classification, ranging from 1.1 to 1 for marginal areas to 1.5 to 1 for extreme areas. Emission reductions to achieve the offsets must be from the same source or other sources in the same nonattainment area. However, the state may allow obtaining such reductions in another nonattainment area if (1) the area has an equal or higher nonattainment classification than the area where the source seeking the permit is located, and (2) sources in another nonattainment area contribute to the nonattainment in the area where the source in question is located. Emission reductions otherwise required by the CAA will not be creditable as emission reductions for purposes of meeting these offset requirements.

By November 1991, the USEPA had intended to propose a regulatory package that would implement these and other changes to the NSR provisions. Final adoption of these revised USEPA regulations was targeted for November 1992.

NO_x

Since 1970, the USEPA's O_3 policy has been based on the premise that VOC control is the optimum approach to O_3 pollution reduction. However, the new CAAA of 1990 require specific provisions for sources of NO_x. First, the Administrator, in conjunction with the National Academy of Sciences, conducted a study on the role of O_3 precursors in tropospheric O_3 formation which examined the role of NO_x and VOC emissions, the extent to which NO_x reductions contribute or were counterproductive to achieving attainment in different nonattainment areas, the sensitivity of O_3 to the control of NO_x, the availability and extent of controls for NO_x, the role of biogenic VOC emissions, and the basic information required for air quality models. The study was submitted to Congress in December 1991.

Secondly, Section 182(c)(2)(C) called for the Administrator to issue guidance concerning the conditions under which NO_x control may be substituted for VOC control or may be combined with VOC control in order to maximize the reduction in O_3 air pollution. This guidance was intended to govern the substitution of NO_x reductions for the required 3% per year VOC reductions beginning in November

1996. In order to substitute NO_x for VOC, the state must demonstrate to the USEPA that the NO_x reductions would result in O_3 concentrations at least equivalent to that which would result from the amount of VOC emission reductions otherwise required. Since in some instances the control of NO_x may detract from the effectiveness of VOC controls, safeguards have been incorporated into the new Act. These provisions would allow states to opt out of additional stationary source NO_x reductions if these are shown to be counterproductive in reducing O_3.

Some research has concluded that when biogenic emissions are mixed with anthropogenic VOC emissions, the VOC-to-NO_x ratio is raised, which lessens the importance of VOCs. Thus, in those urban areas where biogenics are substantial and well mixed with anthropogenic emissions, NO_x control would impact ambient O_3 more effectively.

Some areas may require combined VOC and NO_x controls, while others could attain with NO_x-only or VOC-only strategies.

Table 4 lists the major sources of NO_x pollution.

TITLE II REQUIREMENTS

In many ways, the 1970 CAA was a great success in reducing pollution from automobiles. Today's cars, for instance, emit 60 to 80% less pollution over their lifetimes than their 1970 counterparts. Despite this considerable progress, however, the nation's clean air goals have not been met. Urban sprawl has created new pollution sources and contributed to a doubling of vehicle miles traveled since 1970.

The 1990 Amendments strengthen components of the earlier law. The tailpipe standards now in effect for cars, buses, and trucks will be tightened, for instance, and I/M programs have been expanded to include more areas and allow for more stringent tests.

The 1990 Amendments also present several entirely new concepts with regard to reducing motor vehicle-related air pollution. For the first time, fuel is considered along with vehicle technology as a potential source of emission reductions. In addition, more attention is focused on reducing the growth in vehicle travel. The new provisions are given below.

Emphasis on Fuels

The Amendments mandate improved gasoline formulations be sold in some polluted cities to reduce emissions of carbon monoxide or hydrocarbons. Other programs set low vehicle emission standards to stimulate the introduction of even cleaner cars and fuels.

Non-Road Engines

Under the previous legislation, the USEPA's mobile source pollution control authority applied only to highway vehicles. The 1990 Amendments require the USEPA to control locomotive emissions, as well as study, and regulate if war-

Table 4. Sources of NO_x, 1980

Source	%
Highway vehicles	35
Other transportation	10
Power plants	28
Industrial combustion	19
Other sources	8
Total	100

From NAPAP Interim Assessment, Vol. II, MacKenzie, 1988, 37.

ranted, any category of non-road engines that contributes to urban air pollution. This includes boats, farm equipment, bulldozers, lawn and garden devices, etc.

Alternative Transportation Programs

The law requires areas with serious pollution problems to employ ride-sharing and other transportation alternatives to reduce the amount of vehicle traffic. In areas where smog levels exceed certain criteria, employers of 100 or more employees will be asked to find ways to increase the average number of passengers in each vehicle for commutes to work and during work-related driving trips.

Some of the specific CAA programs under Title II are discussed below.

Tighter Tailpipe Standards

The current tailpipe standards for cars are 0.41 grams per mile (gpm) total hydrocarbons, 3.4 gpm CO, and 1.0 gpm NO_x. Lower standards of 0.25-gpm nonmethane hydrocarbons and 0.4-gpm NO_x will be phased in between 1994 and 1996 (the CO standard does not change). The USEPA is required to study whether even tighter standards are needed, technologically feasible, and economical. If the USEPA determines by 1999 that lower standards are warranted, the standards will be cut in half beginning with 2004 model year vehicles.

O_3 Control

As a complement to stricter tailpipe standards, the new law introduces several programs to minimize "evaporative emissions", as gasoline vapors from fuel evaporation are a major source of hydrocarbons. Devices that trap gasoline vapors from the engine and fuel system will be improved. In addition, systems to capture gasoline vapors during refueling will be required. Most importantly, gasoline volatility will be capped, reducing the propensity for fuel to evaporate in the first place.

Reformulated Gasoline

By 1995, all gasoline sold in the nine worst O_3 areas must contain a minimum oxygen content and a maximum benzene content. It must also reduce (by at least

15%) emissions of toxic and smog-forming compounds, without increasing emissions of NO_x. By 2000, gasoline sold in these cities must be able to reduce emissions of toxic and smog-forming compounds by at least 20 to 25%. Other cities may elect to use this cleaner gasoline.

Clean Fleets

Beginning in 1998, 30% of new vehicles purchased by centrally fueled fleets will be required to use clean fuels and meet tailpipe standards that are lower than those in place for general passenger cars (0.075 gpm hydrocarbons, 3.4 gpm CO, and 0.2 gpm NO_x). The purchase requirements will grow to 70% by 2000. The program, which is intended to stimulate development of new, low-polluting fuel/vehicle combinations, will affect areas in 22 states across the country where pollution levels are high.

I/M Corrections

The 1990 CAA Amendents require that designated marginal areas submit "immediately" after enactment corrected upgraded plans to any previously required basic vehicle emission control I/M program. The USEPA realizes that meeting such a schedule immediately is virtually impossible and thus expects an SIP submittal to implement corrections to the I/M program by November 15, 1992.

As required by the CAA Amendents, the USEPA plans to issue guidance on making corrections to existing I/M programs and procedures for adopting basic and enhanced motor vehicle I/M programs. At a minimum, such guidance will include frequency of inspections, test method and procedures, proof of corrective actions performed on faulty vehicles, etc.

Motor Vehicle Inspection and Maintenance

Under Section 182(b)(4), states with moderate nonattainment areas must immediately after enactment submit revised SIPs with provisions for a vehicle I/M program, even if not required to do so under the 1977 Act. The USEPA recognizes that this may be a first-time requirement for some new moderate areas and will allow at least 1 year for such areas to adopt and implement an I/M program. The CAA Amendents allow the USEPA to conditionally approve basic I/M programs in such areas upon the state's commitment to develop an I/M program within 1 year of conditional approval.

Enhanced I/M

Within 2 years of enactment, the states with *Serious* O_3 nonattainment areas must submit a revised SIP to provide for an enhanced program to reduce hydrocarbon and NO_x emissions from in-use motor vehicles registered in each urbanized area (1980 population of 200,000 or more) in the nonattainment zone. The enhanced I/M programs must include computerized emission analyzers, certain waiver restrictions, enforcement via vehicle registration denial, and, generally, annual centralized testing and inspection. Guidance issued in June 1991 gives additional details on these programs, such as achievable performance standards. States are also required to biennially prepare a report for the USEPA, evaluating emission reductions achieved by the enhanced I/M program based on data collected at the time of inspections and repairs of vehicles.

Gasoline Vapor Recovery

Section 182(b)(3) requires *Moderate* or above areas to revise their SIPS to require that owners or operators of gasoline dispensing systems install and operate gasoline vehicle refueling vapor recovery systems (known as Stage II). Stage II is required at such dispensing facilities which sell more than 10,000 gallons of gasoline per month or 50,000 gallons per month in the case of small independent marketers.

Additional Stage II provisions are required in Title II of the amended CAA, that is, onboard vehicle refueling controls. According to these Title II requirements, the Stage II requirements of Title I are not required in *Moderate* areas if onboard controls were promulgated as required by November 15, 1991, and such controls are in widespread use. Likewise, other O_3 nonattainment areas may waive or revise Stage II requirements after widespread use of onboard vehicle refueling controls.

Clean-Fuel Vehicle Programs

SIP requirements for Clean-Fuel Vehicle Programs in Serious or above O_3 nonattainment areas (population areas of 250,000 or more in these areas) are found in Sections 182(c)(4) and 246 of the Act. Measures described in Title II Part C for implementing the clean-fuel vehicle program for centrally fueled fleets were scheduled to be adopted by May 15, 1991. Also, necessary measures to make the use of clean alternative fuels in these vehicles economical must be included in the SIP revision.

Substitution of the clean-fuel vehicle program is allowed if substituted control measures achieve the same emission reductions of O_3 precursors and toxic substances as the clean-fuel program.

Transportation Control

Beginning 6 years after enactment and each third year thereafter, states with *Serious* areas or above are required to submit a demonstration as to whether current aggregate vehicle mileage, aggregate vehicle emissions, congestion levels, and other relevant parameters are consistent with those used for demonstration of attainment. If the emissions indeed exceed those projected for the attainment demonstration, then a revised SIP (including transportation control measures to bring the area into attainment) must be submitted by the state within 18 months. States should be aware of the importance of TCMs (transportation control measures) so as to adequately plan for the future.

Vehicle Miles Traveled

By November 15, 1992, the State must submit a revision that specifies enforceable transportation control measures (TCMs) to offset any growth in emissions due to an increase in vehicle miles traveled or numbers of vehicle trips in the area and measures to reach the necessary reductions. Within the above timeframe, the state must also submit a revision requiring employers in such *Severe* areas to implement programs to reduce work-related vehicle trips and miles traveled by employees. Specifics of such trip programs are discussed in greater detail in the General Preamble to Title I.

CONCLUSIONS

The 1990 Amendments to the CAA are very complex and represent an attempt to *not* repeat the mistakes of the past. The tiered approach to the nonattainment problem is an attempt to avoid the "one size fits all" programs of the past. The combination of technology-based controls and a long-term planning approach is designed to put together a program that starts quickly but looks towards longer-term problems as well. Finally, this time there are provisions for the possibilities of not meeting the milestones on time and there are mechanisms to add on controls when necessary. The 1990 Amendments are a challenge to everyone involved in meeting the goal of clean and healthy air, a partnership that everyone has a stake in.

CHAPTER 8

THE EFFECTS OF PHOTOCHEMICAL AIR POLLUTION ON PEOPLE

David V. Bates

INTRODUCTION

Studies of the effect of air pollution on people have been complicated by a number of factors:

- The dominance of cigarette smoking in the genesis of lung disease
- The fact that most respiratory disease is multifactorial in origin
- The difficulty of drawing conclusions about causation from demonstrated associations
- The difficulty of comparisons between populations living under different pollution conditions, but with many other differences between them such as climate, living conditions, methods of heating and cooking, etc.
- The uncertainty of the long-term importance of some outcome measurements, such as small transient declines in lung function
- The difficulty of separating the effects of different pollutants when they commonly rise together

After the obvious associations between acute mortality and discrete air pollution episodes in the 1950s and 1960s in the west were lessened by control methods, there was temptation to dismiss continuing evidence of adverse health effects from air pollution as essentially unimportant. Thus, a respected scientist

0-87371-475-X/94/$0.00+$.50
© 1994 by Lewis Publishers

with wide experience in environmental matters, noted in a volume of autobiography published in 1990[1] that:

> "It is now nearly 20 years since I prepared that paper for presentation in England. I am still convinced that the dangers of pollution are greatly exaggerated in the public mind. This is particularly true so far as health effects are concerned... The greatest danger from pollution is the threatened increase in global temperature resulting from the increasing carbon dioxide content of the atmosphere." (See p. 228.)

During the past 2 decades, there have been substantial advances in environmental epidemiology; in addition, other lines of evidence have become available, such as that deducible from controlled human exposures to pollutant mixtures. In retrospect, it is remarkable that it took 28 years after the 1952 London pollution episode before the increased sensitivity of asthmatics to SO_2 was discovered.[2] In the case of ozone (O_3), controlled human exposure experiments began in the mid-1960s;[3] since then, a substantial body of data has accumulated. The controlled exposure data provide a foundation for the interpretation of epidemiological studies and indicate (with some precision) what concentrations of O_3 or SO_2 are likely to impact the normal population.

In this chapter, an attempt is made to give a general overview of the present status of our understanding concerning O_3, with no attempt to conceal areas where we know that our present understanding is deficient.

ACUTE CONTROLLED EXPOSURE EXPERIMENTS

The results of many studies of normal human subjects exposed to O_3 in controlled experiments enable us to make the following general statements.

- The FVC* and FEV_1** is lowered by a 6-h exposure with exercise to as little as 0. 08 ppm O_3 (the current Canadian 1-h standard).[4]
- There is substantial individual variation in response (the reasons for which are not known), but reproducibility is good.[5]
- Under standard conditions, the mean group response is related to concentration, time, and the minute ventilation.[6]
- There is generally satisfactory concordance in results from different laboratories.[7]
- The first effect of O_3 is not to cause airway constriction, but to limit the maximal inspiration the individual can take (probably through a spinal reflex activated by irritant receptor stimulation).[8]
- O_3 increases lung permeability, as indicated by DTPA (^{99M}Tc-labeled diethylene triamine pentacetic acid) clearance.[9]
- When the FVC is lowered, airway inflammation is demonstrable;[11] this was

* FVC is the forced vital capacity and measures a maximal expiration.

** FEV_1 is the forced expiratory volume in 1 second, a simple test of lung function first shown by using bronchoalveolar lavage 18 h after the exposure. More recent data indicates that the polymorphonuclear flux is even greater immediately after exposure has ceased.[1] However, there does not appear to be a close association between the degree of inflammatory response and the decline in FVC and FEV_1.

- The decrement in lung function increases with increasing exposure;[12] however, repetitive exposures produce less of a response (this "adaptation" appears to last for about 10 days). With exposure to 0. 30 ppm for an hour with continuous moderate exercise, the response to the same concentration the next day is enhanced.[13]
- A greater decrement in FEV_1 results if, over a 2-h period, O_3 is breathed at an increasing concentration reaching 0. 24 ppm at 1 h and then declining to zero, than if a constant concentration of 0. 12 ppm of O_3 is breathed.[4] Thus, the response does not seem to be invariably expressible in terms of the total amount of O_3 inhaled.
- O_3 increases nonspecific airway responsiveness (to mecholyl or histamine) in normal subjects; however, asthmatics are not more sensitive to O_3 than normal nonasthmatic subjects. Prior O_3 exposure increases the subsequent response to SO_2 in asthmatics.[15]
- O_3 exposure limits human maximal athletic performance.[16,17]
- A low O_3 exposure of 0. 12 ppm breathed for an hour at rest leads to a reduction in dose of inhaled allergen for a given response. This was shown in six of seven asthmatics tested.[18] If confirmed in larger-scale studies, this would have important long-term implications.

Many of these observations fit well with others. Last[19] has found that albumin in bronchoalveolar lavage of rats after O_3 exposure is directly dose dependent. O_3 induces a general increase in lung permeability, presumably as a consequence of the inflammation induced; this would permit more effective penetration of allergens.

There are, however, a number of important but unexplained observations. These include the reasons for individual variability in response, the mechanism of "adaptation", whether the inflammatory response is reduced in successive exposures, and why there is no concordance between the degree of inflammatory response and the fall in FVC and FEV_1.

These studies, very briefly summarized in the paragraphs above, have contributed a great deal to our knowledge of the effects of O_3.

EPIDEMIOLOGICAL STUDIES

There is great variety in the epidemiological observations made on oxidant-exposed populations. Although the general pattern of symptomatology and impaired lung function that is found in populations exposed to major particulate or high SO_2 pollution is well established (and has been generally known since 1965),[20] it is not yet certain what adverse effects from O_3 exposure should be specifically looked for. For this reason, most of the studies have been of a "hypothesis-generating" type; and few dose-response deductions are yet possible.

Studies of Hospital Admissions or Emergency Visits

In 1966 and 1967, Sterling and colleagues reported a positive association between admissions of Blue Cross patients to Los Angeles hospitals with more than 100 beds and oxidant levels.[21] In 1981, Richards et al.[22] reported finding no association in the same city between emergency visits for asthma at a children's hospital and oxidant levels. The study period extended from August 1979 to January 1980. Goldsmith et al.[23] studied emergency room visits in four southern California communities during 1974 and 1975. Maximum hourly averages of oxidants and temperature were associated with daily visits in the high oxidant area (Azusa). In retrospect, it is surprising that this finding did not trigger follow-up epidemiological studies.

In 1983, Bates and Sizto[24] reported finding an association between acute respiratory admissions to 79 acute care hospitals in Southern Ontario and O_3 levels. A later and more detailed analysis, including data on concomitant sulfate measurements, was published in 1987.[25] Admissions on a given day were compared to the mean admissions for that day of the week (in the same season of the same year only) to correct for day-of-the-week effects and also long-term trends. This conservative approach to the data confirmed the association between summer O_3 levels and hospital admissions for acute respiratory disease; it also showed that the association was as strong for sulphates as for O_3. In this and similar studies, it has proved difficult to separate the effects of these pollutants which, in this region, commonly rise and fall together.[26,27] The association of hospital admissions and sulphate levels in Southern Ontario was confirmed by Plagiannakos and Parker.[28] They analyzed all acute respiratory admissions over a period of 6 years, and reported significant regression associations between SO_2 and SO_4, and hospital admissions expressed as a rate for each county. This study used all the available sulphate and SO_2 data for the region and attempted to ensure that differences in socioeconomic status had not interfered with the analysis. An indirect regional index of tobacco consumption was also used. The importance of this study was that the two analyses in Southern Ontario, though planned and conducted independently, served to reinforce each other. Unfortunately, Plagiannakos and Parker[28] did not analyze O_3 levels.

A significant association between O_3 levels and respiratory admissions was found in Massachussetts by Ozkaynak and associates.[29] Using hospital emergency visits rather than admissions as an index, Bates et al.[30] found no summer relationship between O_3 levels in Vancouver and visits for acute respiratory conditions; O_3 levels there were about half what they were in Southern Ontario. White and colleagues,[32] at the Center for Disease Control, have recently reported a summer study conducted at a children's hospital in Atlanta. It showed a close and highly significant association between attendance for acute asthma and O_3 levels. In 1980, Whittemore and Korn[31] found that O_3 levels were a significant risk factor for worsening of asthma in Los Angeles. At an Air Waste Management Association (AWMA) Meeting (November 1991), preliminary data from an ongoing study in New York state was reported; the data showed a significant association between O_3 levels and hospital admissions for acute repiratory disease in Buffalo, but weaker associations in Albany and White Plains.[33] Preliminary data were also presented from a similar study in New Jersey,[34] which also found a significant association between admissions and O_3 levels, but no association with H^+ levels in the region. As no detailed publications have yet appeared on these three studies, a more detailed analysis is not possible.

All of these studies may be criticized on a number of grounds. It may be objected that the exposure data is very imprecise [(the USEPA Criteria Document[35] noted in connection with the Bates and Sizto study that: "Lack of sufficient exposure analysis limits the quantitative use of this study" (see Table 11–7, pp. 11–37, Vol. V)]. This difficulty is, of course, inherent in a population study involving 6 million people studied over a period of 8 years. It is also difficult to quantify, with any accuracy, the probable exposure profile of acutely ill patients arriving at a hospital — though one can imagine a better approximation than simple dependence on regional monitors. However, imprecise exposure data will probably weaken any association rather than produce a false positive association.

Other critics have not always understood that factors that are important confounders in cross-sectional comparisons (such as smoking, gas stoves, overcrowding, etc.) do not apply to studies showing associations between the population (as it exists) and ambient pollutants.

A second criticism is that all data based on computerized hospital records are too imprecise for analysis. This is a point stressed in USEPA Criteria documents,[35] but there have been no published detailed analyses that show this to be the case. Recently, Delfino[36] has analyzed 669 records of hospital admissions in Quebec, comparing the diagnosis entered on the computerized record, with a diagnosis reached after perusal of the hospital records by experts. This comparison is shown in Table 1. It shows a very good concordance in the general category of acute respiratory disease; for the 151 cases of asthma so coded, in 141 cases the experts agreed; in 4, the diagnosis was considered to be chronic obstructive lung disease; in 1, pneumonia; in 1, an acute large airway infection; and in 4 others, they were classified in other categories. This study indicates that for epidemiological purposes, the coding is precise enough for the data to be used. Such misclassifications as do exist will, in general, have the same effect as exposure misclassification, namely the association with other factors will be weakened.

A third problem with most air pollution studies of this kind, elegantly pointed out by Whittemore,[37] is that it is difficult to draw any precise dose-response conclusions from them. This only becomes possible when comparisons can be made between results of studies from different regions; but such comparisons are always made difficult by possible changes in other factors. If a number of studies in different regions (with different patterns of health care) show significant associations between hospital admissions and O_3 levels, at least it can be inferred that existing levels of tropospheric O_3 are having an adverse health impact (and this is the conclusion to which the regulator's attention should be drawn).

Some of the criticisms of the indirect nature of air pollution studies seem to be little more than a yearning that they could be patterned after studies of animal

Table 1. Analysis of 669 Database Entries in Quebec

Recorded database diagnosis	Diagnosis after reassessment					
	Uri	Ac L A infect	Pneum	COPD	Asthma	Other
URI (462–463)	25	0	0	0	0	7
Ac Large airway infections (464–466)	0	84	1	3	1	6
Pneumonia (480–486)	0	0	217	0	2	11
COPD minus asthma (490–492 494,496)	0	9	6	130	6	10
Asthma (493)	0	1	1	4	141	4
Total	25	94	225	137	150	38

From Delfino, R. Personal communication (1991). With permission.

toxicology, in which all variables can be controlled. This is unsophisticated thinking; it pays little attention to the wiser approach of Hill in 1965[38] to the question of concluding that an association indicates a causal relationship.

Camp and Panel Studies

Studies of FEV_1 in children at summer camps in Canada and the U.S. have indicated that the summer mix of pollutants, including O_3, is associated with measurable declines in function.[39] There is reasonable conformity between the different studies in relation to the relationship between O_3 level and function decrement. Adults jogging during their lunch hour have also been studied.[40] Many of these were found to have ventilation levels in excess of 60 l/min (they had selected their own exercise level).

Lioy et al. found evidence that the effect of O_3 on peak expiratory flow rate might last for several days.[41] In a recent summer camp protocol, Spektor and colleagues[42] studied 46 children aged 8 to 14 years at a camp for a 4-week period. The highest O_3 level encountered was 150 ppb. They noted that regressions of morning function on the previous day's O_3 were significant. No correlations were observed for H^+ (in previous studies, it had proved difficult to be sure that the response might not be due to accompanying acid aerosols). These authors concluded that camp studies generally yield higher effects of O_3 than chamber exposures for four reasons.

- The exposures in the field are longer
- There may be potentiation by other factors such as particulates
- There may be persistence of effects from the prior day's exposure

- There may be persistence of a transient response associated with the daily peak exposure

The authors cautioned: "It follows that projections of likely effects in the real world from controlled chamber studies should either have a large margin of safety, or the judgment of the extent of effects likely to occur among populations should be based directly on the effects observed in field studies."

Lebowitz and co-workers[43] studied panels of asthmatics and others in Tucson, AZ, drawn from 117 white families from a stratified sample. Detailed measurements of air pollutants, pollen, bacilli, fungi, and algae were made in and around a random cluster sample of 41 households; 35 asthmatics provided daily peak flow measurements. The very complex resulting data showed that both O_3 levels and temperature had effects on the peak flow data. Javitz et al.[44] reanalyzed a study of 286 subjects with asthma and chronic bronchitis (probably with and without accompanying emphysema) in Houston, TX. Daily symptom diaries were kept and home spirometry testing was done in about one third of the subjects. Respiratory symptoms and medication use increased as the daily O_3 maxima increased. More recently, Gong[45] studied 83 asthmatics in Los Angeles between February and December 1983. It was reported that 39 of the subjects were "O_3 responders", with symptoms and medication use consistently related to O_3 levels, although the total group results had not shown a significant relationship. The cases were relatively mild, since 77% of the total group had peak flow measurments within the normal range.

In the early 1980s, Linn and colleagues[46] carried out an extensive series of studies in Los Angeles using a mobile van which contained an exposure chamber and pulmonary function testing equipment. These studies are summarized in detail in the USEPA Ozone Criteria Document.[35] After an initial finding that suggested that function decrements in the natural environment with photochemical pollutants was greater than that found in chamber exposures, it was finally concluded that the observations of decreased function were explicable on the basis of the O_3 exposure. Later studies of competitive cyclists in the same mobile facility showed that FEV_1 declines occurred at O_3 levels as low as 0.16 ppm (a result in accord with the later controlled exposure data described above).

Cross-Sectional Comparisons

Hodgkin et al.[47] found some evidence of more respiratory symptoms in nonsmokers in higher oxidant regions, but the difference was not great. Linn and coworkers could not demonstrate consistent differences in either respiratory symptomatology or function between populations of office workers in areas with different levels of oxidant pollution.[48]

Very recently, a most interesting comparison has been published of children in Austria living in relatively high and low O_3 regions.[49] Some 218 children (mean age 11. 6 years) lived in the high O_3 region in which O_3 exceeded 100 ppb 9. 68% of the time, and 281 children lived in a low O_3 region where that value was never exceeded. Other pollutants were similar between the regions. There were no differences in allergic background, as judged by IgE levels or asthma symptoms; no differences in cough or breathlessness, nor in pulmonary function status. However, there was a higher incidence of bronchial hyperresponsiveness to methacholine in the high O_3 group, and differences in the ratios of types of lymphocytes. The "high O_3" children showed a decrease in T-helper lymphocytes and an increase in T-suppressor cells compared to the children in the low O_3 area. Although the significance of these hematological differences is unclear, the authors interpreted their overall findings to indicate that O_3 had caused a significant difference in status.

Longitudinal Studies

Detels and co-workers planned an ambitious longitudinal study of smoking and nonsmoking subjects in different regions of Los Angeles.[50] This showed a faster rate of longitudinal decline in more polluted regions when subjects were retested after an interval of 5 years; however, there was a considerable loss of sample between the two observations, making interpretation difficult. Bresnitz and Rest,[51] in a review of oxidant epidemiologic data, noted difficulties in the interpretation of the UCLA longitudinal study.[50] The question of whether the differences in rate of longitudinal decline could be attributed to sample loss is a difficult one. Lippmann, in a review[52] of O_3 effects, felt that the preliminary evidence from this study could not be dismissed.

It may be noted that on theoretical grounds, longitudinal studies should be more sensitive indicators of long-term adverse effects of oxidant pollutant exposure, but such studies are very difficult to organize, and it is difficult to avoid loss to follow-up of a significant fraction of the surveyed population. If this occurs, the interpretation of any result becomes open to question. Hackney and Linn[53] have recently demonstrated that the level of airway responsiveness changes with the seasons in Los Angeles, being increased during and just after the high O_3 season, and thereafter declining.

Health Interview Survey Data

Other epidemiologic evidence of long-term adverse effects of O_3 exposure has come from the analysis of the NHANES II Survey data in the U.S. provided by Schwartz.[54] The author reported a decrement in FVC in boys in relation to calculated long-term O_3 exposure. Ostro and Rothschild[55] used data from the Health Interview Survey in the U.S. between 1976 and 1981, and reported a relationship between O_3 exposure and minor reduced activity days. This study confirmed earlier observations by Portney and Mullahy,[56] who noted a relationship between O_3 levels and reduced activity days by combining aerometric and survey data.

Autopsy Data

Sherwin and Richters[57] have described changes of severe respiratory bronchiolitis in a high proportion (27%) of autopsy lungs from 107 young adults aged 14 to 25 years who died in the Los Angeles district from nonrespiratory causes. Although this study has not yet been compared to a control region, and the degree of small airway disease has not been measured by quantitative morphometry, it seems likely that the lesion represents an aggravation of the respiratory bronchiolitis known to be a consequence of cigarette smoking. It may be significant that theoretical dosimetric calculations have indicated that O_3 deposition in the human lung would be maximal in this region, and data in O_3-exposed non-human primates indicated that a respiratory bronchiolitis was the principal induced lesion.[58] In Vermont, Adesina et al.[59] have quantitated bronchiolar wall inflammation and fibrosis in autopsy material from 42 smokers and 13 nonsmokers living in a relatively nonpolluted environment. They confirmed the highly significant differences between the two groups, particularly in subjects below the age of 40 years. The question of whether living in an oxidant environment results in a more severe degree of bronchiolar inflammation — either in nonsmokers, or possibly in smokers as well — should be answerable if these data could be compared with similar data from Los Angeles.

Mortality Data

It is remarkable that although Mahoney[60] reported in 1971 that respiratory mortality rates in Los Angeles seemed to follow the contours of oxidant pollution, it was not until 1991 that any definitive study of the relationship between O_3 levels and mortality had been conducted. Kinney and Ozkaynak[61] have recently published a very carefully conducted analysis of such data between 1970 and 1979. It indicates that O_3 levels are associated with increases in mortality (both from respiratory and from cardiovascular disease). Their analysis by modern statistical techniques indicates that O_3 is the important factor and that temperature changes are unlikely to have interfered with the regression analyses.

CONCLUSIONS

This summary of a considerable amount of information relating to the effects of the most irritant gas known to us raises as many questions as it answers. There has been inherent difficulty in visualizing what the consequences of living in a high oxidant environment for some part of each year might be. Epidemiological tools such as questionnaires that concentrate on symptoms of chronic bronchitis, that proved so valuable in analyzing the effects of gross particulate and high SO_2 exposure, might well be inappropriate for studying the effects of oxidants. In

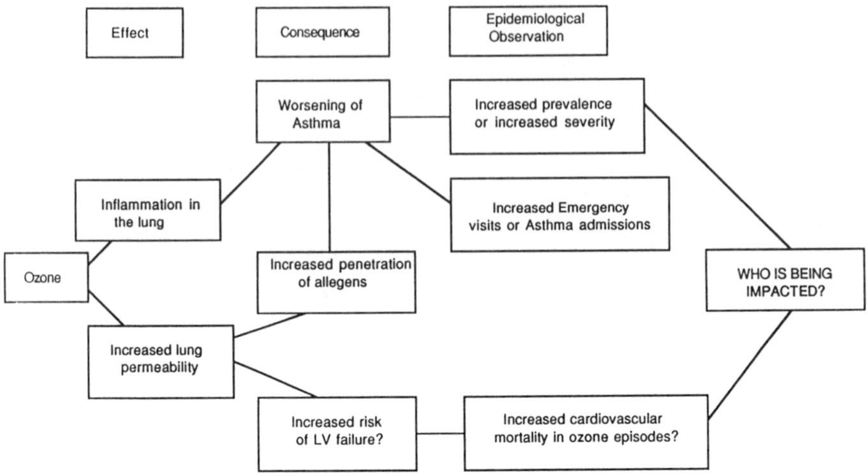

Figure 1. Speculative integration of current information on O_3.

addition, the FEV_1, a simple and well-standardized test suitable for field use, which might be considered an early indicator of the effects of chronic bronchitis and emphysema, is hardly a sensitive enough tool to detect chronic changes in small airways.[62] We know that sensitive epidemiological analyses require as precise data on exposure as is obtainable, yet personal monitors of O_3 exposure have not been generally available.

Figure 1 indicates a preliminary attempt to integrate some of the information available on O_3. Both the induction of inflammation by O_3 and the increased permeability would be predicted to worsen asthma. Also, the heightened airway responsiveness in someone with an already raised level might be predicted to be particularly serious. The increased permeability induced by O_3 might be invoked to explain the relationship between increased cardiovascular mortality and O_3, but more work will be needed to establish a biological basis for this relationship. It might exist because of the danger of increased lung permeability when the left atrial pressure is raised. There is evidence that increased hospital attendances and admissions for acute respiratory disease are associated with O_3 levels when these commonly exceed 60 ppb, but not when they are below this. The data quoted from Austria is the first to suggest that living in a higher oxidant atmosphere may induce increased airway responsiveness in children (though some preliminary data did indicate a higher prevalence of airway responsiveness in children in southern Ontario compared to children in New Zealand tested by the same investigator).[63] There is thus some coherence in these observations.

There is also coherence between the chronic bronchiolitis documented in non-human primates exposed to O_3,[58] and in the preliminary autopsy data from Los Angeles. There was also some indication in the UCLA longitudinal study that a faster-than-predicted fall in mid flow rates (FEF25-75) had occurred in nonsmoking women living in a high oxidant region. If a chronic bronchiolitis had been induced, this is what one might expect to find.

It is clear that what is needed in the further analysis of hospital admission data, as well as mortality studies, is a detailed analysis of exactly who is being impacted. Such information would contribute greatly to our understanding of such data and enable us to attach a proper estimate of significance to it.

We should be unwilling to dismiss repetitive induced falls in lung function as insignificant, as these may well indicate repetitive inflammation. At the very least, the burden of proof should surely be on those who claim that such events are insignificant to demonstrate it. Repetitive attacks of asthma are obviously adverse [though whether they should be appropriately costed at $25.00 (U.S.) is open to question[64]].

It is hoped that many of the present uncertainties will be resolved by work now being planned. Since 86 million Americans — and countless more people in Canada, Europe, and millions more in Third World cities such as Mexico City — now have to live in O_3 concentrations that frequently exceed 80 ppb, one need not apologize for regarding the question of the long-term effect of O_3 as an urgent one.

REFERENCES

1. Eisenbud, M. "An Environmental Odyssey: People, Pollution, and Politics in the Life of a Practical Scientist," (Seattle: University of Washington Press, 1990), p. 264.
2. Sheppard, D., W. S. Wong, C. F. Uehara, J. A. Nadel, and H. A. Boushey. "Lower Threshold and Greater Bronchomotor Responsiveness of Asthmatic Subjects to Sulfur Dioxide," *Am. Rev. Respir. Dis.* 122:873–878 (1980).
3. Young, W. A., D. B. Shaw, and D. V. Bates. "Effect of Low Concentrations of Ozone on Pulmonary Function in Man," *Appl. Physiol.* 19:765–768 (1964).
4. Horstman, D. H., L. J. Folinsbee, P. J. Ives, S. Abdul-Salaam, and W. F. McDonnell. "Ozone Concentration and Pulmonary Response Relationships for 6.6-hour Exposures with Five Hours of Moderate Exercise to 0.08, 0.10, and 0.12 ppm," *Am. Rev. Respir. Dis.* 142:1158–1163 (1990).
5. McDonnell, W. F., III, D. H. Horstmann, S. Abdul-Salaam, and D. E. House. "Reproducibility of Individual Responses to Ozone Exposure," *Am. Rev. Respir. Dis.* 131:36–40 (1985).
6. Hazucha, M. "Relationship Between Ozone Exposure and Pulmonary Function Changes," *J. Appl. Physiol.* 62:1671–1680 (1987).
7. Graham, J. A. NAPAP: National Acid Precipitaton Assessment Program State of Science/Technology. Report No. 22. Direct Health Effects of Air Pollutants Associated with Acidic Precursor Emissions (1991).
8. Hazucha, M. J., D. V. Bates, and P. A. Bromberg. "Mechanism of Action of Ozone on the Human Lung," *J. Appl. Physiol.* 67:1535–1541 (1989).
9. Kehrl, H . R., L. M. Vincent, R. J. Kowalsky, D. H. Horstman, J. J. O'Neil, W. H. McCartney, and P. A. Bromberg. "Ozone Exposure Increases Human Respiratory Epithelial Permeability in Humans," *Am. Rev. Respir. Dis.* 135:1124–1128 (1987).
10. Koren, H. S., R. B. Devlin, D. E. Graham, R. Mann, M. P. McGee, D. H. Horstman, W. J. Kozumbo, S. Becker, D. E. House, W. F. McDonnell, and P. A. Bromberg. "Ozone-Induced Inflammation in the Lower Airways of Human Subjects," *Am. Rev. Respir. Dis.* 139:407–415 (1989).
11. Devlin, R., W. McDonnell, R. Perez, S.Becker, and H. Koren. "Humans Exposed to 0.4 ppm Ozone Exhibit Cellular and Biochemical Changes in the Lung One Hour Following Exposure," *Am. Rev. Respir. Dis.* 141:A71 (1990).

12. McDonnell, W. F., Kehrl, H. R., Abdul-Salaam, S., Ives, P. J., Folinsbee, L. J., Devlin, R. B., O'Neil, J. J., and Horstman, D. H. "Respiratory Response of Humans exposed to Low Levels of Ozone for 6.6 Hours," *Arch. Environ. Health* 46:145–150 (1991).
13. Brookes, K. A., W. C. Adams, and E. S. Schelegle. "0.35 ppm O_3 Exposure Induces Hyperresponsiveness on 24-h Reexposure to 0.20 ppm O_3," *J. Appl. Physiol.* 66:2756–2762 (1989).
14. Hazucha, M. J., E. Seal, and L. Folinsbee. "Effects of Steady-State and Variable Ozone Concentration Profiles on Pulmonary Function in Man," *Am. Rev. Respir. Dis.* 141:A71 (1990).
15. Koenig, J. Q., D. S. Covert, Q. S. Hanley, G. Van Belle, and W. E Pierson. "Prior Exposure to Ozone Potentiates Subsequent Response to Sulfur Dioxide in Adolescent Asthmatic Subjects," *Am. Rev. Respir. Dis.* 141:377–380 (1990).
16. Linder, J., D. Herren, C. Monn, and H. U. Wanner. "Die Wirkung von Ozon auf Die Koerperliche," Leisatungsfaehigkeit. *Schweiz. Z. Sportmed.* 36:5–10 (1988).
17. Bates, D. V. "Effects of Irritant Gases on Maximal Exercise Performance," in *Exercise Bioenergetics and Gas Exchange,* P. Cerretelli and B. J. Whipp, Eds. (Amsterdam: Elsevier/North Holland Biomedical Press, 1980), pp. 337–344.
18. Molfino, N. A., S. C. Wright, I. Katz, S. Tarlo, F. Silverman, P. A. McClean, J. P. Szalai, M. Raizenne, A. Slutsky, and N. Zamel. "Effect of Low Concentrations of Ozone on Inhaled Allergen Responses in Asthmatic Subjects," *Lancet* 338:199–203 (1991).
19. Last, J. A. "Synergistic Effects of Air Pollutants: Ozone Plus a Respirable Aerosol," Health Effects Institute Research Report No. 38. (1990).
20. Holland, W. W., and D. D. Reid. "The Urban Factor in Chronic Bronchitis," *Lancet* i:445–448 (1965).
21. Sterling, T. D., S. V. Pollack, and J. H. Phair. "Urban Hospital Morbidity and Air Pollution: A Second Report," *Arch. Environ. Health.* 15:362–374 (1967).
22. Richards, W., S. P. Azen, J. Weiss, et al. "Los Angeles Air Pollution and Asthma in Children," *Ann. Allergy* 47:348–354 (1981).
23. Goldsmith, J. R., H. L. Griffith, R. Detels, S. Beeser, and L. Neumann. "Emergency Room Admissions, Meteorologic Variables, and Air Pollutants: A Path Analysis," *Am. J. Epidemiol.* 118:749–779 (1983).
24. Bates, D. V., and R. Sizto. "Relationship Between Air Pollutants and Hospital Admissions in Southern Ontario," *Can. J. Public Health* 74:117–122 (1983).
25. Bates, D. V., and R. Sitzo. "Hospital Admissions and Air Pollutants in Southern Ontario: The Acid Summer Haze Effect," *Environ. Res.* 43:317–331 (1987).
26. Bates, D. V. "Air Pollution and Hospital Admissions: Identification of the Causative Agent," *Environ. Health Perspect.* 79:69–72 (1989).
27. Keeler, G. J., J. D. Spengler, P. Koutrakis, G. A. Allen, M. Raizenne, and B. Stern. "Transported Acid Aerosols Measured in Southern Ontario," *Atmos. Environ.* 24A:2935–2950 (1990).
28. Plagiannakos, T., and J. Parker. "An Assessment of Air Pollution Effects on Human Health in Ontario," Energy Economics Section, Economic and Forecasts Division, Ontario Hydro: File No. 706.01 (#260) March 1988.
29. Ozkaynak, H., B. Burbank, K. Van Dyke, F. McConkey, B. Chang, and J. Spengler. "Refinement of a Health-Based Air Quality/Visibility Index for Massachusetts, Phase II Draft Final Report. Submitted to Division of Air Quality Control, Dept. of Environmental Quality Engineering, Boston, MA. From Energy and Environmental Policy Center, Harvard University, Cambridge, MA.
30. Bates, D. V., M. Baker-Anderson, and R. Sizto "Asthma Attack Periodicity: A Study of Hospital Emergency Visits in Vancouver," *Environ. Res.* 51:51–70 (1990).

31. Whittemore, A. S., and E. L. Korn. "Asthma and Air Pollution in the Los Angeles Area," *Am. J. Publ. Health* 70:687–696 (1980).
32. White, M. C., R. A. Etzel, W. D. Wilcox, P. D. T erry, C. Lloyd, and K. Beusterian. "Childhood Asthma and Ozone Pollution in Atlanta," *Abstr. Annu. Meet. SOEH*, Washington, D.C., March 1991.
33. Thurston, G. Personal communication, (1991).
34. Lioy, P. J. Personal communication, (1991).
35. "Air Quality Criteria for Ozone and Other Photochemical Oxidants," Office of Research and Development, U.S. Environmental Protection Agency, Washington, D.C. 20460. USEPA 600/8-84/020dF (August 1986) 5 volumes.
36. Delfino, R. Personal communication, (1991).
37. Whittemore, A. S. "Air Pollution and Respiratory Disease," *Annu. Rev. Publ. Health* 2:397–429 (1981).
38. Hill, Sir Austin B. "The Environment and Disease: Association or Causation?" *Proc. Roy. Soc. Med.* 58:295–300 (1965).
39. Kinney, P. L., J. H. Ware, and J. D. Spengler. "A Critical Evaluation of Acute Ozone Epidemiology Results," *Arch. Environ. Health* 43:168–173 (1988).
40. Spektor, D. M., G. Lippmann, G. D. Thurston, P. J. Lioy, J. Stecko, G. O'Connor, E. Garshick, F. E. Speizer, and C. Hayes. "Effects of Ambient Ozone on Respiratory Function in Healthy Adults Exercising Outdoors," *Am. Rev. Respir. Dis.* 138:821–828 (1988).
41. Lioy, P. J., T. A. Vollmuth, and M. Lippmann. "Persistence of Peak Flow Decrement in Children Following Ozone Exposures Exceeding the National Ambient Air Quality Standard," *J. Air Pollut. Control. Assoc.* 35:1068–1071 (1985).
42. Spektor, D. M., G. D. Thurston, J. Mao, D. He, C. Hayes, and M. Lippmann. "Effects of Single- and Multiday Ozone Exposures on Respiratory Function in Active Normal Children," *Environ. Res.* 55:107–122 (1991).
43. Lebowitz, M. D., C. J. Holberg, B. Boyer, and C. Hayes. "Respiratory Symptoms and Peak Flow Associated with Indoor and Outdoor Air Pollutants in the Southwest," *J. Air Pollut. Control Assoc.* 35:1154–1158 (1985).
44. Javitz, H. S., R. Kransnow, C. Thompson, K. M. Patton, D. E. Berthiaume, and A. Palmer. "Ambient Oxidant Concentrations in Houston and Acute Health Symptoms in Subjects with Chronic Obstructive Pulmonary Disease: A Reanalysis of the HAOS Health Study," in *Int. Symp. Biomedical Effects of Ozone and Related Photochemical Oxidants,* S. D. Lee, M. G. Mustafa, and M. A. Mehlman, Eds. (Princeton, NJ: Princeton Scientific, 1982), pp. 227–256.
45. Gong, H., Jr. "Relationship Between Air Quality and the Respiratory Status of Asthmatics in an Area of High Oxidant Pollution in Los Angeles County," Report to California Air Resources Board. April 1987. Contract number A1-151-33 and A4-135-33.
46. Linn, W. S., J. D. Hackney, E. E. Pedersen, P. Breisacher, C. A. Mulry, and J. F. Coyle. "Respiratory Function and Symptoms in Urban Office Workers in Relation to Oxidant Air Pollution Exposure," *Am. Rev. Respir. Dis.* 114:447–483 (1976).
47. Hodgkin, J. E., D. E. Abbey, G. L. Euler, and A. R. Magie. "COPD Prevalence in Nonsmokers in High and Low Photochemical Air Pollution Areas," *Chest* 86:830–838 (1984).
48. Linn, W. S., M. P. Jones, E. A. Bachmayer, C. E. Spier, S. F. Mazur, E. L. Avol, and J. D. Hackney. "Short-Term Respiratory Effects of Polluted Ambient Air: A Laboratory Study of Volunteers in a High Oxidant Community," *Am. Rev. Respir. Dis.* 121:243–252 (1980).

49. Zwick, H., W. Popp, C. Wagner, K. Reiser, J. Schmoger, A. Bock, K. Herkner, and K. Radunsky. "Effects of Ozone on the Respiratory Health, Allergic Sensitization, and Cellular Immune System in Children," *Am. Rev. Respir. Dis.* 144:1075–1079 (1991).
50. Detels, R., D. P. Tashkin, M. S. Simmons, H. E. Carmichael, J. W. Sayre, S. N. Rokawe, and A. H. Coulson. "The UCLA Population Studies of Chronic Obstructive Respiratory Disease." *Chest* 82:630–638 (1982).
51. Bresnitz, E. A., and K. M. Rest. "Epidemiologic Studies of Effects of Oxidant Exposure on Human Populations," in *Air Pollution, the Automobile, and Public Health*, A. Y. Watson, R. R. Bates, and D. Kennedy, Eds. (Washington, D.C.: National Academy Press, 1988), pp. 389–413.
52. Lippmann, M. "Health Effects of Ozone," *J. Air Pollut. Control Assoc.* 39:672–695 (1989).
53. Hackney, J. D., and W. S. Linn. "Evaluating Relationships Among Personal Risk Factors, Ambient Oxidant Exposure, and Chronic Respiratory Illness," in *Susceptibility to Inhaled Pollutants*, ASTM ST1024, M. J. Utell and R. Frank, Eds. (Philadelphia: American Society for Testing and Materials, 1989), pp. 174–181.
54. Schwartz, J. "Lung Function and Chronic Exposure to Air Pollution: A Cross-sectional Analysis of NHANES II," *Environ. Res.* 80:309–321 (1989).
55. Ostro, B.D., and S. Rothschild. "Air Pollution and Acute Respiratory Morbidi ty: An Observational Study of Multiple Pollutants," *Environ. Res.* 50:238–247 (1989).
56. Portney, P. R., and J. Mullahy. "Urban Air Quality and Acute Respiratory Illness," *J. Urban Economics* 20:21–38 (1986).
57. Sherwin, R. P., and V. Richters. "Centriacinar Region (CAR) Disease in the Lungs of Young Adults. A Preliminary Report," in *Tropospheric Ozone and the Environment*, L. Berglund, D. R. Lawson, and D. J. McKee, Eds. (Pittsburgh: Air and Waste Management Association, 1991), pp. 178–196.
58. Eustis, S. L., L. W. Schwartz, P. C. Kosch, and D. L. Dungworth. "Chronic Bronchiolitis in Nonhuman Primates After Prolonged Ozone Exposure," *Am. J. Pathol.* 105:121–137 (1981).
59. Adesina, A. M., V. Vallyathan, E. N. McQuillen, S. O. Weaver, and J. E. Craighead. "Bronchiolar Inflammation and Fibrosis Associated with Smoking," *Am. Rev. Respir. Dis.* 143:144–149 (1991).
60. Mahoney, L. E. "Windflow and Respiratory Mortality in Los Angeles," *Arch. Environ. Health* 22:344–347 (1971).
61. Kinney, P. L., and H. Ozkaynak. "Associations of Daily Mortality and Air Pollution in Los Angeles County," *Environ. Res.* 54:99–120 (1991).
62. Richards, G. A., A. J. Theron, C. A. Van der Merwe, and R. Anderson. "Spirometric Abnormalities in Young Smokers Correlate with Increased Chemiluminescence Responses of Activated Blood Phagocytes," *Am. Rev. Respir. Dis.* 139:181–187 (1989).
63. Fitzgerald, A. M., M. R. Sears, R. S. Roberts, M. M. Morris, D. A. Fester, and F. E. Hargreave. "Symptoms of Asthma and Airway Hyperresponsiveness to methacholine in a Population of Canadian School Children," *Am. Rev. Respir. Dis.* 137:285 (1988) (Abstract).
64. Krupnick, A. J., and P. R. Portney. "Controlling Urban Air Pollution: A Benefit-Cost Assessment," *Science* 252:522–528 (1991).

CHAPTER 9

A LOCAL PROGRAM'S INITIATIVES TO REDUCE OZONE LEVELS

Roger P. Stewart, Executive Director and Staff

INTRODUCTION

Preamble of the Executive Director

Local environmental programs are a unique and essential player in the three-level (along with federal and state) hierarchy of environmental regulation in the U.S. Unfortunately, very few communities have established the legal framework essential to mount such an effort or are prepared to devote the monetary support on the local level. The USEPA has for many years wisely devoted support funding to develop and maintain local (as well as state) air quality programs (there is, regrettably, no comparable monetary support program in the non-air areas); despite this support, there is little evidence of any great enthusiasm to get involved at the local level. This is probably true in the less-developed or more sparsely populated areas of the nation. This author has long believed that much more could be done with far greater efficiency if there was a state-level incentive program or even a legal mandate insuring effective local participation in the state of Florida. To date, Florida's Department of Environmental Protection (DEP) is doing more to encourage the development of new air programs on the local level. This is good business given the fact that every dollar spent to encourage local authorities to develop local programs elicits a disproportionately large input of local dollars and effort.

0-87371-475-X/94/$0.00+$.50
© 1994 by Lewis Publishers

To this author, the greatest drawback to the effectiveness of local programs is the potential for local political interference which may unduly temper or thwart the mandate of the higher level of government. Political influence can and does manifest itself at all levels; at the local level, we often see significant changes in approach and policy with changes in administration at the federal and state levels. Some of these changes may not always be consistent with the environmental imperatives of the nation. I point out these things to emphasize that the Environmental Protection Commission (EPC) of Hillsborough County (Florida) has, throughout its 23-year existence, remained relatively free of any political influence and currently enjoys a largely environmental-oriented and knowledgable elected policy board.

The remainder of this chapter is a technical review of recent efforts on our part to effect the national and state mandates for clean air in Hillsborough County, Florida. Worldwide environmental trends (which it is not our intent to address herein) as currently emerging will no doubt demand a far greater and more comprehensive effort in the near future.

Purpose

The purpose of this chapter is to describe one local program's initiatives in reducing ambient ozone (O_3) levels. For the most part, the local program is the front line of the regulatory hierarchy. Field work is their top priority and their expertise is practical experience. Usually governed by a board of locally elected officials, local programs are more responsive to their citizens and in a better position to choose effective control strategies. Besides carrying out federal and state control measures, local programs can take additional initiatives as the problem and their resources allow.

EPC's Authority

There are over 200 local programs in the U.S. and the basis of this chapter is the experiences of one such agency — the EPC of Hillsborough County. The EPC was created by legislative mandate in 1967 under Chapter 84-446, Laws of Florida, initially titled "Hillsborough County Environmental Protection Act" — Chapter 67-1504. As of 1991, the agency has 140 full-time employees, with 38 working the Air Management Division. Through various agreements, the EPC acts as the state's and the USEPA's agents on most matters.

Organizational Philosophy

The Air Management Division's philosophy in controlling air pollution is twofold: preventive and corrective. Preventive options include ambient air monitoring and compliance inspections; the corrective option is enforcement. Preventive options are obviously always pursued because avoiding the degradation of air quality is the major goal of any regulatory air pollution program. The EPC annually commits to an ambitious schedule of inspections of air pollution sources. In situations where corrective measures are necessary, the enforcement section is committed to timely corrections of the problem and moving the situation once again into a preventive mode. Some enforcement may require permits, periodic testing, or providing public service-type presentations or pamphlets in lieu of, or in addition to, mitigation and monetary penalties.

Oxides of NO_x/O_3 Ambient Air Network

The Oxides of Nitrogen/Ozone (NO_x/O_3) Ambient Air Monitoring network consists of four ozone-monitoring stations (Figure 1), two of which also have NO_2 monitors. The four stations are Davis Island (Station Number 63), Gandy Station (Station Number 137), Simmons Park (Station Number 113), and Gaither High School (Station Number 138). Davis Island Station is the central-most O_3 monitoring station in the county. This station also has an NO_2 monitor which is not part of the network, but is used as a special-purpose monitor for the Differential Optical Absorption Spectrum (DOAS) unit (also known as OPSIS) feasibility study.

The EPC is the first local program in the U.S. to use the OPSIS. It is being used as part of a grant from the USEPA to determine its feasibility as compared to conventional monitors.

The NO_2 monitor is located at Gandy Station is part of the National, State and Local Air Monitoring (NAMS/SLAMS) network. The USEPA requires one NO_2 monitor based on Hillsborough County's population; however, an additional monitor may be required. NO_2 monitors are very expensive and labor intensive; it is hoped that by the time another NO_2 monitor is required, OPSIS will have received equivalency.

In 1985, the six ozone nonattainment counties in Florida, along with the state, formed an Ambient Monitoring Quality Assurance (QA) Task Force. As required by the USEPA, procedures had to be developed to ensure consistency regarding interpretation of the regulations and to ensure consistency of the data produced. From the task force emerged a QA plan which allows all the counties to verify the data which are produced. The Quality Assurance Plan supports the Ambient Air Monitoring Network by ensuring the data produced are valid. In turn, the Ambient Air Monitoring Network supports the permitting and field staff efforts by recording, over a period of time, the air quality which determines the successes of a program and also determines what controls should be emphasized in the future.

Stationary Source Compliance

A major source (A1) is a source with the potential to emit greater than 100 tons per year of any criteria pollutant. Hillsborough County has some 35 of these facilities. An A2 source has a potential of less than 100 tons, but uncontrolled, will exceed 100 tons; there are another 35 of these. Smaller sources (Class B) in

EPC #	STATION NAME & ADDRESS	PARAMETERS	NAMS/SPM/SLAMS
063	Davis Island 155 Columbia Drive	CO, O_3, NO_2, SO_2, PM_{10}, Wind D/S	NAMS/SPM/SLAMS
113	Simmons Park 2401 19th Ave., N.W.	O_3, Wind D/S	SPM/SLAMS
137	Gandy 5121 Gandy Blvd.	O_3, NO_2, Wind D/S	NAMS/SPM/SLAMS
138	Gaither	O_3, SO_2, CO	NAMS/SPM/SLAMS

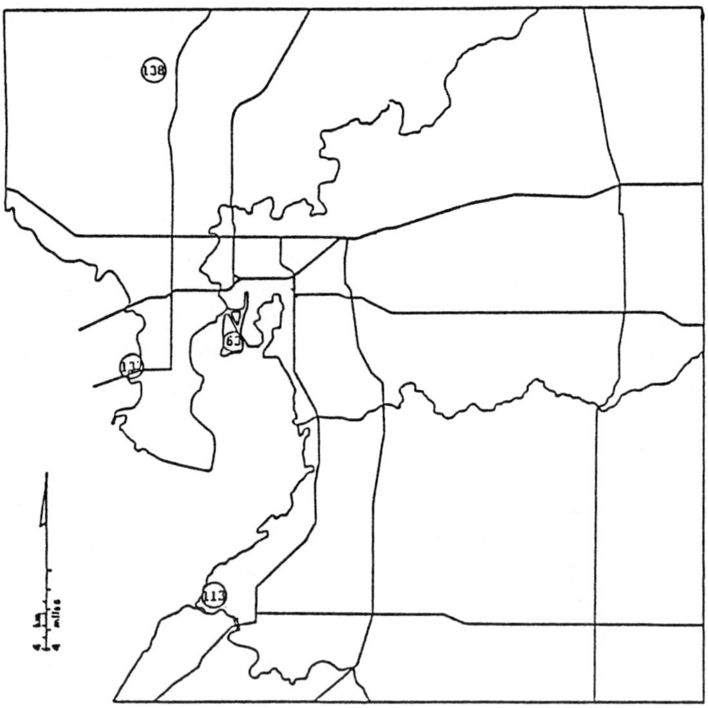

Figure 1. Tampa O_3/NO_x Ambient Air Monitoring Network. (From Environmental Protection Commission of Hillsborough County, Tampa, Florida. With permission.)

Hillsborough County have uncontrolled emissions between 25 and 100 tons. There are several hundred of these, although most are particulate sources. All A1 and A2 sources are permitted and most of the Class B sources are as well. Special emphasis is given to small VOC (volatile organic compound) sources because of the county's O_3 nonattainment classification.

The EPC annually commits to inspecting 100% of all the A1 sources and 50% of all A2 sources. Small source inspections are not a commitment, but are targeted.

Table 1. 1989 A1 and A2 Facility Emissions

Facility	PM	SO_2	NO_x	CO	VOC	Pb
Agrico	36.9	0.0	0.0	0.0	0.0	0.0
Alumax Extrusions Co.	0.8	0.1	16.3	3.3	101.5	0.0
Amoco Oil Co.	0.0	0.0	0.0	0.0	60.4	0.0
Ball Metal Corp.	0.3	0.0	0.4	0.0	134.1	0.0
B. P. Oil Co.	0.0	0.0	0.0	0.0	45.6	0.0
Cargill Inc.	0.3	0.0	0.0	0.0	0.0	0.0
Chevron USA	0.0	0.0	0.0	0.0	72.9	0.0
Citgo Corp.	0.0	0.0	0.0	0.0	51.6	0.0
Colored Metals Co.	0.1	0.0	1.2	0.3	60.4	0.0
COMCO			Did not operate			
ConAgra	8.9	0.0	0.0	0.0	0.0	0.0
Consolidated Minerals Inc.			Did not operate			
Consolidated Minerals Inc.	81.9	0.6	68.0	17.0	1.4	0.0
Couch Construction	6.2	0.5	17.1	4.1	0.3	0.0
Couch Construction			Did not operate			
CSX Transportation Corp.	144.4	0.0	0.0	0.0	0.0	0.0
C. F. Industries	85.2	4645.8	165.5	5.6	0.4	0.0
David J. Joseph Co.	3.1	0.0	0.0	0.0	0.0	0.0
Delta Asphalt	5.9	56.6	9.6	0.7	0.1	0.0
Detsco Terminals			Did not operate			
Eastern Associated Term.	57.5	0.0	0.0	0.0	0.0	0.0
Fina Oil & Chemical Co.	0.0	0.0	0.0	0.0	104.5	0.0
Florida Steel Corp.	140.5	1.2	41.3	377.0	0.6	1.2
GAF Building Corporation	20.2	0.0	4.5	2.9	2.6	0.0
Gardinier Inc.	173.2	2510.3	158.3	6.2	0.4	0.0
Garrison Stevedoring	7.4	0.0	0.0	0.0	0.0	0.0
GATX Terminals Corp.	0.0	0.0	0.0	0.0	74.2	0.0
GNB Inc.	0.6	0.0	0.0	0.0	0.0	2.7
Gold Bond Building Prod.	89.8	0.4	86.3	21.6	1.7	0.0
Graves Enterprises	15.1	0.1	4.6	23.2	22.1	0.0
Gulf Coast Lead Co.	5.0	1321.3	25.7	8.9	0.0	0.2
Gulf Portland Cement	1.0	0.0	0.0	0.0	0.0	0.0
HC Resource Recovery Unit	73.4	1160.9	1774.4	49.8	11.2	0.9
Ideal Basic Industries	2.4	0.0	0.0	0.0	0.0	0.0
Ideal Cement	1.7	0.0	0.0	0.0	0.0	0.0
IMCC Fort Lonesome	79.3	0.0	0.0	0.0	0.0	0.0
IMCC Port Sulton	465.4	0.2	52.5	13.1	1.1	0.0
Johnson Controls	0.0	0.0	0.0	0.0	0.0	0.1
Lafarge Corp.	235.0	0.0	0.0	0.0	0.0	0.0
Lafarge Corp.	5.0	0.0	0.0	0.0	0.0	0.0
Manna Pro. Corp.	1.8	0.0	0.0	0.0	0.0	0.0
Marathon Oil Co.	0.0	0.0	0.0	0.0	304.8	0.0
Master Packaging Inc.	0.0	0.0	0.0	0.0	201.6	0.0
McKay Bay RTE Facility	193.7	432.0	895.0	84.8	10.4	1.2
Mobil Chemical Corp. 1			Did not operate			
Murphy Oil Co.	0.0	0.0	0.0	0.0	29.0	0.0
Nitram Inc.	124.1	0.0	292.4	1.1	0.1	0.0
Paktank Florida	222.2	0.0	0.0	0.0	0.0	0.0
Phillips Pipeline Co.	0.0	0.0	0.0	0.0	23.7	0.0
P-Squared	0.0	0.0	0.0	0.0	6.8	0.0
Reed Minerals	27.3	0.0	1.7	0.4	0.0	0.0
Reynolds Metals Co.	0.1	0.0	3.7	0.9	180.5	0.0
Shell Oil Co.	0.0	0.0	0.0	0.0	84.9	0.0
Tampa Bulk Services	31.0	0.0	0.0	0.0	0.0	0.0
Tampa Shipyard-Mainyard	0.0	0.0	0.0	0.0	115.5	0.0
Tampa Shipyard-Southslip	0.0	0.0	0.0	0.0	23.4	0.0
Tampa Shipyard-Westshore	0.0	0.0	0.0	0.0	4.1	0.0

Table 1. 1989 A1 and A2 Facility Emissions (Continued)

Facility	PM	SO_2	NO_x	CO	VOC	Pb
TECO-Big Band	2823.5	157364.4	45670.2	1285.9	157.1	0.0
TECO-Gannon	2062.5	58548.9	44071.5	781.5	92.4	0.0
TECO-Hooker's Point1			Did not operate			
Trademark Nitrogen	0.0	0.0	35.5	0.1	0.0	0.0
UNOCAL Corp.	0.0	0.0	0.0	0.0	66.0	0.0
Verlite	4.1	0.0	0.2	0.0	0.0	0.0
W. R. Bonsal	1.1	0.0	0.0	0.0	0.0	0.0
Emissions (TPY)	7238.0	226043.4	93395.9	2688.5	2047.4	6.3

From Environmental Protection Commission of Hillsborough County, Tampa, Florida,1989 Air Quality Report, 1989. With permission.

These inspections ensure that the subject sources are in compliance with the regulations and their permit conditions. There are currently 28 small VOC facilities operating in Hillsborough County under permit. Table 1 lists the A1 and A2 sources in the county and their respective 1989 emissions.

Administrative Enforcement Protocols

In order to properly carry out obligations to protect the citizens and the environment from air pollution in Hillsborough County, the EPC must be able to effectively ensure that applicable rules and regulations are followed. Once a violation of a permit condition or a regulation/rule is detected, the Air Management Divisions Enforcement Section takes action. The choice of enforcement action taken depends on the circumstances and the consequences relates to the violation. The various options are described below.

Notification mechanisms for suspected or confirmed violations include:

1. Letter of Notification — The lowest level of agency enforcement action is a letter of notification or information, usually provided by the appropriate staff member responsible for the regulation of the environmental policy at issue. The letter is addressed to an individual or facility which staff has determined may in some minor manner not be in full compliance. The intent of such action is to provide information to the concerned party and alert them to the concern. This action does not initiate an enforcement case.
2. Warning Notice — Where staff believes a violation has in fact occurred based upon all the information available, a Warning Notice (WN) may be issued. The WN is normally prepared and issued in the field, and is intended to advise the responsible party or organization that a violation has possibly occurred. The purpose of the WN is to give the responsible party an opportunity to correct the information or explain the circumstances causing staff to believe a violation has occurred, or to quickly cease and correct the activity causing the violation so as to resolve the matter without enforcement. The WN action is preliminary to initiation of enforcement; if no response from the responsible party is received, enforcement will ensue.
3. Notice of Intent to Initiate Enforcement — For confirmed violations of environmental regulations where a negotiated settlement is reasonably likely (such as a breach of permit conditions), a Notice of Intent to Initiate Enforcement (NOI) is issued. The NOI provides notice that staff has confirmed the violation

and intends to pursue enforcement. The responsible party is requested to provide a written explanation of the violation and identify what actions it would be willing to implement to correct and prevent future similar violations. This information will normally be used to resolve the violation by Consent Order. Usually, 10 days are allowed for the source to respond.

Once a violation has been confirmed but has not yet been resolved, the following enforcement mechanisms may be used.

1. Citation — The Citation is issued by the Director pursuant to Section 14 of the Act. It is a formal administrative finding of violation and order to cease. It may include an order to correct the violation. The Citation gives specific notice that it may be appealed as per Section 9 of the Act. This enforcement action is usually implemented for serious violations, those in which settlement may be difficult to achieve, or in those cases where settlement through the NOI and initial draft Consent Order is at an impasse. Failure to comply with a Citation is a specific and separate violation of the Act.
2. Emergency Order — An Emergency Order is issued by the Director pursuant to Section 15 of the Act and is reserved for violations which create an immediate public health hazard or cause irreparable harm to aquatic life or property. The Order requires immediate cessation of the activities causing the threat of harm. Failure to comply with an Emergency Order is a specific and separate violation of the Act.
3. Judicial Action — Section 18 of the Act authorizes the Commission to seek civil judicial remedies for violations of the Act. Judicial remedies may be sought before, concurrent with, or after administrative enforcement is pursued. Judicial actions may be to enforce compliance with the Act, regulations, and orders; seek an injunction to cease a nuisance or other violation; and to require correction. The commission is entitled to recover its costs and the court can assess a civil penalty of up to $5000 (or $10,000 if state remedies are sought) per day per violation.
4. Criminal Violations — Section 17(3) of the Act provides that a violation of the Act can be a criminal violation. If a violation was intentional, or was in knowing disregard for the likelihood of violation and within the clear control of an individual or individuals, the Director will consider referral to the State or Federal Attorney. Referral will depend upon the seriousness of the harm and our ability to prevent the recurrence by agreement.

Once the circumstances of the violation have been agreed upon, the following "Resolution Mechanisms" are applied.

1. Letter of Agreement — The Letter of Admission or Agreement is a staff-level means of resolving a violation prior to it being formally determined to be a violation. The letter is an agreement signed by an appropriate person accepting responsibility for a circumstance and committing to resolving the circumstance and its effects. It must include the following: the facts of the violation, the actions taken to abate same, the actions taken to prevent a recurrence of the violation, and any agreed-upon settlement terms. These terms may include a settlement amount payable to the Pollution Recovery Fund and reimbursement of EPC investigative expenses.

2. Consent Order — The Consent Order is a legally enforceable settlement document which outlines the facts of a violation and the requirements for a mutually satisfactory resolution. It usually follows staff review of the violator's response to the NOI; however, in certain circumstances, staff may forward a draft Consent Order along with the NOI or other explanatory cover letter. The Consent Order provides for recovery of the costs incurred in investigating and resolving the case and usually includes a settlement amount in lieu of pursuing penalties judicially. This document must be reviewed by the Chief Counsel in draft form prior to mailing. The cover letter normally requires a response from the violator within 10 days and if all parties agree, the Consent Order is executed by the Director.
3. Closeout Letter — A Closeout Letter is mailed when a violator has fulfilled all agency requirements and is in compliance with regulations addressed by the enforcement. It is an acknowledgment by the agency that the enforcement case is closed. This is required for all cases where a Notice of Intent to Initiate Enforcement, Citation, or Emergency Order has been issued.

BACKGROUND

Physical Description of Hillsborough County

Hillsborough County is located on the west coast of central Florida (Figure 2). The total area of the county is 1072 sq. mi. (1048 sq. mi. land area and 21 sq. mi. inland water area). The population (1988) of the county is approximately 826,000 residents and the county is fairly heavily industrialized. There are large tracts of land used for agricultural crops, and there is some surface phosphate rock mining in the eastern part of the county. The climate is semitropical and air stagnations are rare.

A majority of the population resides in the unincorporated portions of the county (504,000). Tampa is one of the three incorporated cities, with a population of 286,000 residents. The other two cities (Temple Terrace and Plant City) are much smaller, with populations of 15,000 and 21,000 residents, respectively. Between 1984 and 1988, the county grew by approximately 14% (100,000 residents), with a majority of the growth concentrated in unincorporated portions of the county (82,000 residents). The unincorporated portions of the county now accounts for approximately 62% of the total county population. As a result of more than half of the county's population living outside the Tampa city limits, the air quality in Hillsborough County and the Tampa Bay area has been greatly affected. The problem is compounded by the fact that approximately 7500 residents from Pinellas County [located on the other side of Tampa Bay (west)] commute to Hillsborough County daily. This figure was taken from traffic counts between 7:30 and 8:30 a.m. (in 1989) at the Courtney-Campbell, Gandy, and Howard Frankland Bridges (the main transportation routes between Pinellas and Hillsborough counties). An average of 125 cars per minute pass the traffic counter within the 1-h peak time period.

The extremely rapid growth in the area has resulted in poor or inadequate infrastructure. This results in very slow-moving or idling cars on congested

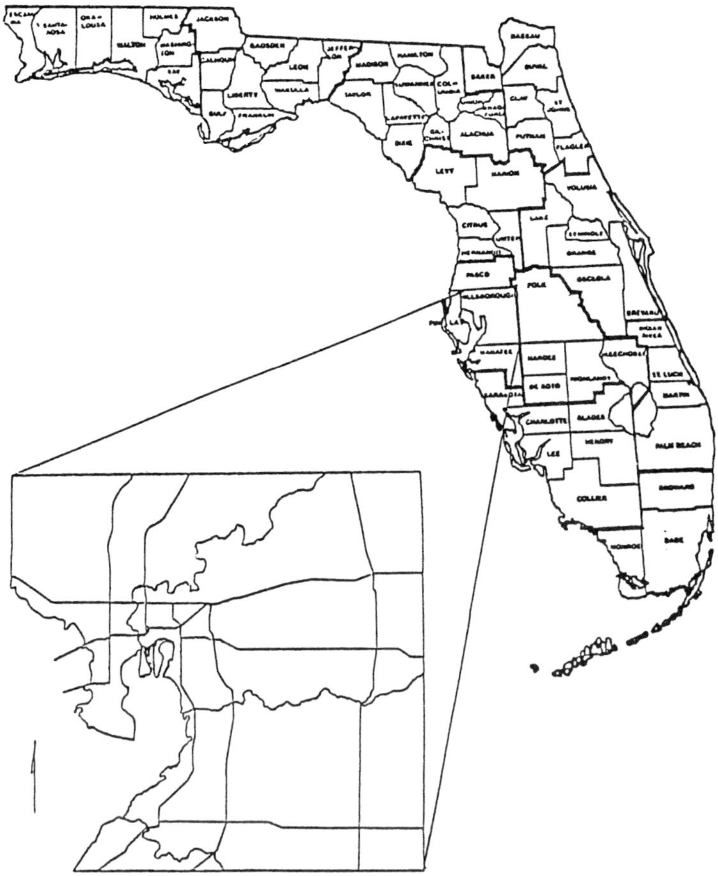

Figure 2. Hillsborough County, Florida. (From Environmental Protection Commission of Hillsborough County, Tampa, Florida. With permission.)

highways. The traffic problem, coupled with a majority of the major industries located along Hillsborough Bay and the Port of Tampa (Florida's largest port, which is within the Tampa city limits), also contributes significantly to the air quality of Hillsborough County and the Tampa Bay area.

Historical Perspective of West Central Florida's O_3 Problem

In 1970, the Clean Air Act (CAA) mandated that the USEPA establish National Ambient Air Quality Standards (NAAQS) for certain air pollutants. The USEPA established a standard of 0.08 ppm for photochemical oxidants (PHOX). In 1974, as a result of the South Florida Ozone Study, the USEPA determined that Hillsborough County and other population centers of Florida needed to install ambient air monitoring devices. The study found that due to Hillsborough County and the state's rapid growth in both population and number of vehicles, it was essential that the air quality be monitored. Since 1974, the ambient air in

Table 2. Photochemical Oxidant (PHOX) Concentrations Meeting or Exceeding the 0.08-ppm Concentration at the Davis Island Station

Year	# Days/year 0.08 ppm met or exceeded	# Hours/year 0.08 ppm met or exceeded
1974	61	244
1975	22	68
1976	55	191
1977	48	197
1978	18	65
1979	28	88
1980	25	114

From Florida Lung Association. With permission.

Hillsborough County has been monitored. Table 2 depicts the number of days the 0.08-ppm standard was exceeded in Hillsborough County.

In 1977, after the Clean Air Act was amended and made public law, the USEPA published a list of nonattainment areas based on ambient air monitoring data and urban population. This list identified seven Florida counties, including Hillsborough. One county, Orange, has since been classified as attainment.

In 1979, the USEPA changed the standard from 0.08 to 0.12 ppm. It also dropped all the chemicals comprising the PHOX group except the primary one, O_3. Figure 3 depicts the number of exceedances in Hillsborough County based on the new standard.

Since 1977, Hillsborough and Pinellas counties, occupying a common air shed, have been classified by the USEPA as ozone nonattainment areas. Each county has its own local program and assists the state in carrying out air quality control measures. The area is classified as a nonattainment area since at least one of the ambient air monitoring sites for O_3 exceeded the hourly O_3 standard of 0.12 ppm greater than once per year over a three year period. As indicated, Figure 3 depicts the number of days in which the NAAQS for O_3 has been exceeded in Hillsborough County.

O_3 continues to be Tampa Bay's most pervasive air pollution problem. The fact that O_3 is not directly emitted into the atmosphere, but is formed from precursor emissions (primarily composed of VOCs and NO_x) and that such precursor emissions can be transported many miles from the original sources, complicates the regulatory equation and makes the attainment of NAAQS much more difficult to achieve.

In Hillsborough County, the main source of oxides of nitrogen (NO_x) is stationary sources. It comes primarily from industrial and commercial boilers. There are 10 large coal-fired boilers in operation for generation of electricity.

The main source of VOCs is from automobiles. Other VOC sources include evaporation of gasoline, solvents, degreasers and surface coatings, and any sources of combustion.

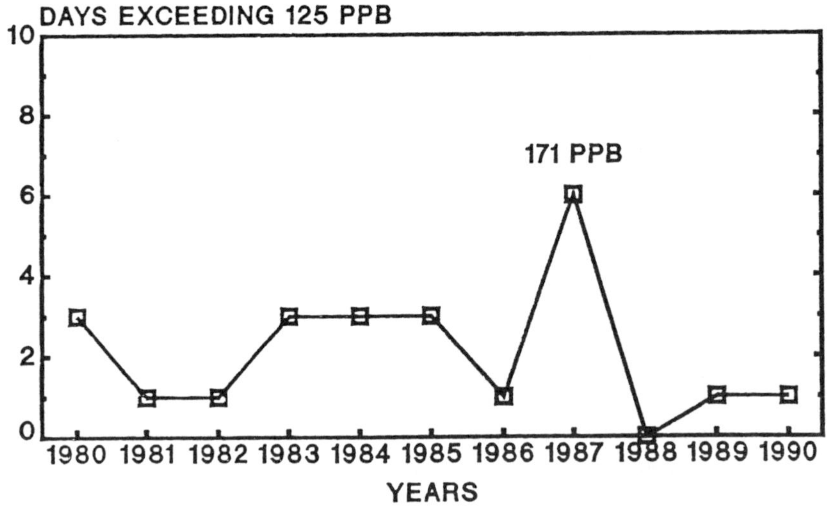

Figure 3. Hillsborough County O_3 levels. NAAQS exceedances in Hillsborough County. (From Environmental Protection Commission of Hillsborough County, Tampa, Florida. With permission.)

CONTROL STRATEGY

State Implementation Plan

In 1970 as a result of the CAA, not only did the USEPA need to develop air quality standards, but states needed to submit State Implementation Plans (SIPs) demonstrating their initiatives to achieve and maintain these standards. Since the USEPA felt the states were not acting fast enough, the CAA was amended in 1977. The Amendments required states to submit to the USEPA revised SIPs. These revised SIPs were required to demonstrate attainment of the NAAQS for PHOX by December 31, 1982. An extension was given until December 31, 1987 for states not able to meet the 1982 deadline, provided other provisions to reduce PHOX were initiated. Florida did not believe it could make the standard by 1982 so it prepared, as did Hillsborough County, to initiate other provisions. However, as stated earlier, in 1979, the USEPA eased the standard from 0.08 to 0.12 ppm, plus the standard was to include only O_3, and not other PHOX chemicals. After further review of computer modeling, Florida now felt it could meet the standard; therefore, some of the new initiatives were put on hold. It soon became apparent, however, that even with the new standards, not all nonattainment areas in Florida could reach attainment by the deadline.

The state of Florida needed an extension of the deadline to 1987 and a revised SIP was submitted to the USEPA. Basically, the 1977 CAA SIPs were to include the following:

1. Regulations to control existing industrial sources and new minor sources through the application of reasonably available control technology (RACT)
2. Procedures to control new major sources through the implementation of nonattainment New Source Review (NSR); this translates to lowest achievable emission rate (LAER)
3. Commitment to implement automobile inspection and maintenance programs
4. Commitment to implement transportation control measures, such as bus and carpool lanes, transit improvement, park-and-ride lots, and traffic flow improvements

Florida chose to implement items 1, 2, and 4 in the west central counties of Hillsborough and Pinellas counties to control this regional problem.

The 1982 revised SIPs included, in addition to the 1977 requirements, plans to:

1. Determine the level of emission reductions needed to reach the standard
2. Demonstrate how the proposed emission reductions would achieve the standard
3. Implement additional transportation control measures
4. Implement regulations to control other sources through additional RACT applications

At this time, the state further committed to use a linear rollback approach on total VOC emissions and implement any applicable new RACT categories.

RACT evolved from the USEPA's Control Technique Guideline (CTG) publications in the late 1970s. CTGs grouped specific sources or similar sources based on VOC use (i.e., bulk gasoline storage, dry cleaners, or degreasers). All 17 CTG categories in Hillsborough County were to be in compliance by 1982, according to the CAA of 1977. However, only four of the categories met the deadline; they were petroleum storage, (internal and external floating roofs), bulk gasoline terminals, and can coating operations. As stated earlier, an extension was granted to the 1982 deadline and some additional categories were brought under RACT requirements — gasoline service stations, graphic arts, and some surface coating facilities. There were still, however, many sources not in compliance.

In 1986, the EPC decided on a strategy which was to correct this situation. Certain facilities suspected of being VOC RACT sources were to be targeted for inspection and enforcement as necessary. A year's effort resulted in 19 facilities, previously unregulated, being inspected. Five were found to be subject to RACT requirements; in addition, four others were determined to be significant VOC sources but were not subject to RACT. This effort was helpful but was too slow. There were still hundreds of potential sources within the county which had not been investigated. To make matters worse, the USEPA had added 12 additional CTG categories for the EPC to look for.

In 1987, the EPC committed to another effort which would hopefully identify all the remaining RACT sources. The USEPA granted the EPC $18,000 to provide a complete list of VOC RACT sources with uncontrolled emissions greater than 25 tons/year in Hillsborough County. The major source book used for contacting potential VOC facilities was the *Contacts Influential Directory*, which conveniently listed sources by Standard Industrial Classification (SIC) codes. SIC codes were looked up for each CTG category except for gasoline service stations and solvent metal cleaning operations. The gasoline stations were already being monitored through the Stage I program and the solvent metal cleaning facilities would be detected through the miscellaneous metal parts inventory forms. Some 12 inventory forms were developed to cover the 21 categories believed to be operating in the county. The inventory forms were a modification of the USEPA forms. EPC simplified the format so that small facilities, not familiar with USEPA language, could understand the forms. The forms basically asked for production information and Material Safety Data Sheets (MSDS). The EPC opted to do the calculations. The purpose of undertaking the calculations for the potential 700 sources was twofold:

1. To improve the quality of data received and consistency of the emissions calculations
2. To relieve the burden on small sources, where they would not need to hire a consultant

Approximately 700 inventory forms were mailed in mid-March 1987. A summary of the forms mailed by CTG category is depicted in Table 3. By May 1, 40% returned completed forms, the remaining 60% were contacted by telephone inquiry to determine why they did not respond. By mid-June, an additional 10% returned completed forms. A second mass mailing was initiated and a 10-day suspense date to return completed forms was given. Due to the lack of response again, telephone inquiries were reinstituted. Some 90% (630) of the forms were eventually completed and returned; the remaining 10% were inspected by the EPC.

Some 300+ facilities responded with no emissions; one third of those sources were inspected. Based on the inspectors finding, 97% of those claiming no emissions were correct. It seemed obvious that the *Contacts Influential Directory* was not site-specific when assigning SIC codes, since 45% of the facilities with no emissions were from sales and distribution centers. The 3% of facilities found to have misrepresented themselves were very small sources who stated they did not understand the inventory.

For each CTG category at a facility, a worksheet showing the EPC emission calculations was completed. The worksheet listed the sources annual uncontrolled emissions, 1986 actual emissions, CTG category, percent reduction necessary to meet the CTG/RACT standard, and ton reduction possible based on 1986 production. For sources using less than 200 gallons of coating, EPC assumed all VOCs evaporated, which is much less than 1 ton VOC emissions. For those sources, 0.1 tons actual emissions and 1 ton uncontrolled emissions were estimated.

The final numbers resulting from the EPC's review of the inventory forms are summarized in Table 4. If all the sources identified in the inventory were complying with the CTG/RACT standards, then VOC emissions in Hillsborough County could be reduced by 228 tons per year. Since 98 tons of this was photochemically nonreactive perchloroethylene, the benefit in terms of reducing ambient O_3 was somewhat less.

Table 3. Summary of the Inventory Mailing and of the RACT Implementation in Hillsborough County

CTG category	Number of inventory forms mailed	Number of sources under permit in 1986
Surface coating of automobiles light-duty trucks	7	0
Surface coating of cans	0	2
Surface coating of metal coils	2	1
Surface coating of paper products	2	0
Surface coating of fabrics	0	0
Surface coating of metal furniture	3	0
Surface coating of magnet wire	2	0
Surface coating of large appliances	0	0
Fixed-roof petroleum storage	0	11
Bulk gasoline plants	31	0
Tank truck gasoline loading terminals	0	10
Gasoline service stations Stage I control	0	618
Miscellaneous refinery sources	0	0
Cutback asphalt	7	0
Solvent metal cleaning	0[a]	0
Leaks from petroleum refinery equipment	0	0
Surface coating of miscellaneous metals, parts, and products	400	5
Surface coating of flat wood paneling	12	0
Manufacture of synthesized pharmaceutical products	1	0
Manufacture of pneumatic rubber tire	0	0
Graphic arts-rotogravure and flexography	98	1
Petroleum liquid storage in external floating roof tanks	0	11
Perchloroethylene dry cleaning systems	130	0
Leaks from gasoline tanks, trucks, and vapor collection systems	0	500
Large petroleum dry cleaners	0[b]	0
Manufacture of high density polyethylene, polypropylene, and polystyrene resins	3	0
Natural gas/gas processing plants	0	0
SOCMI fugitive emissions	0	0
SOCMI — air oxidation	0	0

[a] Solvent metal cleaning operations were to be detected through coating of miscellaneous metal parts forms.

[b] Petroleum dry cleaners were to be detected through perchloroethylene dry cleaners forms.

From Environmental Protection Commission of Hillsborough County, Tampa, Florida. With permission.

Essentially, there were four categories where reactive VOCs could be reduced: metal furniture coating, bulk gasoline plants, metal parts coating, and petroleum dry cleaners. These source types represented 130 tons of potential VOC reductions. Within these four categories, there were 65 sources which contributed to these VOCs, averaging out to approximately 2 tons per source. Since roughly half of the noncomplying sources were so small they made no real contribution to the 130 tons,

Table 4. Summary of Inventory Results by CTG Category

CTG category	# of Class A-2/B sources found	VOC reduction possible from Class A-2/B/NB sources (TPY)
Surface coating of paper products	1	0
Surface coating of metal furniture	4	20
Bulk gasoline plants	1	16
Cutback asphalt	3	0
Surface of miscellaneous metal parts and products	24	48
Perchloroethylene dry cleaning systems	26	98
Large petroleum dry cleaners	11	46
Solvent metal cleaning	6	0
Totals	76	228

From Environmental Protection Commission of Hillsborough County, Tampa, Florida. With permission.

the average per source was more accurately reported at 4 tons. The largest single reduction possible was 11 tons from a metal furniture manufacturing facility.

The overall compliance rate for sources contacted in the inventory was 52%. That means five out of every ten sources subject to a CTG were determined to be in compliance with the applicable standards. When the perchloroethylene dry cleaners were not considered, the compliance rate improved to 59%. The individual categories with the highest compliance rates were paper coating, cutback asphalt, and solvent metal cleaning. On the other hand, the lowest compliance rates were found in bulk gasoline plants, metal furniture coating, and petroleum dry cleaning.

Although the discovery of 76 unregulated CTG sources in one country seemed surprising, it was not unexpected given the history of VOC RACT in Tampa. After the initial push in the early 1980s, VOC RACT took a backseat to other priorities. What was surprising was the small increment of reductions available. The expectations were several orders of magnitude greater than 130 tons. The 1986 strategy of targeting specific facilities was far more successful in terms of tons per year eliminated. The addition of an incinerator on a single coil coater found in 1986 practically matched the entire VOC savings identified in this inventory.

The addition of 50 reactive VOC sources to the regulated community had a significant impact. It more than doubled the number of VOC sources which were either under permit or an enforceable order to become permitted. The actual 1986 emissions from the inventoried facilities was 694 tons, with 150 tons coming from perchloroethylene dry cleaners compared to 3000+ tons of VOC emissions reported for the permitted stationary sources in 1986. Interestingly, only half of the emissions were subject to RACT regulation.

Beyond the USEPA request, the EPC inventoried an additional 200 VOC sources in 11 non-CTG categories. The categories included: fiberglass boat building, aircraft/ship painting, solvent storage, lithographic printing, paint manufacturing, plastic coating, wooden furniture coating, metal drum coating, fabric printing, photographic supplies, and large-scale automotive/truck refinishing.

The implementation of RACT significantly affected the enforcement scheme and provided interesting, but meaningful insight to the enforcement program. Sources found to be over 25 tons per year, based on EPC calculations, were referred to enforcement. The enforcement proceedings in some of these cases were very difficult for two primary reasons.

1. The penalties were very high because in some cases the VOC content of the coatings used were 50 to 200% greater than the standards
2. The smaller sources all claimed ignorance

Between fiscal years 1986 and 1988, the number of VOC enforcement cases increased from 3 to 48% of the total air enforcement caseload. During fiscal years 1988 and 1989, 46% of the VOC cases involved surface coaters of miscellaneous metal parts; 23% of violators were subject to non-RACT VOC regulations, including a styrofoam cup manufacturer and a circuit board manufacturer. Other types of VOC sources under enforcement during those 2 years were dry cleaners, bulk gasoline terminals, a coil coater, a graphic arts operation, and a metal furniture coater.

As noted above, at one time almost one half of the VOC enforcement cases in Hillsborough County involved coaters of miscellaneous parts. Although VOC emissions from these coating operations were high, the sources EPC worked with were small, unsophisticated businesses, some operating out of vacant lots. They were not active in trade organizations, nor did they have an in-house environmental staff; thus these violators were unaware of their contribution to air pollution in Hillsborough County. An integral part of settling the VOC enforcement cases was the education of source owners, and in some cases their legal counsel, about the requirements of state and local VOC regulations. Only after they understood the regulations and how they applied to their operations were they willing to work with the EPC toward settlement.

Enforcement actions initiated against petroleum dry cleaners produced some of the most interesting results. Persistence on the part of some of the affected dry cleaners and their trade organizations drew attention to deficiencies in the state's rule that regulated their operation. The result was an amendment to the rule affecting those dry cleaners. As such, all of the enforcement cases against the dry cleaners were closed without formal enforcement action: some were relieved from previous permitting requirements and others voluntarily agreed to obtain the required permits.

To achieve acceptable settlement in these cases, a Consent Order was the normal means of settlement. A typical Consent Order required a compliance schedule, fulfillment of permitting requirements, reimbursement of staff costs, and payment of a penalty for the violation of environmental regulations. Between fiscal year 1986 and fiscal year 1988, the EPC collected over $120,000 in settlement of VOC violations. More recently, a penalty in excess of $94,000 was collected from New Source Performance Standards (NSPS) procedural and emissions violations at a major VOC source in the county.

The emphasis on VOC enforcement at the local level led to some unforeseen difficulties, which in turn resulted in improved enforcement procedures that can be used in many types of enforcement actions. First, the VOC enforcement cases were some of the longest in our recorded enforcement history, from date of discovery to date of closure. This led to the implementation of a set deadline for staff review of cases and transfer of those cases to our legal department for further action. This 180-day deadline has been adopted as standard agency procedure for all enforcement cases.

The second improvement to the EPC's enforcement procedure resulted from the inordinate amount of staff time and effort expended to settle procedural VOC violations. To address that problem, it was determined that any unpermitted, minor VOC source not in violation of an emissions standard would be notified of the requirement to obtain a permit and would be given 45 days to do so. Noncompliance with those requirements would result in enforcement. The benefits from this change have been twofold: sources more quickly become part of the regulated community and compliance oversight is easier; enforcement staff time has been freed to more actively pursue violators that cause environmental harm.

The Stage I Control Program was developed to help reduce evaporative emissions of VOCs from the storage and marketing of gasoline. This program applied to all gasoline stations with a monthly throughput of more than 20,000 gallons. Chapter 1-8 was revised in April 1991 to include all gasoline service stations. The process of transferring gasoline from tankers to the applicable gasoline stations must include a vapor control system which collects displaced vapors that are created during the refueling.

There are at least 848 Stage I facilities in the county. The EPC inspects these facilities at a rate of approximately 20% per year; thus far, 512 have been inspected. Based on the number of gasoline stations inspected, approximately 15% are found to be out of compliance. Those found out of compliance are required to make appropriate corrections.

Beyond the implementation of RACT, nonattainment New Source Review (NSR) rules were imposed on the area. Their effect has been minimal as no LAER determinations have ever been done for a VOC source in Hillsborough County. On the other hand, many applicants have agreed to more stringent standards to avoid triggering NSR. This has had the desired effect of minimizing emission increases from major stationary facilities.

It should also be mentioned that the USEPA's New Source Performance Standard (NSPS) program has had a measurable impact. Modifications and additions to existing facilities has resulted in NSPS being applied to several of the bulk gasoline terminals and both of the remaining can coaters. This means VOC emissions were lowered.

EPC Initiatives

Much of the air pollution problem in population centers can be directly attributed to the automobile. Studies have shown that 50% of hydrocarbons and 40% of oxides of nitrogen emitted into the atmosphere originate from automobile exhaust. In Hillsborough County, as depicted in Figure 4, the portion of VOCs which are emitted by mobile sources vs. industrial sources is much greater. Mobile sources account for greater than 50% of all man-made VOCs emitted in the county. This rate is continuing to rise due to rapid growth. On the other hand, the NO_x estimates for the county are reversed, with stationary sources contributing three quarters of this O_3 precursor. As such, the environmental community has focused much attention on controlling automobile emissions.

In 1962, the first pollution control devices were installed on vehicles; by 1968, most light-duty vehicles were equipped with some sort of emission control device. Since 1975, the Federal Motor Vehicle Emissions Control Program has required new cars to be equipped with emission control devices. However, this program was not as effective as hoped because of tampering with the devices. The CAA of 1977 made it illegal to tamper (disconnect, render inoperative, or alter) with automobile emission control devices. In 1978, the USEPA began conducting annual tampering surveys nationwide of light-duty motor vehicles to determine the rates and types of tampering which existed. The 1984 survey showed that Hillsborough County had one of the highest tampering rates in the nation. The national average that year was 20%.

In 1987, the Motor Vehicle Emissions Study Commission was created by the Florida Legislature. Its purpose was to study the air quality problems in the six nonattainment counties of the state. Their conclusion was that the only way to meet the NAAQS for O_3 was to establish a mobile source emission control program. As a result of these findings and the inability to meet attainment, in 1988, the Legislature passed Chapter 88-129, Laws of Florida, entitled the Clean Outdoor Air Law (COAL). This law set up a statewide prohibition against anyone tampering with emission control devices (effective January 1, 1991). It also required vehicle emission inspections (effective April 1, 1991) on vehicles registered in the six nonattainment counties, as well as any other counties wishing to participate.

Hillsborough County felt action was needed more quickly after just completing the worst O_3 season (May through October) in 1987. As a result, in September 1987, the county enacted a mobile source rule, Chapter 1-8.

Chapter 1-8 prohibits tampering, visible emissions, misfueling, and the manufacturing, using, or selling of circumvention devices. The rule also authorizes vehicle inspections; facility inspections such a vehicle retainers, repair shops, parts stores, and service stations; and requires record keeping by these facilities. The final aspect of the rule is that it establishes enforcement and corrective procedures.

The EPC has conducted 6500 roadside pullovers since Chapter 1-8 was adopted. As mentioned earlier, Hillsborough County's tampering rate was 38% in 1984 and by 1989, the tampering rate was reduced to 24%. This reduction is believed to be due to fleet turnover, increased public awareness, and implementation of Chapter 1-8.

Besides the mobile source initiative, the EPC has also attempted to control open burning. In 1987, Chapter 1-4 was adopted. Chapter 1-4 Open Burning prohibits anyone from open burning without a permit except when used for cooking food or warmth. The permits are only issued for one-time land clearing activities. Agricultural land burning permits are authorized through the Division

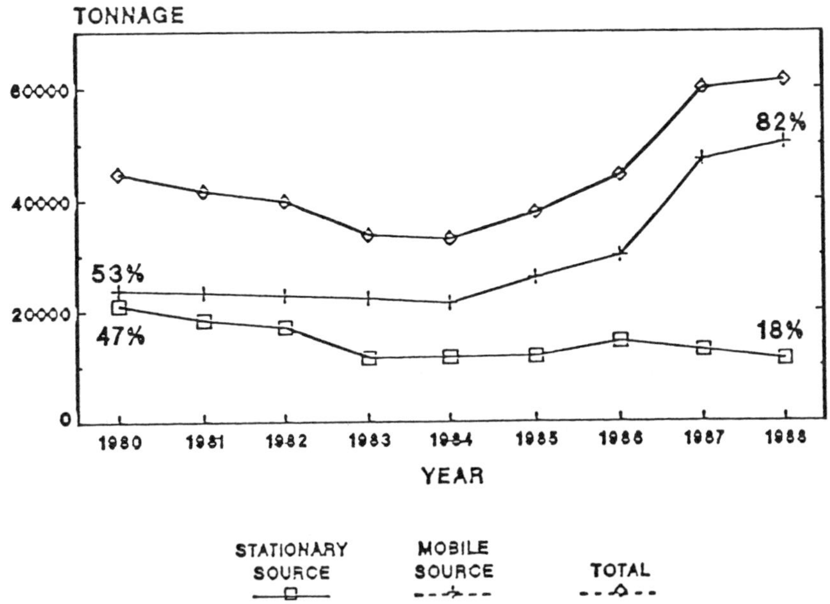

Figure 4. Percent man-made VOC emissions per year by stationary and mobile sources in Hillsborough County. (From Environmental Protection Commission of Hillsborough County, Tampa, Florida. With permission.)

of Forestry; all other open burning permits are issued by the EPC. Since the adoption of Chapter 1-4, the EPC has issued approximately 100 permits annually. Enforcement of this rule has been a high priority.

The EPC has attempted to prevent air exceedances in another manner. That is, if conditions are likely based on meteorological information and current ambient air monitoring data, an advisory is issued to the media. The purpose is to inform the public that pollution levels are elevated and that meteorological condition are such that they could worsen. An advisory may request citizens to drive as little as possible, and usually suspends all open burning permits. This is the agency's attempt to be *proactive* instead of reactive to degrading air quality.

CONCLUSIONS

Failure of Initial SIP

The State Implementation Plan (SIP) for Florida was not as effective as it could have been because only a small group (17) of VOC sources was targeted. These sources potentially made up only 20% of total VOC emissions in the area. The reductions made on this select group of sources in just two of the region's counties was overwhelmed by the area's rapid population growth. Nonattainment counties were the only ones required to identify and evaluate these sources. This was not totally effective because O_3 levels are not confined to political boundaries, but are indicative of air sheds; air sheds can and do cover large geographic areas. The largest emitter of O_3 precursors in the state — the automobile — was not even included in the initial SIP. Although Hillsborough County attempted a modest program to control vehicle emissions in 1987, the surrounding counties did not; therefore, little or no change was seen overall in the Pinellas-Hillsborough O_3 problem. The Mobile source program was, however, very effective in minimizing the tampering rate within the county.

Clear Air Act Mandates

Overall, the enactment of the Clean Air Act Amendments of 1990 will have a significant impact on the regulatory community in Hillsborough County. However, because of the region's designation as a marginal O_3 nonattainment area under Title I, the Act does not mandate much with regard to combating the region's O_3 problem. Perhaps the most significant action is the consideration of expanding the nonattainment area beyond the two counties currently recording exceedances. Although the surrounding counties may not be exceeding the standard, activities within their boundaries do contribute to air quality in Tampa and St. Petersburg. As such, any reasonable resolution to the problem will require their participation. The USEPA Administrator is to act on the boundary issue in the near future.

Another important feature of the Act is the 3-year attainment deadline. If a Marginal nonattainment area such as west-central Florida does not demonstrate compliance with the O_3 standard in 3 years, then the boundaries are reanalyzed and the area is bumped up to the next worse designation — moderate nonattainment. The Act requires these areas to do more than just implement RACT and keep good inventories. In addition to mandatory inspections for vehicles and stage II at gasoline service stations, the Act insists on a 15% reduction in VOC emissions from the area over a 6-year period. This would require some innovative and stringent rules which would go beyond just affecting certain industries. At the moderate level, the individual citizens' way of life could be changed.

It should be mentioned that the state has recognized the fallacy of attempting to correct the O_3 problem without addressing the automobile's contribution. In the six O_3 nonattainment counties in Florida — effective April 1, 1991 — all vehicles must pass an emission test prior to renewing their tags. For Hillsborough and Pinellas counties, this step goes beyond the Act. However, by taking this proactive measure, the state is giving the region a better chance of meeting the 3-year deadline specified in the Act.

Future Initiatives

As stated earlier, the EPC has four existing O_3 monitoring stations. Davis Island is the central-most in the county. Simmons Park, located near Ruskin, is south of Davis Island. The third, Gandy Station, is located west of Davis Island. The final, Gaither High School, was recently installed; it is located north of Davis Island. Another O_3 monitoring site was proposed in fiscal year 1992 and will be placed east of Davis Island. Once the latter site is on-line, all four compass points from the central location (Davis Island) will be covered. The purpose of these sites, geographically, is that O_3 levels can now be tracked across the county. It will also aid in determining how much the outlying counties contribute to the Pinellas-Hillsborough pollution problem. The outlying counties could then be included in any control measure.

Due to the continued rapid growth of the county, more emphasis must be placed on transportation issues. Historically, the reduction of O_3 has focused on the control of VOCs rather than NO_x. While emphasis on the control of oxides of nitrogen may increase in the future in severe or extreme nonattainment areas, it seems highly unlikely that the VOC efforts will be diminished in any program area. Hillsborough County has experienced the usual sequence of stationary source regulation and implementation regarding VOCs since the initial CAA in 1970. Consequently, a very high degree of precursor reduction has been obtained from these sources. In spite of this, the county remained nonattainment, and an evaluation of contribution reasons and sources for this led, quite naturally, to a shift in control focus towards vehicle emissions.

The aspects of the EPC's direct efforts in this area — Chapter 1-8 and its requirements, prohibitions, and subsequent vehicle and facilities inspection programs — have been discussed previously. The most recent mobile source control efforts are the Agency's involvement with transportation planning.

Hillsborough County is considered an urbanized area under the Urban Mass Transit Administration guidelines and, consequently, is required to have a Metropolitan Planning Organization (MPO) for transportation. Although formally a member of the MPO Technical Advisory Committee (TAC) since 1979, the EPC lacked the resources to provide technical analysis on air quality issues to the TAC until 1989, and only in a broad way then.

As mentioned earlier, the tremendous population growth in the county shows little sign of abating. Transportation demand and infrastructure and indirect source increase will place a significant burden on the local air quality. Correspondingly, there is an opportunity now, as never before, to minimize mobile source emissions before the fact.

In response to this challenge, the EPC's Air Management Division has begun the process of resource allocation and acquisition, employee development, and technical training and interagency networking in order to more fully realize its potential in transportation planning. Goals of this effort include continuing the mobile source inventory, technical review of the MPO's needs plan and cost-affordable plan, increasing the adoption and implementation of Transportation Control Measures (TCM), a strong presence on the MPO-TAC, and an air quality analysis of specific projects. Ideally, the EPC would be able to provide conceptual guidance at the MPO level in addition to technical guidance offerable to the MPO-TAC.

The EPC is also in the process of developing a compliance monitoring strategy which will be used by the USEPA as a model for other programs to follow. The strategy is meant to more effectively monitor all air pollution sources. Once completed, the strategy will provide guidance on ranking sources based on size, pollutant, and source type. The purpose is to target those sources which need to be inspected more often from those which do not require the same degree of attention.

Probably, the most significant initiatives in the future will focus on more research in O_3 formation, O_3 transport mechanisms, and the biogenic contribution to elevated O_3 levels. The Southern Oxidants Study, a 4-year $35 million study funded by Congress, will focus on these issues as they relate to the southeastern U.S. Hillsborough County, being a coastal county on a peninsula, does not fit the "usual" regional scenario. This will be researched to more fully understand the O_3 problem within the west-central Florida area. The EPC intends to fully participate in the study, which was scheduled to begin in the fall of 1991.

ACKNOWLEDGMENTS

This chapter represents the collective effort of the Executive Director and the agency staff, in particular the employees of the EPC's Air Management Division. Shiela Luce served as the Project Officier and contributed greatly to the writing of this paper. Assisting her in the writing and editing were Iwan Choronenko, Jerry Campbell, Steve Mlynarek, Kay Strother, and Tom Tamanini. A special recognition is also owed to Carol Ann Green for her typing and proofing of the chapter.

REFERENCES

1. Barton-Aschman Associates, Inc. Vehicle Counts August 23, September 6, and September 12, 1989.
2. Environmental Protection Commission. "VOC Inventory of Select Stationary Sources in Hillsborough County for Calendar Year 1986," Hillsborough County, Tampa, FL.
3. Environmental Protection Commission. "Hillsborough County Air Quality Report 1987," Hillsborough County, Tampa, FL.
4. Environmental Protection Commission. "Hillsborough County Air Quality Report 1988," Hillsborough County, Tampa, FL.
5. Environmental Protection Commission. "Hillsborough County Air Quality Report 1989," Hillsborough County, Tallahassee, FL.
6. Florida Lung Association. "The Status of Ozone Pollution in the State of Florida," (1980).
7. USEPA. "Motor Vehicle Tampering Survey 1986," September 1987.
8. USEPA. "Ozone Its Effects and Control," September 1978.
9. USEPA. "Ozone Policy Summary," May 30, 1985.

CHAPTER 10

COSTS OF ATTAINING OZONE AMBIENT AIR QUALITY STANDARDS

Francis L. Bunyard

INTRODUCTION

This chapter presents a discussion on the cost and impact implications associated with reduction of ozone (O_3) pollution. Several studies have been conducted by Office of Technology Assessment, the U. S. Environmental Protection Agency (USEPA), the South Coast District of southern California, and industry sources in an attempt to determine the magnitude of costs to bring some 95 urbanized metropolitan areas into attainment with the National Ambient Air Quality Standard (NAAQS) for O_3 (0.12 parts per million). Rather than attempt to sort through all these study results and provide a critical assessment of the various study results, this chapter presents an objective assessment of these costs — not so much whether they are low or high estimates, but for their implications in attaining the NAAQS. Now that the Clean Air Act (CAA) has passed, it is time for the implementers to apply their tools. Toward this end, the author attempts to summarize the USEPA's studies conducted for analysis of Title I and II provisions relevant to O_3 nonattainment.

BACKGROUND ON LEGISLATION

On November 15, 1990, the President signed into law the new Clean Air Act. The Administration forged the debate for the passage of the Act by endorsing

0-87371-475-X/94/$0.00+$.50
© 1994 by Lewis Publishers

market-based principles, in conjunction with effective legislation, to help attain cleaner air and better health for the nation's citizens. Several new features of the CAA incorporate provisions for economic incentives, innovative strategies, and cost-effectiveness considerations in developing state implementation plans (SIPs). The cornerstone of market-based principles is illustrated by the Title IV acid rain program, which allows for trading sulfur dioxide emission allowances for utilities and potentially other sources related to emissions of sulfur dioxide (e.g., suppliers of coal). Relative to attainment of the O_3 NAAQS, examples of market-based principles are those dealing with economic incentives, as allowed in Section 110(a)(2)(A) for SIP requirements, and cost-effectiveness guidance, Section 183(d) in evaluating options for control of emissions from stationary sources which contribute to the O_3 nonattainment problem.

The focus of this chapter is on summarizing the USEPA's analysis of the national cost for Title I provisions by categories of control measures, projection of attainment, and the geographic distribution of these costs. Finally, the chapter discusses application of cost effectiveness in the design and implementation of control strategies for attainment of the O_3 NAAQS.

STATUS OF NONATTAINMENT OF O_3 AIR QUALITY

As of October 26, 1991, there were 98 areas in violation of the O_3 ambient air quality standard.[1] Table 1 gives a listing of those nonattainment areas, their respective design values, and classifications. Except as noted in the table, the areas comprise consolidated metropolitan statistical areas (CMSAs) or metropolitan statistical areas (MSAs), as defined by the U.S. Department of Commerce. The areas are ranked according to O_3 design values based on monitoring data over the 1988–1990 time period. In addition, the table lists the classification status of each area based on two factors — current design values and the area classifications referenced in Subpart 2, Section 181(a) of the new CAA. This table gives insight into the level of control for which individual states should strive in designing their SIPs. More specifically, classification indicates the need for emission reductions; that is, in general, increased severity of nonattainment requires greater emission reductions.

SUMMARY OF COST INFORMATION

Various sources have provided cost estimates of requirements to attain the NAAQS for O_3 in our nation's cities, such as those listed in Table 1. Here, we will summarize results of the USEPA studies for reauthorization of the CAA.

In forging the legislative process for reauthorization of the CAA, the USEPA provided technical and economic inputs to the Administration and legislative staff. In this process, the USEPA analyzed the relative costs of the various bills that came to the floor for vote in either the House or Senate. Of particular interest

Table 1. The Status of Nonattainment of O_3 Air Quality

Location	Design value	Class
Los Angeles-South Coast Basin	0.330	Extreme
Southeast Desert Modified, CA	0.240	Severe-17
Houston-Galveston-Brazoria, TX	0.220	Severe-17
New York, NJ-NY-CT CSMA	0.201	Severe-17
Baltimore, MD	0.194	Severe-15
San Diego, CA	0.190	Severe-15
Chicago-Gary-Lake Co., IL-IN	0.190	Severe-17
Philadelphia-Wilm-Trenton, PA-NJ-DE-MD	0.187	Severe-15
Milwaukee-Racine, WI	0.183	Severe-17
Muskegon, MI	0.181	Serious[a]
Sheboygan, WI	0.176	Serious
Greater Connecticut	0.172	Serious
Ventura Co., CA	0.170	Severe-15[a]
San Joaquin Valley, CA	0.170	Serious
El Paso, TX	0.170	Serious
Manitowoc Co., WI[b]	0.167	Moderate[a]
Springfield (Western MA), MA	0.167	Serious
Boston-Lawrence-Worcester, MA	0.165	Serious
Washington, D.C.-MD-VA	0.165	Serious
Portsmouth-Dover-Rochester, NH	0.165	Serious
Huntington-Ashland, WV-KY-OH	0.164	Moderate[a]
Baton Rouge, LA	0.164	Serious
Providence, RI (all RI)	0.162	Serious
Atlanta, GA	0.162	Serious
Beaumont-Port Arthur, TX	0.160	Serious
Sacramento Metro, CA	0.160	Serious
Charlotte-Gastonia, NC	0.158	Moderate
Knox & Lincoln Cos., ME	0.158	Moderate[a]
Cleveland-Akron-Lorain, OH	0.157	Moderate
Cincinnati-Hamilton, OH	0.157	Moderate
St. Louis, MO-IL	0.156	Moderate
Portland, ME	0.156	Moderate
Parkersburg, WV	0.152	Moderate
Greensboro-WS-H Point, NC	0.151	Moderate
Pittsburgh-Beaver Valley, PA	0.149	Moderate
Kewaunee Co., WI	0.147	Moderate
Louisville, KY-IN	0.149	Moderate
Atlantic City, NJ	0.145	Moderate
Detroit-Ann Arbor, MI	0.144	Moderate
Grand Rapids, MI	0.143	Moderate
Salt Lake City, UT	0.143	Moderate
Jefferson County, NY	0.143	Marginal[a]
Salt Lake City, UT	0.143	Moderate
Dayton-Springfield, OH	0.143	Moderate
Richmond-Petersburg, VA	0.142	Moderate
Phoenix, AZ	0.141	Moderate
Reading, PA	0.141	Moderate
Raleigh-Durham, NC	0.141	Moderate
San Francisco-Bay Area, CA	0.140	Moderate
Dallas-Fort Worth, TX	0.140	Moderate
Edmonson Co., KY[b]	0.140	Marginal[a]
Santa Barbara-Santa Maria-Lompoc, CA	0.140	Moderate
Memphis, TN-AR-MS	0.140	Marginal[a]
Toledo, OH	0.140	Moderate
Miami-Fort Lauderdale-W. Palm Beach, FL	0.138	Moderate
Monterey Bay, CA	0.138	Moderate

Table 1. The Status of Nonattainment of O_3 Air Quality (Continued)

Location	Design value	Class
Charleston, WV	0.138	Moderate
Nashville, TN	0.138	Moderate
Lewiston-Auburn, ME	0.137	Moderate
Allentown-Bethlehem-Easton, PA-NJ	0.137	Marginal
Owensboro, KY	0.137	Marginal
Harrisburg-Carlisle-Lebanon, PA	0.136	Marginal
Canton, OH	0.135	Marginal
Knoxville, TN	0.135	Marginal
Poughkeepsie, NY	0.134	Marginal
Youngstown-Warren-Sharon, OH-PA	0.134	Marginal
Birmingham, AL	0.133	Marginal
Hancock & Waldo Cos., ME[b]	0.133	Marginal
Johnstown, PA	0.133	Marginal
Cherokee Co., SC[b]	0.132	Marginal
Buffalo-Niagara Falls, NY	0.131	Marginal
Columbus, OH	0.131	Marginal
Kent & Queen Anne's Co., MD[b]	0.131	Marginal
Lake Charles, LA	0.131	Marginal
Reno, NV	0.131	Marginal
Seattle-Tacoma, WA	0.131	Marginal
Norfolk-Virg. Beach-Newport N, VA	0.130	Marginal
Sussex Co., DE[b]	0.130	Marginal
York, PA	0.129	Marginal
Tampa-St. Petersburg-Clear, FL	0.129	Marginal
Walworth Co., WI[b]	0.129	Marginal
Scranton-Wilkes-Barre, PA	0.129	Marginal
Altoona, PA MSA	0.129	Marginal
Erie, PA	0.129	Marginal
Portland-Vancouver, OR-WA	0.128	Marginal
Manchester-Nashua, NH	0.128	Marginal
Albany-Schenectady-Troy, NY	0.128	Marginal
Jersey Co., IL[b]	0.128	Marginal
Essex Co., NY[b]	0.127	Marginal
Door Co., WI[b]	0.126	Marginal
Lexington-Fayette, KY	0.126	Marginal
Lancaster, PA	0.125	Marginal
Smyth Co., VA[b]	0.125	Marginal
Evansville, IN	0.124	Marginal
Paducah Co., KY[b]	0.124	Marginal
Indianapolis, IN	0.121	Marginal
South Bend-Elkhart, IN	0.121	Marginal
Kansas City, MO-KA	0.120	Submarginal

[a] Indicates 5% classification change.
[b] Indicates an area not a CMSA/MSA.

From "Designation of Areas for Air Quality Planning Purposes," U.S. EPA, 40 CFR Part 81, November 6, 1981. With permission.

are the estimated costs for the Title I and Title II provisions in the revised CAA (1990) since these are relevant to reduction of O_3 pollution across the country. In summary, the Title I, Subpart 2 (O_3), provisions require for moderate, serious, severe, and extreme nonattainment areas: (1) 15% across-the-board emission reductions by 1996, and (2) reasonable available control technology (RACT) based on CTGs (Control Technique Guideline) for 11 new source categories,

Table 2. Summary of Nationwide Costs and Emission Reductions for Titles I and II for O_3 Precursors (By Projection Year)

Projection year	Nationwide emission reductions (1000 tons/yr)	Low estimate of costs ($ milion/yr)	High estimate of costs ($ millions/yr)	Residual nonattainment areas
1995	5,400	5,600	5,900	25
2000	6,900	8,400	10,500	5
2005	7,800	9,500	13,900	4
2010	8,900	11,100	19,400	0

From Pechan, E. H. "Ozone Nonattainment Analysis, Clean Air Amendments of 1990" (Draft), Prepared for U.S. Environmental Protection Agency, September 1991. With permission.

gasoline vapor recovery at gasoline refueling stations (Stage II), and motor vehicle inspection and maintenance. Serious, severe, and extreme nonattainment areas are required to adopt, in addition to these measures, enhanced motor vehicle inspection and maintenance and clean fuels programs for mobile sources (e.g., reformulated gasoline, methanol, natural gas) that will foster use of alternative fuels to replace conventional gasoline in specific nonattainment areas. These are not all inclusive because there are also considerations regarding definition of major sources and offset requirements attendant to new sources that are classification dependent. Title II provisions include onboard vapor recovery, lower emission standards for conventional motor vehicles, lower fuel volatility (e.g., 7.8 psi Reid vapor pressure) for nonattainment areas, and alternative fuels requirements for specified nonattainment areas.

Point estimates of costs for controls for VOC (volatile organic compound) and NO_x (nitrogen oxides) are expected to increase from an incremental $5.7 billion in 1995 to $14.2 billion in 2010.[2] A complete summary of costs, emission reductions, and residual nonattainment areas (i.e., those areas out of attainment for the specific projection year) are presented in Table 2. A breakdown of costs and emission reductions by major categories — national stationary measures, area measures, motor vehicles, and additional progress requirements — for full attainment (the year 2010) is presented in Figure 1. Mandatory measures as specified in Titles I and II are captured in the area measures and motor vehicles categories. The category responsible for the largest single component of costs (45%) represents additional emission reductions for O_3 attainment in more problematical nonattainment areas.

What do these costs buy in terms of emission reductions and more importantly, attainment of the O_3 air quality standard? The following conclusions were reported based on modeling for the House Bill (HR 3030), which eventually became the final legislation voted into law.

1. Projection of VOC emission reductions in 1995 are about 45% from 1990 levels in nonattainment areas and about 35% nationwide.
2. After the 15% reasonable further progress requirements are implemented (1996), modeling indicated that there still remain some 16 residual nonattainment areas.
3. Projection of VOC emission reductions in 2005 will be approximately 48% in

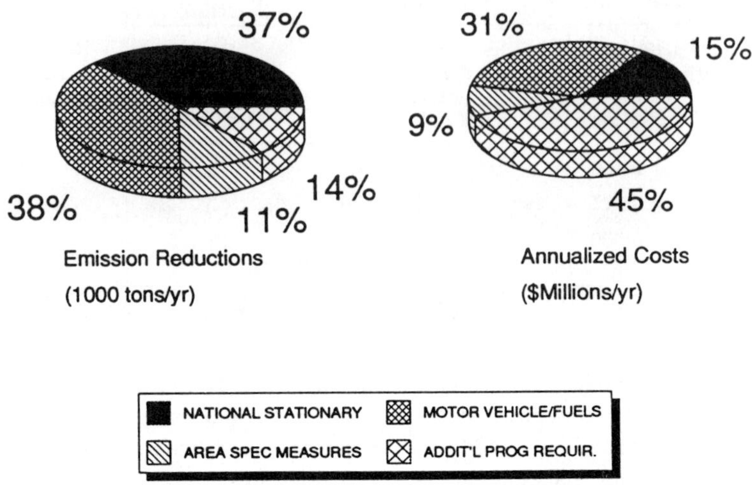

Figure 1. Summary of costs for O_3 attainment. Clean Air Act Amendments; projection year 2010.

nonattainment areas and about 40% nationwide.

4. The remaining 16 serious, severe, and extreme areas will require annual RFP emission reductions beyond the 15%, plus alternative fuels, reformulated gasoline, and other emission reductions from consumer products, solvent clean-up, and similar measures to gradually come into attainment by 2010.

DISCUSSION OF MODELING RESULTS

Modeling Approach

National stationary measures are those controls applied nationwide, mutually exclusive of SIP requirements. For purposes of modeling the various bills in the legislative process of drafting new CAA Amendments, these measures included controlling air emissions from treatment, storage and disposal facilities handling hazardous wastes, municipal landfills, commercial/consumer solvents, architectural coatings, and marine vessel loading/unloading operations. The former two source categories will be regulated post-1990, but under different the USEPA legislation.

Mobile source measures are those measures that apply to on-highway vehicles, excluding categories such as motorcycles. Measures affecting tailpipe emissions, evaporative and running losses, and fuels are regulated under Title II of the CAA. Fuels would include reformulated gasoline, oxygenated gasolines, and alternative fuels such as methanol, ethanol, and natural gas.

Area specific measures are those that are nonattainment area classification dependent (Title I provisions would apply) which would include RACT, new CTGs, I/M (inspection/maintenance) requirements, Stage II vapor recovery (gasoline marketing), and O_3 transport regions. These essentially would be the essence of a control program to achieve the 15% emission reductions from 1990 baseline emissions, which is a Title I requirement for 1996. Since Stage II emission reductions and cost are linked with other motor vehicle/fuels measures, this control measure was modeled with the motor vehicle/fuels category (Title II). Stage II effectiveness would have its greatest significance in the early years (e.g., 1996 through 2000) and onboard would have the greatest impact in the outward years as the motor fleet of 1990 experiences a 100% turnover (i.e., 2010).

Additional progress requirements are considered as a special category. By definition, these requirements represent an annual 3% reduction until the area attains the ambient air quality standard. From a modeling perspective, these emission reductions and control requirements are difficult to quantify for the following reasons: (1) they are dependent on SIP demonstration requiring a photochemical modeling analysis, (2) the control technology appears limited at this point in time, and (3) growth rates in terms of economic activity cast some doubt on any creditable analysis. For modeling purposes, this category was broken into two components: (1) discretionary measures, or identifiable controls for which costs have been somewhat defined, at least for the political debates involving enactment of the CAA Amendments, and (2) assumed controls, such as technology transfer, that could be applied but at high costs. Hence, this is the reason for a wide range of assumed costs.

Results

Figure 2 presents a distribution of costs for reduction of VOC emissions with motor vehicle/measures for the projection year of 2010. The figure shows that fuel volatility has the most efficacious result of available measures, in terms of costs and reduction of emissions. Conversely, reformulated gasoline is the most expensive measure. Although Figure 2 shows a small impact relative to other measures, reformulated gasoline will be an important measure in the near-time years in order for nonattainment areas to meet progress requirements. In the out years, reformulated gasoline has a smaller impact as alternative fuels, low emission vehicles, and the general turnover of the automobile fleet produce greater emission reductions.

Figure 3 presents a similar distribution of VOC emission reductions and cost for area specific measures, or what may be considered as Title I mandatory measures. The most important of these measures is the costs and emission reductions from new CTGs. Other measures include extending existing RACT regulations to smaller point sources, basic and enhanced inspection and maintenance, and "O_3 transport regions". The latter category represents extending other control measures, such as Stage II, inspection and maintenance, CTGs to attainment areas in a region designated as a transport region, such as the northeastern U.S.

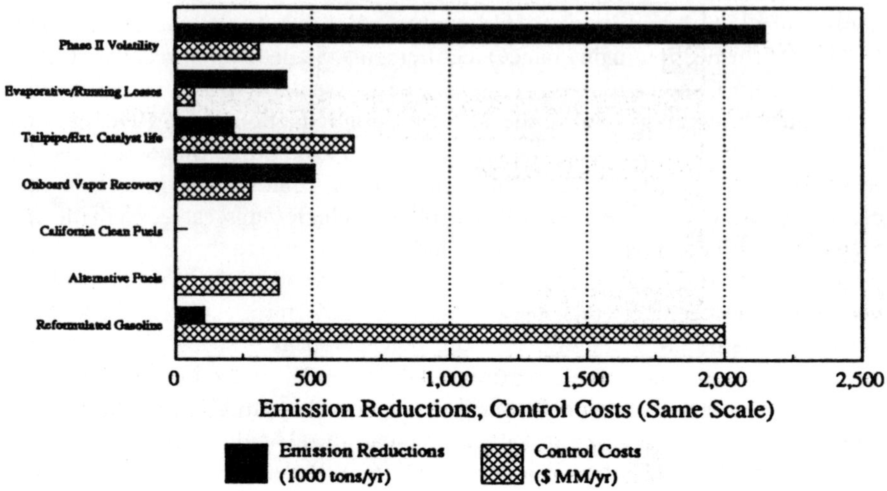

Figure 2. Distribution of VOC emission control costs for motor vehicle/fuel measures; projection year 2010.

FACTORS AFFECTING COSTS

Technology

Traditionally, major stationary point sources have been the focus for implementing rules limiting VOC emissions with such well-known techniques as carbon adsorbers, thermal and catalytic incinerators, and flares. Technology has made some inroads into "command-and-control" by improvements in economy of scale manufacturing (e.g., packaged carbon adsorber units for small sources such as dry cleaners). Technology has also benefitted by engineering improvements in enhanced heat recovery for thermal incinerators. Despite these innovations, end-of-the-pipe control technology has not made as significant advances as hopefully anticipated.

Major inroads have been made in process change or introduction of substitutes for raw materials in the manufacturing process derived from petroleum. The USEPA's experience in working with the coating manufacturers to develop more acceptable coatings from a consumer perspective was a key in the successful development of a series of coating regulations (i.e., New Source Performance Standards and Control Technology Guidance) for VOC Coatings Rules during the late 1970s and 1980s. Adoption of rules based on high solids coatings and waterborne coatings has lowered the cost of emission reductions for a category (namely, coating operations) that possibly could not apply feasible incineration systems. This experience illustrates that incentives can lead to innovative and less-expensive controls.

As indicated in the Introduction, economic incentives were mentioned as an important element of the new CAA. Some of these incentives could directly or

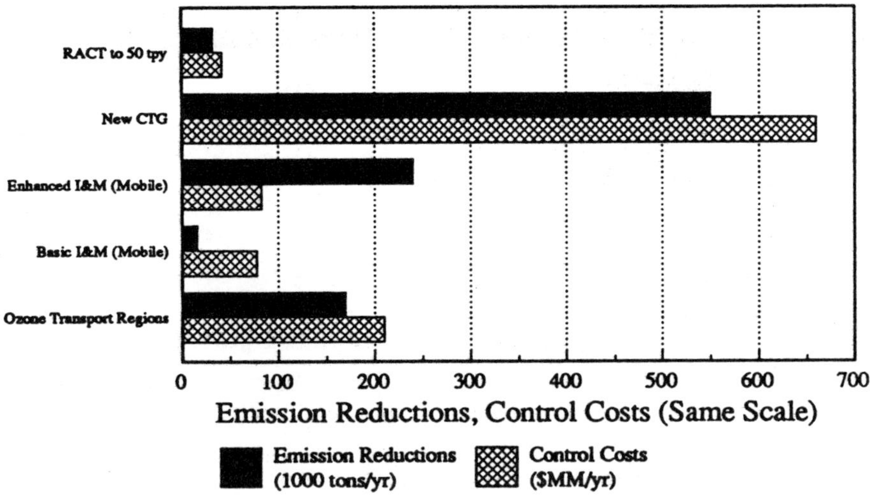

Figure 3. Distribution of VOC emission control costs for area-specific measures; projection year 2010.

indirectly foster the development of innovative technologies to reduce O_3 precursors. Some examples where economic incentives are cited in the CAA under Section 182(g)(4)(B) for revenues generated from fees adopted as part of SIPs and Section 183(e)(4) to enhance regulatory development of consumer and commercial products. For the latter, explicitly allowed are marketable permits and auction of emission rights that would be concerned with the manufacture, processing, distribution, use, consumption, distribution, or consumption of the products.

Implementation Issues

Rule Effectiveness

Expected costs and emission reductions for a given control strategy to attain and maintain the O_3 NAAQS may not be the same as the realized costs and emission reductions. More often than not, when the expectations for a control strategy are not realized, the emission reductions are *less* than anticipated.

Rule effectiveness reflects the ability, or lack thereof, of a regulatory program to achieve all the emission reductions possible through full compliance by all sources all the time. For stationary sources, the the USEPA presumes a rule effectiveness of 80% for SIP rules unless the state demonstrates a higher figure is appropriate for a source category.[5]

By calculating cost-effectiveness numbers assuming 100% rule effectiveness when rule effectiveness is less, the amount of emissions reduced will be overestimated, resulting in an under-estimate of the cost per ton of emissions reduced. This potential effect is illustrated in Table 3.

Table 3. An Illustrative Sensitivity Analysis of Rule Effectiveness

Control efficiency (%)	Controlled emissions (tons/yr)'	Rule effectiveness (%)	Emission reductions (tons/yr)	Cost effectiveness ($ per ton)
95	125	100	2375	632
95	363	90	2138	702
95	481	85	2019	743
95	600	80	1900	789

Note: Basis for analysis: (1) uncontrolled source category emits 2500 tons per year; (2) control cost for source category is $1.5 million per year.

As an example, suppose a control agency determines that a particular source category has uncontrolled emissions of 2500 tons per year. The agency believes that an objective of 90% emissions reduction is possible and specifies some allowable rate based on some output parameter, such as pounds of VOC emitted per pound of high solids coating applied. The source category installs control devices that are supposed to control at 95% control efficiency. With 100% rule effectiveness, emissions are reduced by 2375 tons per year (2500 tons/year × 0.95). However, rule effectiveness of less than 100% may result for a variety of reasons, including equipment leaks and failure to maintain specified operating conditions (e.g., flame temperature). Using the USEPA's default value of 80% rule effectiveness, the estimated emissions reductions are only 1900 tons per year [2500 tons/year × (0.95 × 0.80)]. Improved monitoring and enforcement of presently regulated sources, more inspections, improved record keeping and reporting, and corrective actions should be examined for enhancement of rule effectiveness, emission reduction potential, and cost effectiveness.[3] This is not to say that rule effectiveness is the only way in which to achieve additional emission reductions. Enhanced rule effectiveness should be compared to other methods of achieving reductions.

HOW COSTS MAY VARY GEOGRAPHICALLY

The question may arise as to how costs vary geographically. Figure 4 presents an insight into this question by presenting a distribution of cost requirements for attainment of the NAAQS by classification of nonattainment area. For example, nonattainment areas classified as severe or extreme may need more expensive controls at the margin — for each additional unit of emission reduction — than marginal or moderate nonattainment areas. Similarly, variability in the average cost of control among nonattainment areas is likely to be the norm. Figure 4 presents the modeling results of a control strategy study of 81 nonattainment areas using 1987–1989 O_3 monitoring data and illustrates this variability.[4]

Another approach is to present a profile of the modeled nonattainment areas in the form of an accumulative distribution, such as presented by Figure 5. More specifically, the figure represents the plot of average cost effectiveness of

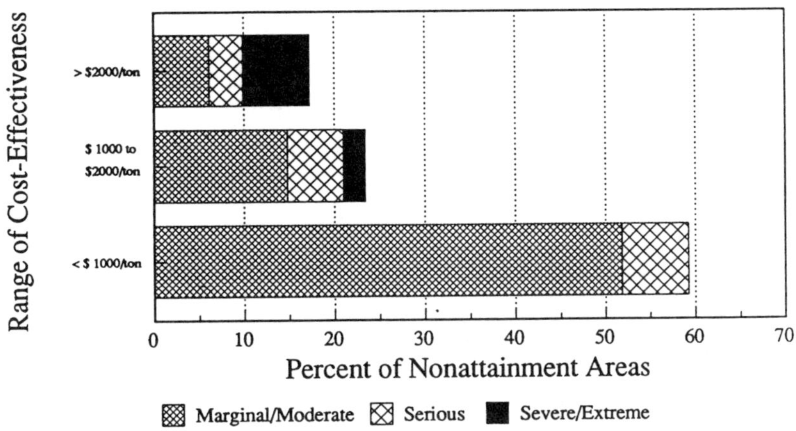

Figure 4. Cost effectiveness for nonattainment areas: a profile of CMSAs/MSAs by average cost per ton (projection year 2010). (From Pechan, E. H. "Ozone Nonattainment Analysis, Clean Air Act Amendments of 1990," E. H. Pechan, Inc. for USEPA. September 1991. With permission.)

mandatory measures for mobile, area-specific measures and post-1996 "progress requirements" for attaining the ambient air quality standard. The "progress requirements" are those requirements beyond the 15% reduction by 1996. These measures are the results of a study modeling 81 nonattainment areas using 1987–1989 O_3 monitoring data. Some of the high-cost areas in which modeled average costs exceed $3000 per ton will need control measures beyond what is currently technically available. These areas were modeled using available controls costed out on the basis of Agency documents and "progress requirements", costed by using a range of $2000–$10,000 per ton, or a point estimate of $5000 per ton, for the residual emission reductions. The USEPA recognizes that technology needed to achieve those residual reductions represent a major uncertainty in the analysis. Notwithstanding, the figure provides insight into control requirement needs. Control requirement needs vary across nonattainment areas; controls for the most problematic nonattainment areas will most likely cost more. This can be illustrated by Figure 6, which illustrates the correlation of nonattainment classification for 16 of the highest-cost nonattainment areas modeled for the U.S. As we would expect, areas ranked as severe/extreme (5) would incur the highest costs at the margin and, thus, the highest average costs. However, areas with lower classification (3) and (2) do appear in this modeled set and would have to be explained by other reasons, such as growth rates, unfavorable mix of NO_x and VOC concentrations from a photochemical modeling perspective, or source mix which may limit application of low cost controls.

The marginal cost per ton of reduced emissions is likely to vary for the following reasons

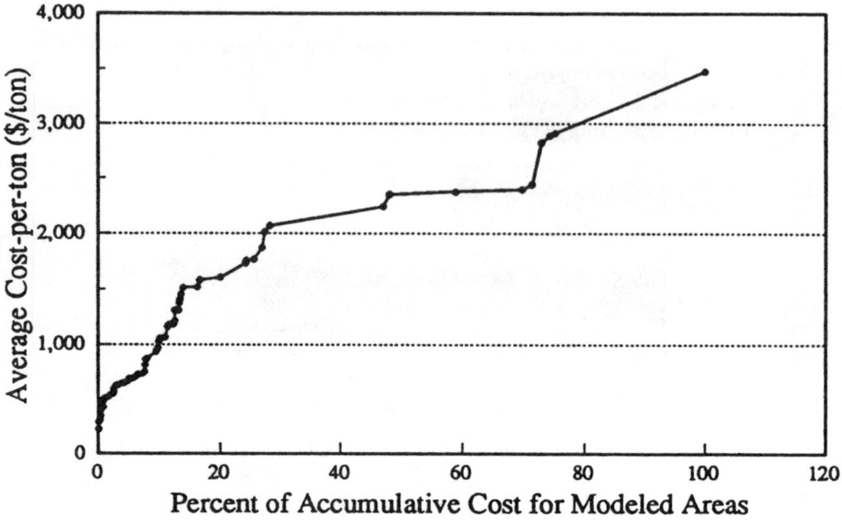

Figure 5. Cost effectiveness of attainment: a ranking by average cost-per-ton per nonattainment.

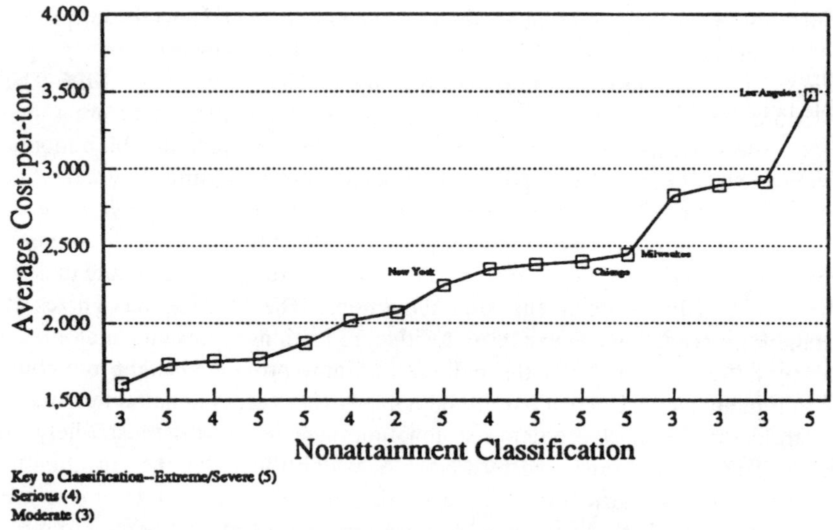

Figure 6. Ranking selected nonattainment areas: by average cost per ton.

- Sources available and selected for control — For example, within a given nonattainment area, there may be more reductions available from mobile sources rather than large point sources. In modeling the final legislation enacted for the CAA Amendments, we found that the marginal cost of control for the entire nonattainment area would vary as a result of the mix of source categories and various control measures available to reduce emissions. For example, some areas had a higher composition of mobile sources; other areas, area or small point sources (i. e., those emitting less than 100 tons per year of VOC).

- Baseline control levels — Some nonattainment areas may have already achieved the lower cost emission reductions available. Higher cost control measures might be required to reduce any additional units of emissions.
- Degree of control required — The amount of emissions reductions necessary to achieve attainment varies across nonattainment areas and therefore affects the relative marginal costs of control. These varying amounts of control are explained by differences in such factors as size and location of sources, as well as daily and seasonal fluctuations in temperature, emission rates, and wind patterns.
- Control techniques — The marginal cost of control is dependent upon the control measure selected to achieve additional emission reductions. In some instances, process change may be less costly than add-on controls, and rule-effectiveness enhancement may be less costly than greater rule penetration.

Table 4 illustrates various VOC control measures and relative cost effectiveness. These costs are national averages and represent current estimates.[5] Again, it should be noted that the marginal costs of VOC control measures for a given nonattainment area may differ from the national averages for these source categories. It should also be emphasized that some of these measures are mandatory, while others may be discretionary in terms of combining various measures for an overall control strategy. In general, process changes are lower in cost than end-of-pipe incineration controls on small sources (including small marine vessels). Rule effectiveness has been added as a *source category* to the table because improving rule effectiveness may help to achieve emission reductions. More inspections, improved record keeping and reporting, and corrective actions represent some of the elements identified in the March 31 Rule Effectiveness Study Protocol.[6] It should be noted that emission reductions resulting from rule effectiveness improvements occurring before 1990 and that are built into the emission inventory baseline are not creditable to the 15% progress requirements. Additionally, rule effectiveness is not without cost. Greater enforcement and/or inspection and maintenance procedures cost resources. Finally, transportation control measures that achieve actual emission reductions are also available, such as employer-based ride-sharing programs, mass public (rail or bus) transit, van pooling, and parking restriction ordinances in centralized business sections of metropolitan areas. A more comprehensive list is included in Section 108(b) of the CAA.

The control measures listed for NO_x emission reductions in Table 5 represent an illustration of various combustion sources to which process changes, such as low NO_x burners, staged air combustion, and add-on controls (namely, selective catalytic reduction) could apply. The range, in costs per ton, is due to factors such as flue gas flow rates, fuel, boiler configuration (tangential, wall), and application.

As described above, control requirement needs and marginal costs and the anticipated environmental quality improvements vary across nonattainment areas; therefore, setting control limits based on single dollars per ton ($/ton) values may not be appropriate.

Table 4. Illustrative VOC Control Measures and Cost Effectiveness

Source category	Control measure	Cost-effectiveness ($ per ton)
Architectural coatings	Application of high solids coating technology	Savings
Autobody refinishing	High solids coatings technology	2,000–7,500[a]
Offset lithography	Decreased alcohol content and magnets for fountain solutions	220–270[a]
Stage II refueling	Vapor balance fuel recovery	770–1,350
Treatment, storage, and disposal facilities (RCRA) air emissions	Tank covers, controls on aerated treatment and storage tanks	190
Enhanced inspection and maintenance	Higher performance standards	1,400–5,300
Volatility rules	Reid vapor pressure 7.8 psi	140
Marine vessel loading/unloading	Ventilation system and incineration	1,000–50,000[b]
Small source coating operation	Ventilation system and incineration	10,000–20,000
Rule effectiveness	More inspections, corrective actions	May lower the cost of control[c]
Consumer products	Substitute stick applicators for aerosol propellants	>400

[a] Draft Control Technology Guidance, mailed to National Air Pollution Control Technology Committee, October 4, 1991, Emissions Standards Division, Office of Air Quality Planning and Standards, U.S. Environmental Protection Agency.

[b] Refer to Table 2.

[c] U.S. Environmental Protection Agency, Office of Mobile Sources, Enhanced Inspection & Maintenance Briefing, October 1991.

From E. H. Pechan, Inc. "Ozone Nonattainment Analysis Clean Air Amendments of 1990," Sept. 1991. With permission.

ROLE OF COST EFFECTIVENESS IN STATE IMPLEMENTATION PLANS

After the the USEPA promulgates NAAQS, the CAA requires states to develop and submit implementation plans for the USEPA approval. SIPs contain enforceable regulations that provide for attainment and maintenance of the NAAQS.

To select a control strategy, states must initially identify mandatory control measures that are required by the CAA, such as the reasonable further progress requirements, RACT for stationary sources, volatility rules for fuels, and inspection and maintenance (I/M) for mobile sources. In addition, the amended CAA requires that control measures adopted or required to be adopted under the preamended CAA remain in effect [Section 193]. Therefore, these mandatory control measures must be adopted and retained for certain nonattainment areas. Beyond these constraints, states may select cost-effective, discretionary measures to attain and maintain the O_3 NAAQS.

Table 5. Sample NO_x Control Measures and Cost-Effectiveness

Source type	Control measure	NO_x emission reduction (%)	Cost-effectiveness ($ per ton)
Utility wall or tangential coal-fired boiler	Low NO_x burners	50	70–830
Utility residual oil-fired boilers	Staged combustion air	42	310–920
Utility tangential natural gas or coal-fired boiler	Selective catalytic reduction	80	3,900–5,300
Utility natural gas-fired boiler	Selective catalytic reduction	80	2,200–2,860
Industrial coal-fired boiler	Staged combustion or low excess air	36	Savings–380
<100 MM Btu/h natural gas	Flue gas recirculation	31	4,200–4,700
>100 MM Btu/h natural gas	Flue gas recirculation	31	1,000–1,100
<100 MM Btu/h natural gas	Selective catalytic reduction	80	12,700–17,400
>100 MM Btu/h natural gas	Selective catalytic reduction	80	2,100–3,200
Gas turbines	Water injection	70	1,000–1,700
Gas turbines	Selective catalytic reduction + water injection	70	2,400–3,900
Internal combustion engines	Change air-fuel ratio	30	140–930
Internal combustion engines	Selective catalytic reduction	80	120–910
Process heaters	Staged combustion air	45	Savings
Process heaters (oil-fired)	Selective catalytic reduction	90	200–4,500

From E. H. Pechan and Associates, under contract with the U.S. Environmental Protection Agency, 1991. With permission.

Figure 7 illustrates the process of selecting a cost-effective control strategy. As the figure shows, the first step in the selection of discretionary control measures is the determination of required emission reductions. Two inputs for determining these reductions are the following.

- A well-defined emission inventory that includes (1) an understanding of the relationships between emission factors (e.g., amount or rate of emissions) and the parameters (i.e., inputs used in the production process such as labor and materials) affecting production of marketable goods and services in the economy; (2) speciation of VOCs in terms of photochemical reactivity; (3) the implications of economic growth on projection of quantities; and (4) the implications of geographical distribution of future emissions for a nonattainment area.
- Air quality modeling for the relevant emissions inventory. Modeling tropospheric O_3 as a criteria pollutant involves a complex set of relationships. These relationships characterize the atmospheric chemical reactions that occur between those emissions that function as precursors — primarily VOCs and NO_x. When the linkage between the emissions inventory and air quality (design value) has been defined, the emission reductions required to meet attainment can be determined. The result is an environmental objective or target. The Urban Airshed Model is available to states to calculate the spatial and temporal concentrations of ground-level O_3 within urbanized areas or regional urbanized areas, such as the northeastern U.S. (see Reference 8).

The second step in the process of selecting a cost-effective control strategy is to catalog all the control possibilities by some measure of cost vs. environmental improvement. The proxy of cost-per-ton ratio is widely used in the USEPA

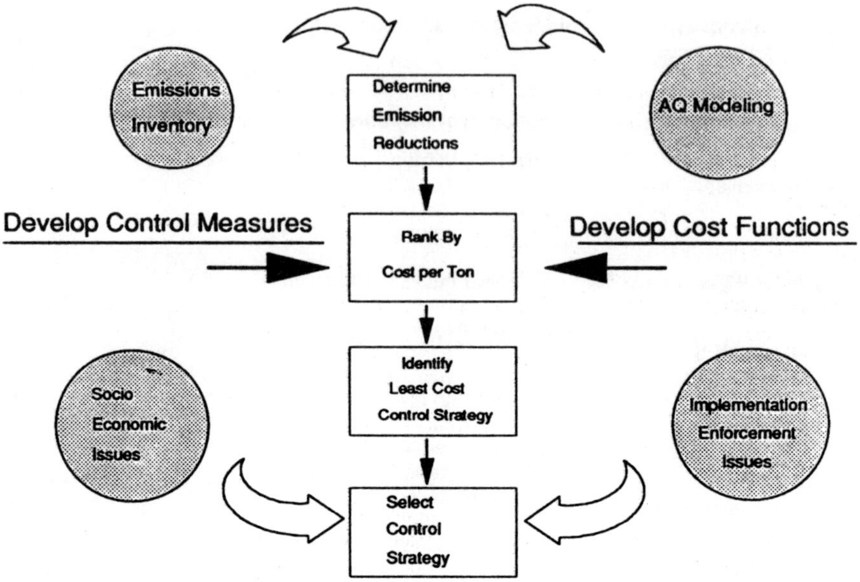

Figure 7. Process for selection of cost-effective control strategy.

analyses for developing regulations for individual source categories. The required inputs for this measurement call for the development of (1) a measurement that tracks control performance such as control efficiency or emission reductions per unit of time or production, and (2) cost (engineering cost) algorithms — mathematical expressions of the relationships between capital and operating costs and engineering parameters, such as size and production rates. Based on a technical assessment of performance and costs, costs per ton of emissions reduction are calculated for each control measure.

The third step is to identify several control strategy options, including the least-cost control strategy for the target emission reductions. Identification of control strategy options is performed by combining various control measures and evaluating the emission reductions and incremental cost for each measure to derive a total incremental cost for implementation of the entire strategy. Different strategies are developed iteratively in this manner to ensure that the least-cost strategy is identified. Mathematical programming techniques are sometimes appropriate to make this determination. It is important to note that the cost effectiveness of a given control strategy may be sensitive to the order in which individual control measures are applied. For example, if add-on control measures controlling 90% of emissions are applied to a stationary source before, after, or simultaneously with reformulated production inputs, the cost per ton of emissions reduced would vary between the three scenarios.

To this point, the process of identifying the least-cost control strategy is straightforward. However, there are policy (growth vs. environmental tradeoffs) and socioeconomic issues (employment dislocation and household sector impacts) that may not be quantifiable, or not readily quantifiable, in a least-cost mathematical programming structure. In addition, there may be implementation and enforcement issues — including the division of certain monitoring and certification responsibilities among various governmental entities and the regulated sources — that may not be quantifiable in this context. Control strategy selection is therefore a multi-attribute decision. In addition to costs, policy, socioeconomic effects, and certain implementation and enforcement considerations may also factor into the decision.

As a further caveat, there are other issues affecting cost effectiveness that have not been mentioned in this chapter. Baseline emission level, specification of emission reductions, rule effectiveness, and rule penetration are important factors that may influence the cost-effectiveness calculation and, possibly, the outcome of the control strategy selection. A discussion of these concepts is presented further in this document. Additionally, speciation may be important in the reactivity of various compounds and how those reactive compounds relate to O_3 formation. The agency position on reactivity is that all VOCs, except for those designated in the *Federal Register* as being negligibly reactive,[7] are of equal importance insofar as the mandatory 15% reductions for all nonattainment areas classified as moderate or above. Reactivity, however, becomes important in modeling for demonstration of attainment and maintenance of the NAAQS.

SUMMARY

This chapter presented an overview of the costs to implement various provisions of the new CAA for attaining the O_3 NAAQS. Meeting these provisions will be expensive; however, as other authors bear out in this book, the environmental goals are worth it. However, there are ways to reduce these costs and this chapter has attempted to address some of the ways that cost-effective principles can be applied to reduce these costs.

REFERENCES

1. "Designation of Areas for Air Quality Planning Purposes," 56 FR 56694 , November 6, 1991, U.S. Environmental Protection Agency.
2. *Ozone Nonattainment Analysis, Clean Air Amendments of 1990*, Draft Report, E. H. Pechan and Associates, Inc., prepared for U.S. Environmental Protection Agency, Office of Air and Radiation, Research Triangle Park, NC, September 1991.
3. Workshop for Implementation of Clean Air Act Provisions Relating to Ozone and Carbon Monoxide Emission Inventories, U.S. Environmental Protection Agency, Office of Air Quality Planning and Standards, Research Triangle Park, NC, June 4–6, 1991.
4. *Ozone Nonattainment Analysis, Clean Air Amendments of 1990*, E. H. Pechan, September 1991.

5. *Ozone Nonattainment Analysis, Clean Air Amendments of 1990*, E. H. Pechan, September 1991.
6. Memorandum from John Seitz, Director of Stationary Source Compliance Division, Office of Air Quality Planning, and Standards, to USEPA Regional Directors, "Implementation of Rule-Effectiveness Studies," March 31, 1988.
7. "Requirements For Preparation, Adoption, and Submittal of Implementation Plans," 56 FR 11387, March 18, 1991, U.S. Environmental Protection Agency.
8. Yocke et al. "Methodologies for Applying the Urban airshed Model to Determine the Effectiveness of Measures to Reduce Ozone Level in the Los Angeles Air Basin," A presentation at the 82nd National Meeting of the Air and Waste Management Association, Anaheim, CA, June 1989.

CHAPTER 11

THE FEDERAL ENFORCEMENT ROLE IN CONTROLLING OZONE

Daniel S. Jacobs*

INTRODUCTION

As discussed in Chapter 7, a broad array of federal statutes and regulations, as well as federally-approved plans and programs, governs the emission of substances which are precursors to ozone. Any regulatory control scheme can only be effective, however, if its requirements are enforced, that is, if those who flout the scheme are subjected to punishment. This chapter discusses the authority under which the federal government may enforce emission limits and related requirements intended to control ozone, as well as the process through which that authority is exercised.[1]

* A.B., Middlebury College; M. Phil., Cambridge University; J. D., Duke University. Mr. Jacobs is presently a trial attorney in the Environmental Enforcement Section, Environment and Natural Resources Division, U.S. Department of Justice, Washington, D.C. He has also served as Visiting Professor in the Department of Political Science at Middlebury College, Middlebury, VT, and Adjunct Professor in Residence in the Department of Government, School of Public Affairs, at American University, Washington, D.C. Mr. Jacobs wrote this chapter in his personal capacity, and therefore is solely responsible for its content.

[1] A discussion of Clean Air Act provisions which have little or no bearing on ozone control is beyond the scope of this chapter, as is an examination of state enforcement powers and citizen suits. For a comprehensive review of the enforcement of the Clean Air Act following the 1990 Amendents, see James Miskiewicz and John S. Rudd, "Civil and Criminal Enforcement of the Clean Air Act After the 1990 Amendments," 9 *Pace Environmental Law Review* 281 (1992) (hereinafter "Miskiewicz and Rudd").

0-87371-475-X/94/$0.00+$.50
© 1994 by Lewis Publishers

THE ENFORCEMENT PROVISIONS OF THE CLEAN AIR ACT

Under the Clean Air Act,[2] federal and federally-approved regulatory requirements[3] may be enforced under the U.S. Environmental Agency's (EPA) administrative authority, through the filing of civil judicial actions in the federal courts, and through criminal prosecution. The vast gamut of penalties, significantly strengthened under the 1990 Amendments to the Act,[4] ranges from an EPA administrative field citation of up to $5000 for relatively minor civil violations, to imprisonment of up to thirty years for the repeated commission of certain serious criminal offenses.

EPA Civil Administrative Enforcement

The Administrator of the EPA[5] has the authority to issue both administrative compliance orders and administrative penalty orders for a variety of violations of the Clean Air Act and requirements and prohibitions issued or approved under the stationary source provisions of the Act.[6] The authority is not mutually exclusive. EPA may order any person[7] found to be violating any such requirement or prohibition to comply, and/or may assess an administrative penalty against such a person.[8] Under the 1990 Amendments to the Act, EPA may assess up to $25,000

---

2 All references to the Clean Air Act ("Act" or "CAA"), 42 U.S.C. §§ 7401 to 7671q, include parallel citations to the Act (sections 101 to 618) and to the United States Code.

3 In terms of ozone control, the most important example of federally-approved regulatory requirements are state implementation plans (SIPs), which are promulgated by the states and must be submitted to EPA for federal approval. Once approved by EPA, they are federally enforceable.

4 Pub. L. 101-549.

5 Under the Act, enforcement authority is formally placed in the Administrator; in practice, the authority is delegated to other senior EPA officials. "Administrator" is used interchangeably here with "EPA."

6 CAA §§ 113(a)(1), (a)(3), and (d); 42 U.S.C. §§ 7413 (a)(1), (a)(3), and (d).

7 The Act defines "person" very broadly, see CAA § 302(e), 42 U.S.C. § 7602(c), but for purposes of civil administrative and judicial enforcement actions, "the term 'a person' shall not include an employee who is carrying out his normal activities and who is acting under orders from his employer," except where a knowing and willful violation is involved. CAA § 113(h), 42 U.S.C. § 7413(h). Inferentially, all other employees in all other circumstances may be subject to both administrative orders and civil judicial actions.

8 The following penalty assessment criteria apply to both administrative and civil judicial cases:

> In determining the amount of any penalty, ... the Administrator or the court, as appropriate, shall take into consideration (in addition to such other factors as justce may require) the size of the business, the economic impact of the penalty on the business, the violator's full compliance history and good faith efforts to comply, the duration of the violation as established by any credible evidence (incluing evidence other than the applicable test method), payment by the violator of penalties previously assessed for the same violation, the economic benefit of noncompliance, and the seriousness of the violation.

CAA § 113(e)(1), 42 U.S.C. § 7413(e)(1).

per day of violation, up to a $200,000 maximum,[9] and may assess up to $5000 for minor violations under a field citation program.[10]

In the event a violator fails to obey a compliance order or administrative penalty order, EPA may enforce the order through the filing of a civil judicial action.[11] In actions to enforce administrative penalty assessments, the Administrator is entitled to recover the assessed penalty with interest, nonpayment penalties, and the government's enforcement expenses.[12]

In addition to these and other options available, the Administrator may also assess and collect a noncompliance penalty under section 120 of the Act,[13] against stationary sources which violate an emission limitation or standard or a compliance schedule under an applicable implementation plan. This particular type of penalty, which dates back to the 1977 amendments to the Clean Air Act,[14] was designed to recover the economic advantage which might otherwise accrue to a source by reason of its failure to comply with air pollution control standards after receipt of a notice of noncompliance.[15] Collection of noncompliance penalties may be enforced through the filing of a civil judicial action under section 113 of the Act.

Civil Judicial Enforcement

Clean Air Act violations may be prosecuted not only administratively but also through the filing of a civil judicial action in federal district court. The Act authorizes the Administrator to bring suit against the owner or operator of a stationary source[16] and any other person who violates any requirement or prohibition of an applicable implementation plan, permit, or permit program.[17] The Administrator is also authorized to bring suit against any person who attempts to construct or modify a major stationary source in a state which is out of compliance with a requirement or prohibition of the Act relating to construction or modification of such sources.[18] In all such cases, the government may seek and obtain as

---

9 The maximum may be exceeded if the Department of Justice concurs. CAA § 113(d) U.S.C. § 7413(d).

10 CAA §§ 113(d)(1) and (3); 42 U.S.C. §§ 7413(d)(1) and (3).

11 CAA § 113(b)(2), 42 U.S.C. § 7413(b)(2).

12 CAA § 113(d)(5), 42 U.S.C. § 7413(d)(5). The expenses include, but are not limited to attorney's fees, collection proceeding costs, and a fixed nonpayment penalty. Such expenses, which can be substantial, ordinarily are not recoverable by the government in federal enforcement actions brought under the Clean Air Act, while the defendant in such actions may win his costs of litigation (including reasonable attorney and expert witness fees), if the court finds that the action was unreasonable. *See* CAA 42 § 113(b), 42 U.S.C. § 7413(b).

13 42 U.S.C. § 7420.

14 Pub. L. 95-95.

15 *See* 40 C.F.R. § 66.1.

16 Defined at section 302(z) of the Act, 42 U.S.C. § 7602(z).

17 CAA §§ 113(b)(1) and (2); 42 U.S.C. §§ 7413(b)(1) and (2).

18 CAA §§ 113(b)(3), 42 U.S.C. §§ 7413(b)(3).

relief, a permanent or temporary injunction[19] and/or up to $25,000 per day in penalties for each violation.

The Administrator is also authorized to file suit for injunctive relief and for civil penalties for violations of the mobile source emission standards provisions of the Act.[20] The court may impose penalties of up to $25,000 for violations of most motor vehicle emission provisions.[21] Violators of the motor vehicle fuels provisions are subject to penalties of $25,000 per day; in addition, the government may recoup the amount of any economic benefit or savings resulting from the violation.[22]

Criminal Enforcement

There is a wide range of offenses punishable as crimes under the Clean Air Act. Any person[23] who knowingly violates any requirement or prohibition of an applicable implementation plan or any compliance order is subject, upon conviction, to imprisonment for up to five years.[24] False statements and the like are punishable by imprisonment for up to two years.[25] The negligent release into the ambient air of certain hazardous pollutants,[26] which negligently places another person in imminent danger of death or serious bodily injury, is punishable by

---

[19] An injunction can take two forms: it can (1) prohibit the defendant from taking a proscribed action; and/or (2) compel a defendant to take a prescribed action. In the context of a Clean Air Act civil action, for example, a defendant can be prohibited from operating in violation of a SIP provision, and/or a defendant can be ordered to take specific measures to correct violations (for example, the installation of add-on controls). The violation of an injunction can lead to coercive court-imposed sanctions, including fines and imprisonment.

[20] CAA §§ 204, 205; 42 U.S.C. §§ 7524, 7525.

[21] Persons other than manufacturers and dealers who tamper with emission control systems are subject to a $2,500 penalty per motor vehicle. Persons who manufacture, sell, or install parts which are intended to bypass or defeat emission control systems are subject to a $2,500 penalty per part.

[22] CAA § 211(d)(2), 42 U.S.C. § 7545(d)(2).The penalty assessment criteria in mobile source and fuels cases, which largely resemble the criteria applicable in stationary source cases, are as follows:

> In determining the amount of any civil penalty to be assessed under this subsection, the court shall take into account the gravity of the violation, the economic benefit or savings (if any) resulting from the violation, the size of the violator's business, the violator's history of compliance with this title, action taken to remedy the violation, the effect of the penalty on the violator's ability to continue in business, and such other matters as justice may require.

CAA § 205(b), 42 U.S.C. § 7524(c); *see also* CAA § 211(d)(1), 42 U.S.C. § 7545(d)(1).

[23] The 1990 Amendments to the Clean Air Act made clear that for criminal enforcement purposes, the term "person" specifically includes "any responsible corporate officer." CAA § 113(c)(6) 42 U.S.C. § 7413(c)(6).

[24] CAA § 113(c)(2), 42 U.S.C. § 7413(c)(2).

[25] CAA § 113(c)(1), 42 U.S.C. § 7413(c)(1). Additional offenses, including recordkeeping violations, are punishable under the same statute.

[26] These include the hazardous air pollutants listed in section 112(h) of the Act, 42 U.S.C. § 7412(h), as well as the extremely hazardous substances listed by the Administrator pursuant to 42 U.S.C. § 11002. *See* 40 C.F.R. Part 355, App. A.

[27] CAA § 113(c)(4), 42 U.S.C. § 7413(c)(4).

imprisonment for up to one year.[27] The most serious type of offense, the knowing release of such a pollutant, accompanied by knowledge that the release places another person in imminent danger of death or serious bodily injury, is punishable by imprisonment of up to fifteen years.[28]

All these offenses are also punishable by fines, including a fine of up to $1,000,000 for an organization convicted of the most serious type of offense. The maximum penalty, both fine and imprisonment, is doubled if the offender has a prior conviction for the same type of offense.[29]

ENFORCEMENT IN PRACTICE

Civil Administrative Actions

While historically less common and less daunting than civil judicial actions, EPA administrative orders can be expected to be issued with greater frequency with the strengthening of the Agency's administrative powers by the 1990 Amendments. Prior to the amendents, EPA did have the authority to issue compliance orders under section 113(a) and assess noncompliance penalties under section 120, but for a number of reasons, it perceived its authority as weak and seldom exercised it.[30] With passage of the 1990 Amendents came both the advent of administrative penalty orders and a clearer congressional mandate on the duration of administrative compliance orders. EPA is already more actively exercising its new administrative penalty authority under section 113(d)(3), averaging just over one penalty order per month for excess VOC emissions in the 10-month period after formal procedures governing issuance of such orders were put into place.[31] These numbers alone suggest that administrative enforcement under the Clean Air Act seems to have taken on a new-found vitality, and that administrative orders are likely to be a more common and perhaps more effective enforcement tool than in the past. With the emergence of administrative penalty orders, noncompliance penalty assessments under section 120 of the CAA seem destined to become *de facto* defunct.

By law, an administrative order shall not take effect until the person to whom it is issued has had an opportunity to confer with the EPA concerning the alleged violation.[32] Additionally, before the Agency may assess an administrative penalty, the person to be assessed is entitled to participate in a full adversarial proceeding

---

28 CAA § 113(c)(5), 42 U.S.C. § 7413(c)(5).

29 42 U.S.C. § 7413(c).

30 *See* CAA § 113(c); Miskiewicz and Rudd at 308–310.

31 EPA, unpublished data. EPA formally promulgated these procedures in February 1992, applying preexisting consolidated rules of practice governing the administrative assessment of civil penalties and the revocation or suspension of permits procedures (*see* 40 C.F.R. part 22) to its new authority, adopting only a few "supplemental rules of practice". *See* 57 *Fed. Reg.* 4316 (February 4, 1992).

32 CAA § 113(a)(4), 42 U.S.C. § 7413(a)(4).

33 A person whom the Agency intends to assess a penalty is entitled by law to notice and a hearing, CAA § 113(d)(2)(A), 42 U.S.C. § 7413(d)(2)(A), as well as to review of any assessment in federal district court, CAA § 113(d)(4), 42 U.S.C. § 7413(d)(4).

conducted by the Agency.[33] Such proceedings are governed by formal Agency procedures which replicate, in certain fundamental respects (albeit on a far smaller scale), the principles of the federal court rules applicable to civil judicial actions.[34]

Civil Judicial Actions

Prior to passage of the Clean Air Act Amendents, the most common means of enforcement of statutes and regulations governing the emmissions of ozone-causing substances was the filing of a civil judicial action in federal district court. In all civil actions brought under the Clean Air Act, the Administrator shall request the Attorney General to represent him.[35] Under a memorandum of understanding between EPA and the Justice Department, the Department is supposed to file civil suits within sixty days of the Administrator's request. This time limit is particularly ambitious in view of the complexity of the cases often referred and the level of review they generally undergo before a suit is filed. In practice, for a variety of reasons, cases often take far longer than sixty days to be filed.

Under an Executive Order issued by President Bush in 1992, government attorneys generally must notify defendants before filing a lawsuit and provide them with an opportunity to settle.[36] Although few cases settle between the time EPA refers the case to the Justice Department and the time suit is filed, eventually most civil judicial actions do settle. In fact, an actual trial for civil violations of the Clean Air Act, particularly SIP violations, is a rarity. It is far more common for the violator and the government, at some point in time, to come to an agreement, which is memorialized in a formal, legally-enforceable, document called a "consent decree." The principal components of a consent decree are payment of a negotiated civil penalty, performance of any necessary compliance measures, and payment of additional stipulated penalties if the decree itself is violated. After all parties have signed the decree, it must be approved by the court in order to take effect.

As a matter of practice, all suits are brought in the name of the "United States", that is, with the "United States" as plaintiff.[37] The legal document which actually

---

[34] *Compare* Consolidated Rules of Practice Governing the Administration Assessment of Civil Penalties and the Revocation or Suspension of Permits, 40 C.F.R. Part 22, *with* Federal Rules of Civil Procedure.

[35] CAA § 305, 42 U.S.C. § 7605.

[36] Executive Order 12778, 57 *Fed. Reg.* 3640 (January 30, 1992). Since notice is provided by the Justice Department *after* receipt of the Administrator's request, the sixty-day "clock" runs while the Justice Department must both evaluate and ready the case for filing *and* attempt to negotiate a settlement.

[37] The mobile source injunctive enforcement provisions actually specify that such suits are to be brought in the name of the United States. Other Clean Air Act enforcement provisions are silent on this technical legal issue, occasionally tempting crafty lawyers to seek dismissal of the suit on the grounds that the Act purportedly requires that the suit name the Administrator as plaintiff. While one such argument did prevail over a decade ago [*see United States v. Associated Electric Cooperative, Inc.*, 503 F. Supp. 92 (E.D. Mo. 1980)] the better and more recent view is that such a hypertechnical reading of the Act is meritless. *See United States v. Zimmer Paper Products, Inc.*, 733 F. Supp. 1265 (S.D. Ind. 1989); *United States v. R.E.A.G.*, 730 F. Supp. 482 (D. Conn. 1989).

begins the lawsuit is known as a "complaint." It is generally no more than a streamlined statement of the allegations against the defendant. Once the complaint has been filed, the defendant has the opportunity to file an "answer." The answer is generally no more than a streamlined response to the complaint, which, more often than not, merely denies most substantive allegations while asserting most conceivable defenses.

From the point in time that the lawsuit has been filed, until a deadline is imposed by the court, both parties are entitled to embark on a lengthy, expensive, and often intensely probing process known as "discovery." Discovery is the ordered exchange of information (testimony and documents) potentially relevant to the lawsuit. Its most common forms are oral depositions, document requests, written interrogatories, admission requests, and inspections. It is not at all uncommon for discovery to last for a matter of years in environmental enforcement cases.

As discovery progresses, there are usually flurries of motions by one party or another relating to contested points of discovery or other matters. Towards the end of the discovery process, one or more parties commonly files what is known as a "summary judgment motion," asking the court to decide the case, or certain parts of it, on the evidence existing to date, without the necessity of a trial. Inasmuch as they generally involve fewer genuine issues of material fact[38] than most types of lawsuits, summary judgment motions are an important tool in efficiently resolving environmental enforcement cases. If a party wins its summary judgment motion, the case is over, at least with respect to any issue raised in the motion.[39]

If a case cannot be resolved by the court in the context of a summary judgment motion, it must be tried. As in all civil trials, the burden is on the plaintiff (in these cases, the government) to show by a preponderance of the evidence that it is entitled to relief. In any Clean Air Act trial in which the government seeks a penalty, the defendant has the right to a jury.[40]

From 1980 through 1992, the federal government brought a total of 127 civil judicial actions for violations by stationary sources of VOC emission limitations in state implementation plans (SIP/VOC cases).[41] During the same period, the federal government also brought a total of 110 civil judicial actions for violations of motor vehicle emission and fuel standards. For a number of reasons, the

38 The court will grant the motion if there is no genuine issue as to any material fact. *See* Rule 56, Federal Rules of Civil Procedure. In layman's terms, this means that if the judge is persuaded that all the relevant facts have already been established through discovery, he/she can decide the case (or issue) on the law and no trial is necessary.

39 It is common for the government to move for summary judgment on the issue of liability and reserve for trial the issue of relief (i.e., penalties and/or required compliance measures).

40 *See Tull v. United States*, U.S. 412 (1987).

41 This statistic, and the statistic which follows, were compiled from unpublished Justice Department computer data, and should be read with the caveat that the data is compiled by the Department in such a fashion that it is not possible to discern from it all cases which relate to enforcement of statutes and regulations which bear on ozone control. Since I have counted only SIP/VOC and mobile source cases, I have likely somewhat underreported the number of cases actually filed which arguably relate to ozone control.

stationary source cases may be viewed as the more significant cases. First, a given stationary source may emit hundreds of tons of VOCs annually, while a single mobile source typically emits less than 0.1 ton VOCs per year.[42] Second, a stationary source which is out of compliance for a matter of years easily accrues millions of dollars in penalty "exposure." For that reason alone, the stationary source defendants tend to have more at stake. Third, stationary source defendants tend to be far larger companies than mobile source defendants, with the resources and resolve to defend against an action.

A recent Clean Air Act enforcement action against Mobil Oil Corporation serves to illustrate a number of the points discussed above. On April 30, 1987, the federal government filed suit against Mobil, alleging that it had violated the Act by exceeding the limitation in the applicable California SIP for emissions of VOCs from Mobil's polystyrene foam manufacturing plant in Bakersfield, CA. The government contended that Mobil had unlawfully exceeded the applicable VOC emissions limit by hundreds of tons over a period of some two years. Mobil vigorously denied the allegations, asserting a plethora of legal and technical defenses. Discovery progressed on and off, contentiously at times, for a period of over four years. A barrage of motions were filed. There were dozens of depositions. A number of experts were retained, at least one of whom was world-renowned. Literally tens of thousands of pages of documents were exchanged as a result of several sets of document requests. Scores of written interrogatories were propounded. At the end of 1991, after discovery was closed by the court, the federal government filed an elaborate summary judgment motion, on which the court held oral argument in May 1992. Trial was set for October 1992, but was delayed because the parties were near settlement. Finally, in November 1992, the parties submitted to the court, which ultimately approved, a consent decree in which Mobil agreed to pay $950,000 in penalties, the third largest penalty at the time for a SIP/VOC enforcement action.[43]

By comparison, a mobile source case was settled in the same year against Economy Muffler and Tire Center, Inc. of Richmond, VA, after much less fanfare, for $35,000. Notwithstanding their lower profile, mobile source violators cannot be overlooked, however, since motor vehicles are generally considered to account for as much as one third of the ozone "inventory" in most nonattainment areas.[44] In light of the EPA's increased administrative authority under the 1990 Amendments, enforcement action against mobile source violators in particular, is increasingly likely to be handled administratively, freeing up government resources to pursue the larger stationary source violators in civil judicial actions.

---

[42] *See* U.S. Environmental Protection Agency, *National Air Quality Emissions Trends Report, 1990* (EPA-450/4-91-023) (1990) (hereinafter "Trends Report").

[43] The largest civil penalty to date in a Clean Air Act case was $11,100,00, in a 1993 case brought against Louisiana-Pacific, Inc. for numerous violations, including the failure to report VOC emissions, at several of its plants.

[44] *See Trends Report.*

Criminal Prosecution

The task of prosecuting Clean Air Act violations as federal crimes also falls to the Justice Department. Once criminal violations of the Act are referred by EPA's Criminal Investigation Division (CID), either Washington-based Justice Department attorneys and/or attorneys in the Justice Department's U.S. Attorneys Offices across the country supervise any further investigation, present the case to a federal Grand Jury for indictment if a felony is involved, or charge the defendant by means of a simpler "information," if a misdemeanor is involved. As in any criminal case, the defendant has the right to a trial by jury.

It should be noted that there is substantial overlap between the civil and criminal enforcement authority under the Clean Air Act. The best practical example is probably a SIP violation. Most SIP violations are handled as a civil enforcement matter, even though under the law if the violation is "knowing" — as most presumably are — it may be prosecuted as a crime.

According to EPA statistics covering fiscal years 1983 through 1992, there has been only one criminal case brought for SIP violations and only a handful of cases for crimes relating indirectly to motor vehicle emissions.[45] The conclusion is inescapable that, for whatever reason, stationary and mobile source cases, with few exceptions, have historically been handled as civil matters.[46]

CONCLUSION

There can be little doubt that the Clean Air Act, particularly since passage of the 1990 Amendments, provides a potent source of enforcement authority. The federal government historically has used that authority more modestly than it might. If used aggressively and effectively, both in the civil and criminal enforcement arenas, federal enforcement conceivably could have a greater impact in curbing violations of statutes and regulations designed to control the emission of substances which are precursors to ozone.

---

45 *See* U.S. Environmental Protection Agency, Office of Enforcement, Summary of Criminal Prosecutions Resulting from Environmental Investigations, May 31, 1992.

46 The federal government's record in prosecuting environmental crimes has been the subject of some close scrutiny. *See*, e.g., *The Washington Post*, "Pollution Leniency Alleged," September 10, 1992, page 1. This scrutiny has focused primarily on political interference as the reason for the lack of more aggressive criminal prosecution. Other possible explanations include the complexity of the cases, their lack of perceived jury and prosecutorial appeal, proof problems, and resources. With a new administration now in place, time will tell whether an increased emphasis will be placed on criminal prosecution.

CHAPTER 12

OZONE HEALTH EFFECTS — A PUBLIC HEALTH PERSPECTIVE

Ronald H. White

INTRODUCTION

The 1970 Amendments to the Clean Air Act[1] [hereafter referred to as the Clean Air Act (CAA)] embody the heart of the public health provisions of the nation's effort to control air pollution emissions. Section 108 of the CAA requires the U.S. Environmental Protection Agency (USEPA) to identify all air pollutants that, in the judgment of the USEPA Administrator, cause or contribute to air pollution which "... may reasonably be anticipated to endanger public health",[2] and to issue air quality criteria for such pollutants that reflect "... the latest scientific knowledge useful in indicating the kind and extent of all identifiable effects on public health and welfare which may be expected from the presence of such pollutants in the ambient air".[3] Based on the information generated under Section 108, Section 109 of the CAA requires the USEPA Administrator to promulgate National Ambient Air Quality Standards (NAAQS)"... which in the judgement of the Administrator, based on such criteria and allowing an adequate margin of safety, are requisite to protect the public health".[4] Conspicuous by its absence is the lack of any reference in this section of the CAA allowing for consideration of economic cost or technological feasibility in setting these standards. In fact, the legislative history of the CAA[5] indicates that the U.S. Congress made a deliberate decision that protection of public health and welfare was to be the sole determinant in deciding an acceptable

0-87371-475-X/94/$0.00+$.50
© 1994 by Lewis Publishers

level of air pollution. Political, economic, and technological factors were then to be integrated into the process of selecting and implementing pollution control strategies for attaining these standards.

The legislative history for Section 109 describes in further detail the concept of setting the NAAQS at a level that is low enough to protect the health, not only of the general healthy population but also the people who are particularly sensitive to air pollution such as people with asthma and emphysema.[6] The Congress also included their definition of protecting public health in the legislative history, noting that "... ambient air is sufficient to protect the health of such [sensitive] persons whenever there is an absence of adverse effect on the health of a statistically related sample of persons in sensitive groups from exposure to the ambient air. An ambient air quality standard, therefore, should be the maximum permissible ambient air level of an air pollution agent or class of such agents (related to a period of time) which will protect the health of any group of the population".[7] The 1977 Amendments to the CAA[8] reinforced the concept of protecting the health of sensitive populations when it added a requirement to Section 108 that the USEPA Administrator publish and make available to other agencies information on "...other measures which may be employed to reduce the impact [of air pollution] on public health or protect the health of sensitive or susceptible individuals or groups".[9] The definition of sensitive populations has been broadened over the years by the USEPA to also include people who demonstrate a greater than average response to an air pollutant or who have an inherent predisposition to such a response. Thus, for ozone (O_3), the USEPA includes in their definition of sensitive populations otherwise healthy people who exercise outdoors, as well as those people who respond with particular intensity to even relatively low O_3 levels. The USEPA also considers people with preexisting respiratory disease, including people with asthma, allergy, and chronic obstructive pulmonary disease (such as chronic bronchitis and emphysema) to be particularly at risk to O_3 exposure.[10]

Another mechanism employed by Congress to ensure that public health would be protected by attainment of the NAAQS was the concept of margin of safety. As noted above, Section 109 includes in its definition of the NAAQS the requirement that the primary (health-based) standard protect public health with an adequate margin of safety.[11] This margin of safety was to account for uncertainty in the interpretation of scientific information as well as for emerging data on currently unanticipated adverse effects of these pollutants. Thus, the Congress expected that the USEPA Administrator would select a level for the health-based NAAQS that was significantly below the concentration of pollution found to cause an adverse health effect even in sensitive populations.

ACUTE HEALTH STUDIES — COMPELLING LOW LEVEL EFFECTS

As discussed in Chapter 8, the USEPA revised the photochemical oxidant NAAQS of 0.08 parts per million (ppm) averaged over 1 h in 1979,[12] weakening the standard by 50% to 0.12 ppm and restricting the regulated pollutant to O_3 rather than the entire group of photochemical oxidants. The primary scientific bases for this change were clinical and epidemiological studies published in the 1970s of children and asthmatics exposed to O_3 levels of between 0.15 and 0.25 ppm.[13] These studies found increased cough, wheezing, shortness of breath, as well as small loss of lung function in the test subjects. However, as early as 1975, studies of the effect of photochemical smog on school children in Japan found lung function reductions at oxidant levels as low as 0.10 ppm.[14] Other studies found increased susceptibility to respiratory infection in animals at O_3 levels of 0.10 ppm.[15] Upon reviewing this data, a USEPA expert advisory panel concluded in a 1978 report that "...there is no compelling reason to suggest a change [in the ozone NAAQS] from the concentration defined by the existing primary air quality standard, namely 0.08 ppm."[16] The panel also noted that even this standard "...provides very little margin of safety."[17] Despite these conclusions of the USEPA's own expert panel, the USEPA Administrator revised the O_3 NAAQS at the 0.12 ppm level.

Beginning in the early 1980s, studies published on the acute respiratory effects of O_3 on healthy exercising children and adults at or just above the 0.12-ppm level found an increase in the acute respiratory symptoms described above as well as a loss of lung function.[18,19] More recent studies have found similar effects at O_3 concentrations below the 0.12-ppm NAAQS.[20-24] Chapter 3 contains a more detailed discussion of these studies. Toward the end of the decade, a number of clinical studies had found increases in acute respiratory symptoms, loss of lung function, sensitivity to bronchoconstricting agents, and biochemical indicators of inflammation occurring in healthy, exercising subjects at O_3 levels as low as 0.08 ppm when the exposure duration was increased to approximately 7 h.[25,26] This O_3 cumulative effect is of particular concern because the heavily populated areas of the eastern U.S. frequently experience O_3 peaks between 0.08 and 0.12 ppm for 6 to 8 h at a time.[27]

Epidemiological data linking O_3 air pollution to increased hospitalization and hospital visits for respiratory disease (especially asthma) have emerged as another important public health concern.[28-30] The acute respiratory symptoms that accompany O_3 exposure, while unpleasant for the general healthy population, can be particularly devastating to the person with asthma or chronic obstructive lung disease. A 1991 clinical study of people with atopic asthma found increased responsiveness to common allergens with a 1-h exposure to 0.12-ppm O_3 while at rest.[31] These allergens often serve as the "trigger" for asthma attacks, and this finding might help to explain the increase in hospital visits and admissions for asthma-related problems during the higher O_3 summer months.

A 1993 study by the American Lung Association (ALA) estimated that over 6 million adults and children with asthma and over 8 million people with chronic obstructive pulmonary disease (COPD) lived in areas that exceeded the federal primary (health) standard for O_3 of 0.12 ppm.[32] A 1991 ALA study estimated that over 7.5 million adults and children with asthma, amounting to more than three quarters of the total number of people with asthma in the U.S., and over 7 million people with COPD lived in areas that experienced O_3 levels above 0.08 ppm averaged over 8 h for the 1987–1989 period.[33]

To many in the scientific and public health communities, the preponderance of clinical, epidemiological, and field study data was sufficient to question the adequacy of the 1-h 0.12-ppm O_3 NAAQS in protecting public health from the acute effects of O_3. In late 1988, approximately one-half of the members of the Ozone Review Committee of the USEPA's Clean Air Scientific Advisory Committee had recommended to the USEPA Administrator that the primary 1-h O_3 NAAQS be revised to a level below 0.12 ppm. In addition, a vast majority of the Ozone Review Committee members, including several of the experts who had not recommended tightening the primary standard, noted that "...at 0.12 ppm there was little or no margin of safety.[34] By 1991, the convergence of data from the different scientific disciplines regarding the acute adverse effects of O_3 at levels well below the existing health standard had become overwhelming. This accumulation of data was ignored by the USEPA, which twice missed CAA-mandated deadlines to review and, if necessary, revise the O_3 standard in 1985 and 1990.

To ensure that the NAAQS reflected the most up-to-date scientific information, Congress amended Section 109 of the CAA in 1977 to require that the air quality criteria used by the USEPA to set the NAAQS and the NAAQS themselves be reviewed and, if necessary, revised at no more than 5-year intervals beginning in 1980.[35] Responding to the new health study data and to the USEPA's failure to carry out the required O_3 NAAQS reviews, the ALA, joined by two environmental organizations and several northeast states, filed a lawsuit against the USEPA in October 1991 to require that the USEPA quickly complete its review of the adequacy of the 0.12-ppm 1-h NAAQS.[36] This litigation was resolved by a settlement agreement requiring the USEPA to propose a decision on the O_3 NAAQS review by August 1, 1992 and to issue a final decision by March 1, 1993. On August 10, 1992, USEPA published its proposed rule in the *Federal Register*,[37] indicating that the current O_3 NAAQS would not be revised. Publication of the USEPA's final rule on this issue was pending at the time this book was written.

While it is impossible at this time to state the final outcome of this review, the USEPA's history of reluctance to even visit the O_3 NAAQS issue does not bode well for a decision that will be made on the basis of public health protection. The CAA Amendments of 1990 set out an O_3 NAAQS attainment program that utilizes a classification scheme based on the severity of O_3 levels above the 0.12- ppm NAAQS.[38] The practical implications of a revised O_3 NAAQS that reflects the most recent scientific data include expanding the number of areas that would be classified as nonattainment areas, as well as increasing the amount of O_3 precursor pollution reduction required to achieve the NAAQS.

Based on actual monitoring data from 1987 to 1989, a 1991 ALA analysis of an alternative O_3 NAAQS level of 0.08 ppm averaged over 8 h found that almost 32 million people live in 68 areas in the U.S. that would be classified as unhealthy despite meeting the 0.12-ppm NAAQS.[39] O_3 pollution control programs will not be developed for these areas until such time as they are classified as "nonattainment" based on a revised O_3 NAAQS. Until then, people living in these areas will falsely believe that they enjoy healthful air quality.

Another significant finding of the July 1991 ALA study was that the number of O_3 exceedances increased dramatically, often by a factor of ten or more, when comparing the number of violations of the 0.12-ppm 1-h O_3 NAAQS with the number of exceedances of a 0.08-ppm level averaged over 8 h.[40] These large exceedance ratios were particularly prevalent for the medium-sized metropolitan areas of the southeast U.S., which frequently experience O_3 levels between 0.08 and 0.12 ppm for extended durations.

The major remaining controversy regarding the acute effects of O_3 exposure revolves around the significance for public health of relatively small transient, reversible reductions in lung function; mild levels of unpleasant but not incapacitating symptoms such as coughing, wheezing, and shortness of breath; and biochemical evidence of lung tissue inflammation. The American Thoracic Society[41] and the USEPA[42] itself have attempted to put this issue in a structured context. Given the current state of the science regarding the relationship between reversible, transient health effects to the development of more severe, nonreversible chronic effects, it is difficult to fully understand the significance of this relationship for public health. However, a prudent public health policy demands that the public be protected against repeated insult and injury to the delicate lung membrane.

CHRONIC HEALTH EFFECTS — A NASTY CONVERGENCE OF DATA

At a press conference held in conjunction with the May 1990 American Lung Association/American Thoracic Society World Conference on Lung Health, Dr. David Bates, Professor Emeritus at the University of British Columbia, completed his review of the emerging body of scientific evidence regarding the health effects of atmospheric acid air pollution by noting that despite the lack of a single "smoking gun" study, there was a "nasty convergence of data" from a number of studies in the different scientific disciplines that leads to significant concern regarding the potential public health impact of these pollutants. There is a parallel between the current status of knowledge of the health effects of atmospheric acid air pollution and the chronic effects of O_3. Perhaps the most important public health issue related to exposure to O_3 air pollution is the concern regarding chronic effects from long-term exposure.

The health effects that provided the basis for the establishment of the O_3 NAAQS in 1971 and for the 1979 revision were primarily acute effects, such as the occurrence of respiratory symptoms and reduced respiratory function. However, a number of animal toxicology studies have been published since the mid-1980s that find evidence of changes in lung cell type, formation of lesions, and evidence of loss of lung elasticity and inflammation for chronic exposure to O_3

levels that typically occur in the ambient air.[43] Coincident with these findings was the publication of a number of studies that allow for increased confidence in extrapolating the animal studies to human populations.[44]

The findings of epidemiological studies that residents of high O_3 areas in southern California have an accelerated loss of lung function in comparison to residents of areas with lower O_3 levels in southern California[45] and Michigan[46] adds to the mounting evidence of important adverse health effects occurring from long-term exposure to O_3. A 1991 study of children in Austria[47] found that long-term exposure to high O_3 levels may lead to persistent bronchial hyperresponsiveness. These studies and the results from a pilot autopsy study[48] of Los Angeles teenagers and young adults that found a high occurrence of lung lesions of the type and location in the lung normally associated with chronic O_3 exposure in animals add to the "convergence of data" that strongly suggests that chronic exposure to high O_3 levels may be associated with a premature aging of the lungs and the development of chronic lung disease.

An additional source of concern regarding ozone's chronic effects are findings that high levels of O_3 enhance the formation of tumors in mice[49] and enhance the ability of normal cell transformation to cancerous cells, both alone and concurrent with exposure to a carcinogen *in vitro*.[50] More studies of this type at lower O_3 levels that more closely relate to human ambient air exposures will be necessary before these very serious effects can be accounted for in the decision-making process regarding the need for a new public health approach to protecting the public from chronic O_3 exposure.

The implications of this growing body of scientific evidence for public health are immense. Lung disease, including lung cancer, is the third leading cause of death in the U.S. and is the fastest growing cause of death among the top ten causes of death. Nearly 10% of the U.S. population currently suffers from chronic lung disease, and the prevalence of these frequently devastating diseases is increasing rapidly. The cost to society in terms of direct expenditures for health care, lost productivity, restriction of daily activity and a reduced quality of life, and suffering of acute symptoms and premature death is likely in the billions of dollars each year for O_3 alone.[51]

O_3 CONTROL PROGRAMS — TWO DECADES OF FAILURE

In the 1970 Amendments to the CAA, Congress required that the NAAQS be met by 1975. However, by this deadline it was apparent that the majority of urban areas in the U.S. would not meet the NAAQS for O_3 or carbon monoxide. The 1977 Amendments to the CAA revised the NAAQS attainment schedule, requiring attainment by 1982 with a possible extension to 1987. Yet, by the close of the second decade after O_3 control programs had been initiated, 133 million people living in 96 metropolitan areas were still potentially exposed to unhealthy levels of O_3.[52]

Having established that O_3 represents a seriously underestimated public health problem, one might reasonably ask why there has been so little progress in curbing this environmental hazard after more than 20 years of effort and the expenditure of billions of dollars for controls. The answer, I believe, involves a combination of factors, including unrealistic deadlines for accomplishing the objective, poor information on the sources of ozone's precursor pollutants, overestimating the potential pollution control benefits of many of the control measures, underestimating the public's reluctance to support pollution controls that would significantly modify their driving habits and lifestyle, and the frequent lack of the political process at the state/local level to follow through on pollution control commitments made in state implementation plans (SIPs). The implications of this last factor for future pollution control efforts is particularly troubling in light of the tremendously divisive, decade-long debate that led to the reauthorization of the CAA in 1990. To accomplish the objective inherent in the 1990 CAA Amendments of achieving clean air for *all* people by the early 21st century will require an almost unprecedented level of commitment and cooperation at the federal, state, and local levels of government to the aggressive development and implementation of a broad range of air pollution control strategies. After the failures of the 1970 and 1977 CAA efforts to achieve a significant reduction in the nation's O_3 problem, public pressure is greater than ever for the programs developed under the 1990 legislation to succeed. Public opinion polls continue to document that a vast majority of the American public believes that achieving healthful air quality is a high-priority issue. However, the nation's past experience with O_3 control programs underscores the need for enhanced vigilance of these efforts by Congress and the public. A 1988 ALA report[53] on state and local implementation of pollution controls included in a sample of 1979 and 1982 SIPs found numerous cases where implementation of pollution control commitments were delayed, never completed, or significantly weakened. The 1991 report of the National Research Council[54] confirms many of these observations, supporting recommendations that include: improved verification of the implementation of O_3 controls and the emission reductions actually achieved by the implementation of these strategies; improving the accuracy of the VOC and NO_x emissions inventory; shifting the emphasis from VOC controls to NO_x controls for those areas where NO_x is the limiting factor for O_3 production; and improving the ability of air quality models to accurately predict the effect of proposed emission controls on ambient O_3 levels.

RECOMMENDATIONS AND CONCLUSIONS

- The current 1-h O_3 NAAQS of 0.12 ppm should be either revised or augmented with a new 6- or more hour short-term standard in order to protect public health from the acute adverse effects that have been demonstrated with multihour exposures to O_3 levels as low as 0.08 ppm. The current O_3 NAAQS does not protect public health with an adequate margin of safety as required by the CAA.

- An expanded research program on the chronic health effects of long-term exposure to O_3 should be funded and implemented on an accelerated schedule. It is imperative that the preliminary findings of potentially serious health effects from long-term exposure to O_3 be replicated and expanded upon so that a sufficiently robust scientific database from credible studies is available to determine the need for a seasonal or annual O_3 standard. The emerging scientific information on the chronic effects of O_3 should be given significant weight in the regulatory decision regarding an adequate margin of safety for a new or revised short-term standard.
- Tropospheric O_3 remains the most widespread, intractable, and potentially the most damaging to health and the environment of the air pollution problems facing the U.S. and many other parts of the world. Many of the control strategies that will be necessary to finally eliminate the O_3 problem will have more than insignificant costs and lifestyle impacts. However, the cost to human health and the natural environment of not aggressively grappling with this problem is likely to be far greater.
- Aggressive oversight by Congress and the public of federal, state, and local efforts to develop and implement O_3 control measures is essential to accomplishing the O_3 reduction objective of the CAA Amendments of 1990. The increased complexity of the legislative requirements, coupled with the opportunity for undermining aggressive pollution control programs via the political process, presents an immense challenge that will require increased cooperation and coordination between industry, government at all levels, and the public interest/public health community.

REFERENCES

1. Public Law 91-604, 84 Stat. 1676 (1970).
2. 42 U.S.C. Section 7408 (a)(1).
3. 42 U.S.C. 7408 (a)(2).
4. 42 U.S.C. Section 7409 (b)(1).
5. S. Rep. No. 91-1196, 91st Cong. 2d Sess. 11 (1970).
6. Id. pg. 10.
7. Id.
8. Public Law 95-95, 91 Stat. 685 (1977).
9. Section 108 (f)(1)(C), 42 U.S.C. Section 7408 (f)(1)(C).
10. U.S. Environmental Protection Agency. Review of the National Ambient Air Quality Standards for Ozone: Assessment of Scientific and Technical Information OAQPS Staff Paper. Research Triangle Park, NC, pp. VI-8 to VI-13 (1989).
11. 42 U.S.C. Section 7409 (b)(1).
12. 44 F.R. 8220 February 8, 1979.
13. U.S. Environmental Protection Agency. "Summary Statement from the EPA Advisory Panel on Health Effects of Photochemical Oxidants." (1978) pp.18.
14. Kagawa, J., and T. Toyama. "Photochemical Air Pollution: Its Effect on Respiratory Function of Elementary School Children," *Arch. Environ. Health* 30:117–122 (1975).
15. Coffin, D. L., E. J. Blomer, D. E, Gardner, and R. Holzman. "Effect of Air Pollution on Alteration of Susceptibility to Pulmonary Infection," *Proc. 3rd Annu. Conf. Atmospheric Contaminants in Confined Spaces,* Wright Patterson Air Force Base, Aerospace Medical Research Lab., OH, pp. 71–80 (1968).

16. U.S. Environmental Protection Agency. Summary Statement from the EPA Advisory Panel on Health Effects of Photochemical Oxidants. (1987), p. 18.
17. Id. p. 19.
18. McDonnell, W. F., III, D. H. Horstman, M. J. Hazucha, E. Seal, Jr., E. D. Haak, S. Salaam, and D. E. House. "Pulmonary Effects of Ozone Exposure During Exercise: Dose-Response Characteristics," *J. Appl. Physiol.: Respir. Environ. Exercise Physiol.* 54:1345–1352 (1983).
19. McDonnell, W. F., III, R. S. Champan, M. W. Leigh, G. L. Strope, and A. M. Collier. "Respiratory Responses of Vigorously Exercising Children to 0.12 ppm Ozone Exposure," *Am. Rev. Respir. Dis.* 132:875–879 (1985).
20. Spektor, D., M. Lippmann, P. Lioy, G. Thurston, K. Citak, D. James, N. Bock, F. Speizer, and C. Hayes. "Effects of Ambient Ozone on Respiratory Function in Active Normal Children," *Am. Rev. Respir. Dis.* 137:313–320 (1988).
21. Spektor, D., M. Lippmann, G. Thurston, P. Lioy, J. Stecko, G. O'Connor, E. Garshick, F. Speizer, and C. Hayes. "Effects of Ambient Ozone on Respiratory Function in Healthy Adults Exercising Outdoors," *Am. Rev. Respir. Dis.* 138:821–828 (1988).
22. Kinney, P., J. Ware, J. Spengler, D. Dockery, F. Speizer, and B. Ferris. "Short-term Pulmonary Function Change in Association with Ozone Levels," *Am. Rev. Respir. Dis.* 144:1385–1390 (1989).
23. Higgins, I., J. D'Arcy, D. Gibbons, E. Avol, and K. Gross. "Effect of Exposures to Ambient Ozone on Ventilatory Lung Function in Children," *Am. Rev. Respir. Dis.* 141:1136–1146 (1990).
24. Krzyzanowski, M., J. Quackenboss, and M. Lebowitz. "Relation of Peak Expiratory Flow Rates and Symptoms to Ambient Ozone," *Arch. Environ. Health* 47:107–115 (1992).
25. Horstman, D. H., L. J. Folinsbee, P. J. Ives, S. Salaam, and W. F. McDonnell, III. "Ozone Concentration and Pulmonary Response Relationships for 6.6-hour Exposures with Five Hours of Moderate Exercise to 0.08, 0.10, and 0.12 ppm," *Am. Rev. Respir. Dis.* 142:1158–1163 (1990).
26. Devlin, R. B., W. F. McDonnell, III, R. Mann, S. Becker, D. E. House, D. Schreinemachers, and H. S. Koren. "Exposure of Humans to Ambient Levels of Ozone for 6.6 Hours Causes Cellular and Biochemical Changes in the Lung," *Am. J. Respir. Cell Mol. Biol.* 4:72–81, (1991).
27. Rombout, P. J. A., L. van Bree, S. H. Heisterkamp, and M. Marra. "The Need for an Eight Hour Ozone Standard," presented at: *3rd U.S.-Dutch International Symposium on Atmospheric Ozone Research and its Policy Implications*, May, 1988, Nijmegan, The Netherlands.
28. Bates, D. V., and R. Sizto. "Air Pollution and Hospital Admissions in Southern Ontario: The Acid Summer Haze Effect," *Environ. Res.* 43:317–331 (1987).
29. Cody, R., C. P. Weisel, G. Birnbaum, and P. J. Lioy. "The Effect of Ozone Associated with Summertime Photochemical Smog on the Frequency of Asthma Visits to Hospital Emergency Departments," *Environ. Res.* 58:184–194 (1992).
30. Burnett, R., R. Dales, M. Raizenne, D. Krewski, P. Summers, G. Roberts, and T. Dann. "The Relationship Between Hospital Admissions and Ambient Air Pollution in Ontario, Canada: A Preliminary Report," Presented at the *85th Meeting of the Air and Waste Management Association*, Kansas City, MO, June 1992.
31. Molfino, N. A., S. C. Wright, I. Katz, S. Tarlo, F. Silverman, P. A. McClean, J. P. Szalai, M. Raizenne, A. S. Slutsky, and N. Zamel. "Effect of Low Concentrations of Ozone on Inhaled Allergen Responses in Asthmatic Subjects," *Lancet* 338:199–203 (1991).

32. American Lung Association. "Breath in Danger II: Estimation of Populations-At-Risk of Adverse Health Consequences in Areas not in Attainment with National Ambient Air Quality Standards (NAAQS) of the Clean Air Act," March, 1993.
33. American Lung Association. "National Estimates of the Populations-At-Risk Exposed to Adverse Health Consequences in Areas with More than One Exceedance of an Eight-Hour Daily Maximum >.08 ppm Ozone Concentration Averaged Over Three Years (1987–1989)," October, 1991.
34. Letter from Roger O. McClellan, D.V.M., Chairman, Clean Air Scientific Advisory Committee, to William K. Reilly, Administrator, U.S. Environmental Protection Agency. May 1, 1989.
35. 42 U.S.C. Section 7409(d), (1977).
36. *American Lung Association et. al. v. William K. Reilly;* United States District Court, 2nd C.; Civ. Action No. 91 Civ. 4114; October 22, 1991.
37. 57 FR 35542.
38. Public Law 101-549 (1990) Section 181 42 U.S.C. 7511.
39. American Lung Association. Metropolitan Statistical Areas (MSAs) With >1 'Exceedance' of an Eight-Hour Daily Max (DM) >0.08 ppm Averaged Over Three Years (1987–1989)," July, 1991.
40. Id. "Ratio of 8-Hr/1-Hr Exceedances (1987–1989)".
41. American Thoracic Society. "Guidelines As to What Constitutes an Adverse Respiratory Health Effect, with Special Reference to Epidemiologic Studies of Air Pollution," *Am. Rev. Respir. Dis.* 131:666–668 (1985).
42. U.S. Environmental Protection Agency. Review of the National Ambient Air Quality Standards fo Ozone: Assessment of Scientific and Technical Information. OAQPS Staff Paper. Research Triangle Park, NC, pp. VII-51 to VII-54.
43. U.S. Environmental Protection Agency, Office of Health and Environmental Assessment. "Summary of Selected New Information on Effects of Ozone on Health and Vegetation: Draft Supplement to Air Quality Criteria for Ozone and Other Photochemical Oxidant," (Review Draft) EPA/600/8-88/105A pp.54–57.
44. Id. pp. 58–61.
45. Detels, R. et al. The UCLA Population Studies of Chronic Obstructive Respiratory Disease. IX. Lung Function Changes Associated with Chronic Exposure to Photochemical Oxidants—A Cohort Study Among Never Smokers," *Chest* 92:594–603 (1987).
46. Kilburn, K. H., R. Warshaw, and J. C. Thornton. "Pulmonary Function Impairment and Symptoms in Women in the Los Angeles Harbor Area," *Am. J. Med.* 79:23–28 (1985).
47. Zwick, H., W. Popp, C. Wagner, K. Reiser, J. Schmoger, A. Bock, K. Herkner, and K. Radunsky. "Effects of Ozone on the Respiratory Health, Allergic Sensitization, and Cellular Immune System in Children," *Am. Rev. Respir. Dis.* 144:1075–1079 (1991).
48. Sherwin, R., and V. Richters. *Tropospheric Ozone and the Environment*; R. L. Berglund, D. R. Lawson, and D. J. McKee, Eds. (Pittsburgh: Air and Waste Management Association,1991), pp. 178–196.
49. Kobayashi, T., T. Todoroki, and H. Sato. "Enhancement of Pulmonary Metastasis of Murine Fibrosarcoma NR-FS by Ozone Exposure," *J. Toxicol. Environ. Health* 20:135–145 (1987).
50. Borek, C., M. Zaider, A. Ong, H. Mason, and G. Witz. "Ozone Acts Alone and Synergistically with Ionizing Radiation to Induce *in vitro* Neoplastic Transformation," *Carcinogenesis* 7:1611–1613 (1986).

51. American Lung Association. "The Health Costs of Air Pollution: A Survey of Studies Published 1984–1989," January 1990, pp. 41–48.
52. U.S. Environmental Protection Agency. National Air Quality and Emissions Trends Report 1989. EPA/450/4-91-003 (Research Triangle Park, NC: Office of Air Quality Planning and Standards, 1991).
53. American Lung Association, "Compliance of Selected Areas With Clean Air Act Requirements," November, 1988.
54. National Research Council, Committee on Tropospheric Ozone Formation and

CHAPTER 13

OZONE POLICY FROM A PETROLEUM INDUSTRY PERSPECTIVE: LESSONS LEARNED, FUTURE DIRECTIONS

Terry F. Yosie, Howard Feldman, Tim Hogan, and Will Ollison

INTRODUCTION

Environmental policy in the U.S. has enjoyed a number of success stories during the past 2 decades, leading to public health and environmental improvements for the American people. Policies to reduce the problem of tropospheric ozone (O_3), however, have not been successful despite significant reductions in O_3 precursor emissions. A significant explanation for the persistence of the O_3 problem lies neither in the absence of political will nor the unwillingness to spend large sums of money by government or business; rather, it stems in large part from a poor understanding of the characteristics of the O_3 problem and an overly narrow application of policy tools and control strategies appropriate for reducing O_3 levels. "The fact that a fundamental change in the nation's ozone reduction strategy could be necessary after two decades of costly efforts," concludes a recent National Academy of Sciences report, "indicates that there could be an even more fundamental flaw in the overall design of the strategy."[1]

A comparison of ambient O_3 with other criteria pollutants demonstrates some of the unique characteristics of O_3 as a public health and environmental problem. For example, while the USEPA reported that the 1989 composite average of peak ambient O_3 levels was 14% lower than the 1980 average,[2] many other criteria

0-87371-475-X/94/$0.00+$.50
© 1994 by Lewis Publishers

pollutants experienced far greater, in some cases dramatic, decreases during a period of economic growth. During this period, peak O_3 concentrations also exhibited far greater year-to-year variability than the other pollutants. Second, the greater reductions of other ambient pollutants testify that control strategies had appropriately targeted the most significant sources. The fact that ambient lead levels declined almost 90% following greater controls on smelters and the lead content of gasoline demonstrates the degree of success resulting from installing controls on these particular sources. Similarly, the carbon monoxide (CO) emission reductions resulted primarily from better design and maintenance of automobile engines and control technologies and occurred during a time when vehicle miles traveled (VMT) rose by 39%. Continued success in reducing CO emissions depends upon such factors as achieving a slower growth rate in VMT; continued improvements in engines, fuels, and control technologies; and reducing in-use vehicle emissions through such techniques as enhanced inspection and maintenance procedures at the federal, state, and local levels and by identifying and repairing cars that are relatively high emitters.

In contrast, strategies to reduce ambient O_3 concentrations have been much less successful in targeting all major emitting sources. Control strategies employed to date, which have focused primarily on reducing anthropogenic hydrocarbon emissions from large stationary sources or automobiles, have proven ineffective in many areas of the country from an environmental and economic standpoint. The Atlanta metropolitan area is a case in point. From 1980 to 1987, industry and government (and therefore consumers) spent over $700 million to reduce anthropogenic VOC (volatile organic compound) emissions in the region by over 50%. Yet Atlanta's O_3 levels in 1987 exceeded recorded levels at the beginning of the decade.[3]

As Congress debated the efficacy of current control strategies prior to passage of the 1990 CAA Amendments, new and powerful concerns emerged from the scientific community about the health effects of O_3 and whether or not the current 1-h O_3 National Ambient Air Quality Standard (NAAQS) of 0.12 ppm sufficiently protects public health. While evidence that questions the adequacy of the standard to protect against acute effects has accumulated, an increasing number of scientists see a need for a longer-term standard to protect against repeated O_3 exceedances in a number of metropolitan areas.[4] The debate continues as to whether sufficient scientific evidence exists to establish a new suite of primary O_3 standards. The USEPA's Clean Air Scientific Advisory Committee was itself divided on these issues.[5]

Finally, the costs of O_3 precursor controls relative to the benefits continue to escalate. Whereas during the 1980s, the USEPA regarded several thousand dollars per ton as a reasonable option for the cost of reducing a ton of pollution, today's cost per ton for certain O_3 abatement options (such as methanol fuels) can exceed $60,000.[6] In view of the ongoing debate about the effectiveness of existing O_3 control strategies, the rising costs of O_3 control are a major reason why the nation's O_3 policies have become more contentious over the past decade.

The 1990 CAA Amendments empower regulators at various governmental levels to tighten air pollution controls on already regulated source categories (e.g., cars, chemical plants, refineries, etc.), while introducing broad new authority to control area sources (e.g., bakeries and dry cleaning establishments) and consumer products such as gasoline. Implementation of the Amendments will, in some cases, lead to major changes in industrial processes, such as refinery technology, or products such as gasoline or paints.

This chapter will address these and other issues bearing upon the ability of metropolitan areas to achieve O_3 attainment status. Specifically, this chapter will discuss:

- Lessons learned about O_3 emissions, measurements, and meteorological trends during the past decade
- Measuring compliance with ambient O_3 standards
- Public health and environmental effects associated with O_3 exposures
- Evaluating control strategies in place

The chapter will close with an assessment of the petroleum industry's role and responsibility to assist in the effort to reduce O_3 levels and a discussion of future regulatory issues and strategies.

LESSONS LEARNED ABOUT O_3 EMISSIONS, MEASUREMENTS, AND METEOROLOGICAL TRENDS

Emissions

Problem is Different in Different Regions

Most of the major urban areas in the U.S. fail to meet the NAAQS for O_3. This occurs, in part, because of varying emissions and meteorological characteristics across the country. Conditions that favor high O_3 concentrations include:

- The availability of the precursors (VOCs and NO_x) to form O_3
- Transport of O_3 from upwind areas

The relative importance of each of the above factors varies across the nation.

O_3 is formed by the reaction of oxygen, VOCs, and nitrogen oxides (NO_x) in the presence of sunlight. Figure 1 shows the pathway to O_3 in a simplified manner. In the absence of other gases, equilibrium would be reached. Unfortunately, the equilibrium is not maintained due to the presence of compounds known as peroxy radicals, which form from VOC emissions. These radicals alter the steady state and lead to the increased O_3 levels that are currently found.

The anthropogenic emissions inventories of hydrocarbons and NO_x are displayed in Figures 2 and 3. The sources of these emissions are ubiquitous, resulting in quantities of anthropogenic O_3 precursors sufficient to form additional incre-

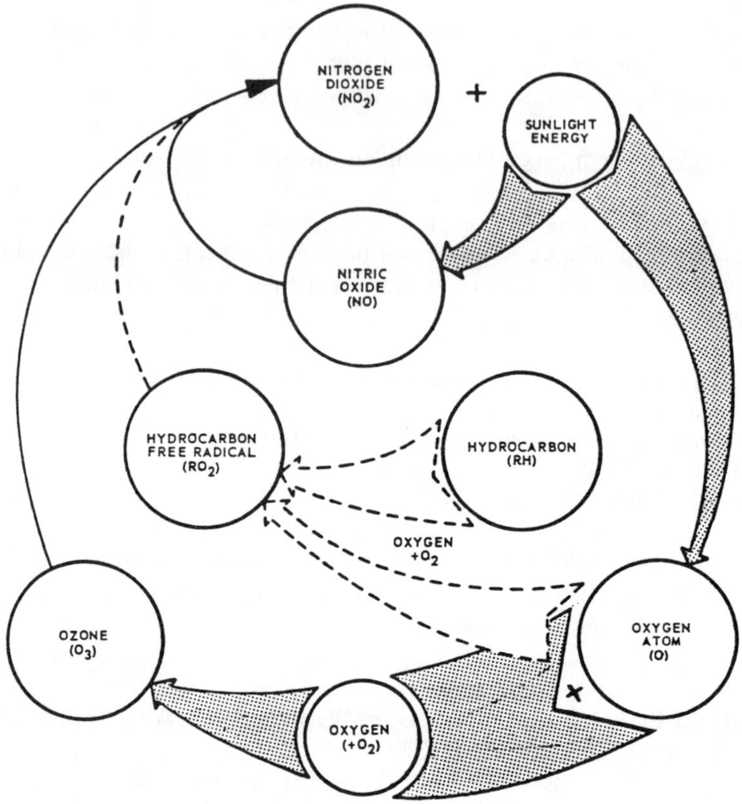

Figure 1. O_3 formation. (From "Air Quality Criteria for Photochemical Oxidants," U.S. Department of Health, Education, and Welfare, Washington D.C., March 1970. With permission.)

mental O_3 in most nonrural areas. Even rural areas that contain a sufficiently dense biosphere may contain sufficient precursors to produce high O_3 levels; a further discussion on this is included under the biogenic emissions section.

Given the availability of sufficient emissions of O_3 precursors, local production of O_3 will be influenced by the local meteorological potential to form O_3. Generally, higher O_3 concentrations are produced locally on days with higher temperatures.[10] Higher temperatures speed the chemical reactions and are correlated with higher levels of solar radiation. Other meteorological parameters that affect O_3 production include both limited wind speeds and vertical mixing. (Note that changing the speed of the reaction will also change the location of the maximum concentration — slower reactions result in maximum concentrations further downwind.)

Attempts have been made to quantify the propensity of meteorological conditions to promote the formation of O_3. One approach combines the Holzworth summer dispersion coefficient of a location with the location's latitude. The

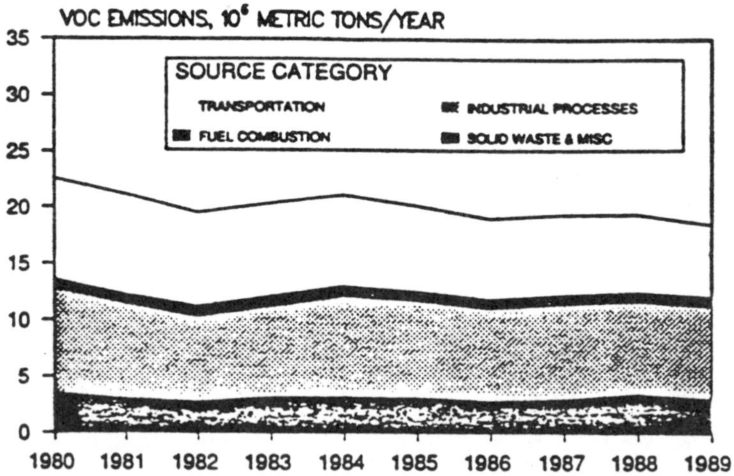

Figure 2. National trends in VOC emissions.

Holzworth summer dispersion coefficient is a function of summer wind speeds and vertical mixing, while the latitude serves as a surrogate for the duration and intensity of sunlight. Figure 4 shows the summer dispersion potential for the nation.

In addition to local production, O_3 may also be transported into an airshed from an upwind location. The degree to which O_3 is transported varies between locations as a function of meteorology and geography. Transport may play a crucial role in some O_3 nonattainment areas. For example, monitors in Acadia National Park in coastal Maine have shown exceedances of the O_3 standard, despite the low level of local anthropogenic emissions. In this case, a large fraction of the O_3 is probably transported into Maine from upwind cities to the southwest. Conversely, there are nonattainment cities that are generally isolated from other urban nonattainment areas and probably receive a minimal amount of transported O_3. An example might be Salt Lake City, where little or no transport from upwind occurs. There are also many cities that are both "transporters" and "transportees", meaning that those cities are the recipients of transported O_3 and precursors from upwind and they also transport O_3 and

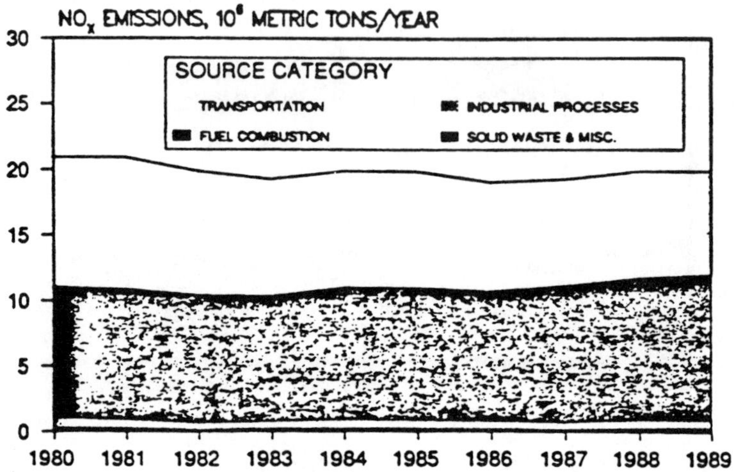

Figure 3. National trends in NO_x emissions.

precursors to cities further downwind. An example of this type of city would be Philadelphia.

NO_x/VOC Ratio

Both VOCs and NO_x contribute to the formation of O_3. Generally, a given unit of VOC reacts most quickly when the ratio of VOC/NO_x in the atmosphere is approximately 5–6 and a given unit of VOC results in the maximum production of O_3 at a ratio of approximately 7.[12] In the simplest sense, when ambient concentrations of VOC and NO_x compounds are outside these ratios, decisions on control strategies should differ. For example, if the VOC-to-NO_x ratio is high (13–20 or more), control strategies which include hydrocarbon emissions reductions may prove fruitless. In such a location, it may prove to be more environmentally and cost effective to reduce concentrations of NO_x in order to reduce O_3. An example of this type of city might be Port Arthur, TX. Conversely, in areas with low ratios of VOC/NO_x, NO_x reductions may not lead to reductions in O_3.

Unfortunately, the real world is far more complex than is implied by the previous discussion. The atmospheric concentrations of various compounds are

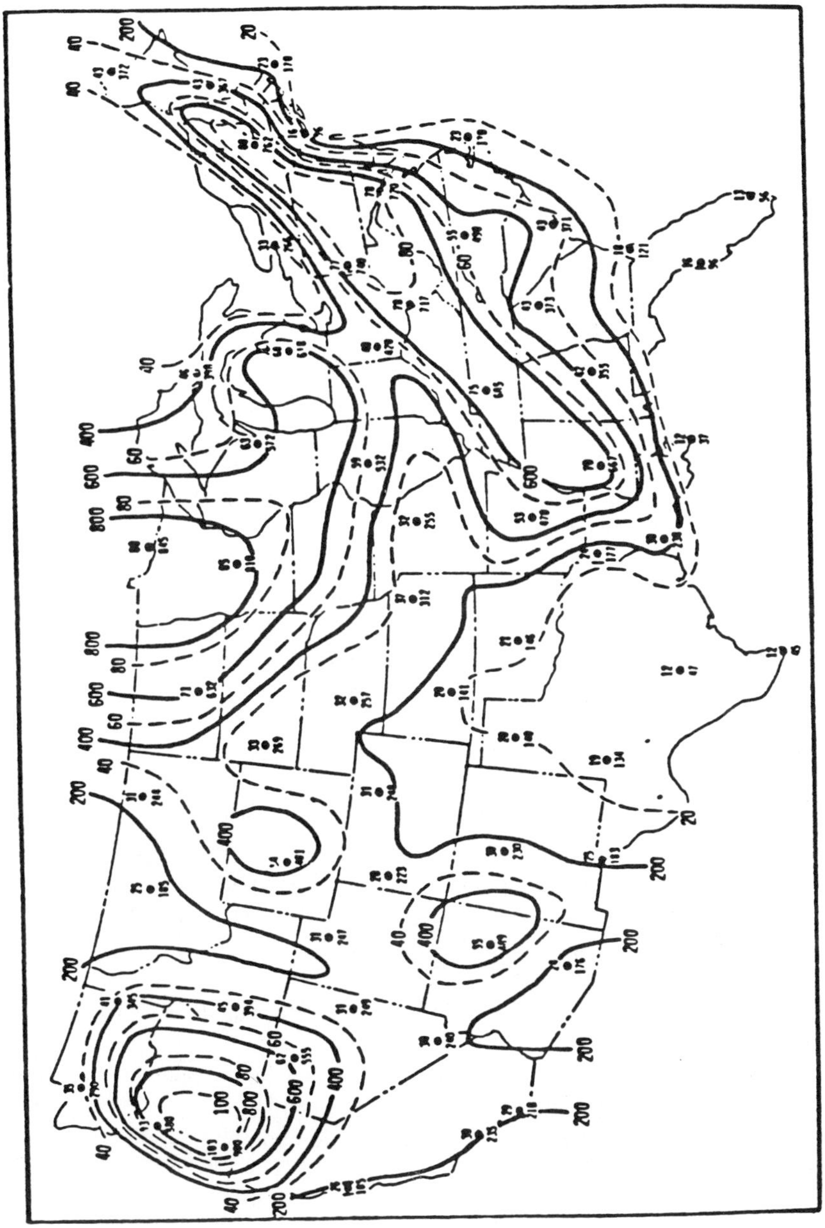

Figure 4. Summer dispersion potential for the U.S.

not homogenous. The concentrations result from different sources in different locations, emitting at different times of the day. Therefore, attempts to assess O_3 formation before or after controls is a complex procedure requiring the use of sophisticated modeling techniques.

Uncertainties in the Emissions Inventory

Emissions of NO_x result from fuel combustion. Major point source NO_x emissions have been measured regularly, some even continuously. There is a high level of confidence in those measurements. With respect to mobile sources, on-road studies have corroborated laboratory measurements of automotive NO_x emission rates, even though the transportation models remain open to further question.

NO_x emissions estimates contrast significantly with the current understanding of anthropogenic hydrocarbon emissions. Sources of the latter are more numerous and widespread. The largest category of hydrocarbon emissions is *mobile sources*, but scientific understanding of hydrocarbon emissions from mobile sources remains quite limited. Recently, as part of the Southern California Air Quality Study (SCAQS), the Southwest Research Institute estimated the mobile source emissions factors from vehicles passing through a tunnel.[13] That study concluded that emissions factors for NO_x appear to be correct, but that current emissions factors from automobiles may underestimate hydrocarbon emissions by a factor of three to five. The findings in the SCAQS tunnel study have confirmed previous on-road emissions studies.[14] Generally, all other on-road emissions studies have found that the mobile source emission models significantly underestimate hydrocarbon emissions.

Role of Natural Emission Sources

Background maximum ambient O_3 concentrations are believed to be at or below 0.06 ppm.[15] Maximum concentrations above this level have, until recently, been attributed to anthropogenic emissions of hydrocarbons and NO_x. More recent information indicates that, in some regions, naturally emitted hydrocarbons may be available in sufficient quantities to promote formation of O_3 in excess of the NAAQS.[16,16a] Natural (or biogenic) emissions have been estimated to be as much as 60% of the northeast regional total hydrocarbon emissions inventory.[17]

Control programs to reduce O_3 concentrations that fail to include biogenic emissions do not address the full emissions inventory and are unlikely to succeed. Unfortunately, data on the concentration of biogenic hydrocarbon emissions are limited because ambient hydrocarbon measurements are usually collected within the urban core, thereby failing to include biogenic emissions. Furthermore, the compounds emitted by the biosphere are often pinenes and terpenes, both of which react quickly in the atmosphere to form other compounds or O_3. Therefore, by the time the upwind biogenic emissions would reach the urban core, they may have partially reacted and would not be measured. Still, intermediate transformation

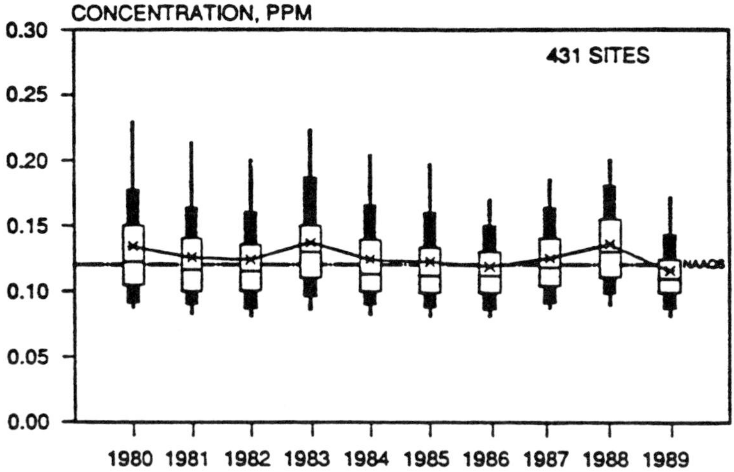

Figure 5. Trends in annual second-highest daily maximum 1-h O_3 concentrations at 431 sites.

Trends: Urban and Rural

Currently, the extent of rural O_3 nonattainment is not well known due to the limited number of monitors found in these locations. Figure 5 shows the trend in O_3 concentrations averaged across the nation's urban areas. No clear trend exists. The high O_3 years tend to be highly correlated with higher than normal summer temperatures.

The O_3 data presented in Figure 5 are, of course, based on the standard monitors. The accuracy of those monitors has been questioned by some researchers since the chemiluminescent and ultraviolet techniques may be subject to a positive interference from relative humidity and aromatic organic species.[21] Many high-ozone episodes in the eastern half of the country are strongly correlated with humidity and, therefore, the high readings may in part be due to humidity interference.

Meteorologically-Adjusted O_3 Concentrations

It is difficult to determine the change in O_3 concentrations that have resulted from emissions reduction efforts during the last 20 years due to the strong correlation between O_3 concentrations and meteorology. Ideally, O_3 concentrations would decline so dramatically that progress could be measured despite year-to-year meteorological variation. Since such progress has not occurred, the interannual weather changes need to be considered when assessing progress towards attainment of the O_3 standard.

Consider an example in which City X measured the following O_3 concentrations:

monitors. The accuracy of those monitors has been questioned by some researchers since the chemiluminescent and ultraviolet techniques may be subject to a positive interference from relative humidity and aromatic organic species.[21] Many high-ozone episodes in the eastern half of the country are strongly correlated with humidity and, therefore, the high readings may in part be due to humidity interference.

Meteorologically-Adjusted O_3 Concentrations

It is difficult to determine the change in O_3 concentrations that have resulted from emissions reduction efforts during the last 20 years due to the strong correlation between O_3 concentrations and meteorology. Ideally, O_3 concentrations would decline so dramatically that progress could be measured despite year-to-year meteorological variation. Since such progress has not occurred, the interannual weather changes need to be considered when assessing progress towards attainment of the O_3 standard.

Consider an example in which City X measured the following O_3 concentrations:

Year	Peak O_3 (ppm)	No. of exceedances
1986	0.15	4
1987	0.14	3
1988	0.16	5

Without considering changes in the weather, one might conclude that City X exhibited no progress or that O_3 was even getting worse. As indicated previously, however, temperature is strongly correlated with O_3 concentrations. Reconsider the example with the addition of the following temperature information:

Year	Peak O_3 (ppm)	No. of exceedances	Days above 90°F
1986	0.15	4	4
1987	0.14	3	6
1988	0.16	5	12

It now appears that concentrations may be constant or declining when the impact of the weather is included in the analysis. This should not be construed that City X now meets the O_3 standard; instead, it implies that progress toward the standard may be underway. The best way to quantify that progress is through a statistical analysis that controls for temperature.

Several researchers have attempted to account for the temperature variable when examining O_3 trends. They have generally found that, during the 1980s, O_3 concentrations have declined when adjusted for variations in meteorology.[22]

O_3 HEALTH EFFECTS DEBATE

Although the USEPA has not complied with the CAA requirement to review the primary and secondary O_3 standards every 5th year a vigorous process of scientific research and review has occurred since the O_3 standards were last revised in 1979. In contrast to the limited scientific basis provided for the 1979 rulemaking, the current review of the standards is characterized by a more robust (though far from complete) database bearing upon the principal policy issues facing decision makers. The USEPA's reevaluation of the O_3 NAAQS also differs from the 1979 standard revisions by expanding beyond the traditional concern over acute effects. Indeed, two of the principal policy issues confronting the USEPA Administrator concern the possibility of cumulative lung damage from chronic O_3 exposure that may accelerate lung aging effects, and a concern over chronic vegetative effects following prolonged O_3 exposures. The convergence of human clinical evidence and the availability of more suggestive epidemiological data place directly before the Administrator the issue of whether or not to establish a longer-term primary O_3 standard to protect against repeated peak exposures, while retaining current safeguards against acute effects expressed in the form of a 1-h standard.

The policy issues before the Administrator occur against the backdrop of a newly reauthorized CAA that provides the USEPA and the states with many more legal and policy tools with which to require reductions in O_3 precursor emissions using different criteria (such as technology-based standards, consumer product reformulation, or the use of incentives such as emissions trading) than those applied to set primary O_3 standards (which are required to be established using public health as the sole criterion). Thus, while the O_3 standard is the benchmark by which to measure attainment, in the future it will no longer serve as the exclusive centerpiece, or driver, of clean air implementation measures. This is because other important provisions of the amended act (e.g., air toxics and cleaner fuels) will also bear directly upon O_3 levels, but the legal and policy rationale for their adoption by Congress is much broader than the health-based focus for establishing an O_3 primary standard. In short, attainment of the health-based O_3 standard is no longer the nation's exclusive O_3 policy goal. And yet, because of the emissions reductions that will occur from implementation of this broader number of policy goals, the air should become healthier.

As a result of this development, scientists, regulators, and the general public should be receptive to an even broader array of information to inform this multiple number of O_3 policy goals. For example, information is emerging in the field of exposure assessment that can be utilized to establish future primary O_3 standards. Detailed population exposure analysis is an integral part of realistic risk assessment, and sophisticated exposure models have been developed by the USEPA during recent years for use in O_3 standard impact assessments. Current versions of the probabilistic NAAQS exposure models provide more realistic simulations of human microenvironmental behavior apportioned by age and work status that include minute-by-minute home/work/indoor/outdoor/roadway locations on seasonal weekday/weekend types of specific breathing rates. These activity patterns are weighted by the census tract demographics for a particular city and then combined with local O_3 data, using indoor/outdoor and roadway/outdoor reduction ratio, to produce more realistic estimates of actual population exposure.[23]

The detailed exposure information generated by such models is crucial for optimization of several aspects of the O_3 regulatory program. For example, by realistically simulating the actual frequency and duration of susceptible subpopulation exposure to specific O_3 levels at given breathing rates, the regulatory relevance of clinical exposures conducted at specific combinations of O_3 concentration, exercise level, and duration can be determined. Thus, specific exposure conditions employed in laboratory clinical studies may or may not be judged sufficiently realistic for use in standard setting. By realistically simulating the frequency and duration of exposure to specific O_3 levels for the susceptible group, the protection afforded by revised forms of the standard may be assessed.

Finally, by using measures of population exposure generated in current probabilistic models, *all* available monitoring data, weighted by local population density, are used to assess regional air quality. Population exposure values thus provide both direct measures of the benefits afforded by proposed control programs and are more robust indicators of continued reasonable progress towards eventual attainment of standards. Emission inventory and design value data currently used to judge adequate progress are at best indirect measures of improvement that are inaccurate. In sum, exposure assessment models should become an integral component of the USEPA regulatory process; sufficient resources should be allocated to refine existing models for eventual routine use. The private sector should provide resources, in addition to those allocated by the USEPA, to further this work.

CONTROL STRATEGIES

The SIP Process

The state implementation plan (SIP) process has failed to produce attainment of the O_3 standard in most of the nation's major metropolitan areas. This section will discuss reasons for this failure.

In general, the SIP process works as follows

1. The current emissions inventory for an airshed is estimated.
2. The inventory is projected for the future year in which attainment is desired.
3. Air quality modeling is undertaken to estimate the need for controls to reduce future emissions.
4. Air quality modeling is repeated, including the emissions controls.
5. Step 4 is repeated until the development of an acceptable strategy contains sufficient emissions reductions to demonstrate attainment.

Some have alleged that continued nonattainment was due to the submission of "cheater" SIPs that complied with the USEPA guidance but failed to take steps that were obviously necessary. A retrospective review of seven 1982 SIPs conducted for API (American Petroleum Institute) indicates that states did follow USEPA guidance but could not reach attainment because of underestimations in future emissions inventories *and* the use of a modeling approach that was insufficient to fully assess the magnitude of the O_3 problem.[24] All seven cities designed control strategies that should have achieved sufficient emissions reductions to attain the standard based on the modeling analyses conducted — yet none met the standard by 1987.

Two examples of problems with the emissions inventories are:

- The USEPA mobile source emissions model used for the 1982 SIPs was MOBILE2. Hydrocarbon emissions estimates from MOBILE4.0, the current model, are approximately double those of the earlier version for typical peak O_3 days.
- Some source categories were entirely omitted from previous inventories. For example, municipal wastewater treatment facilities were not included in SIPs prepared for 1982, and biogenic emissions inventories were either omitted or drastically underestimated.

In most locations, the modeling approach utilized to estimate necessary emissions reductions was insufficient. The Empirical Kinetic Modeling Approach (EKMA) was used for the 1982 SIPs for O_3 in all cities in the API study except one. EKMA is a trajectory model that is applicable to isolated urban areas where emissions are generated in an urban core and the maximum O_3 concentration occurs downwind in that afternoon. The output from EKMA is a set of O_3 isopleths that can be used to determine the percent emissions reduction necessary to meet the standard. The model does not include the effects of precursor emissions and O_3 that are transported from other areas. Also, EKMA only describes the impacts that occur due to the release of emissions on a single day and cannot represent the impacts of multi-day episodes of high O_3 concentrations. Unfortunately, most O_3 nonattainment areas are subject to transport and multi-day episodes.

Better Planning for Control Strategies Development

A more in-depth review of the previous SIP cycle will provide insight into problems and pitfalls that should be avoided in the future. With few exceptions, each of the seven cities studied adopted the minimum stationary source Reasonably Available Control Technology (RACT) established by the 1977 CAA Amendments and required by the USEPA policy for 1982 SIPs. Only three of the cities included stationary source emissions reductions beyond RACT in their SIPs. Stationary and mobile source control measures were less effective than planned because of delays in adopting and implementing these measures, and overly optimistic estimates of their effectiveness.

With respect to delays in adoption and implementation, numerous compliance deadline extensions, variances, and waivers were granted. Furthermore, there are often political and institutional barriers to the adoption of controls. For example, motor vehicle inspection and maintenance programs were included in the seven cities' SIPs, but were delayed to various degrees by the failure to pass enabling legislation in some of the states.

In the 1982 SIPs, emission control effectiveness was assumed to be 100% in many cases. More recent studies have indicated that many rules that have been adopted are only 70 to 80% effective. The result is that additional reductions are needed to achieve the O_3 standard.

There is an additional need for SIP planning to more accurately reflect economic factors. Forecasts of regional economic growth are an important element in SIP development as they affect the quantity of emissions generated. Economic constraints on state and local agencies can also result in reduced resources for planning and enforcement. These factors were not found to play a significant role in the planning process employed in the seven cities studied.

Better planning will be needed to attain the O_3 standard. Emissions inventories will need to be as complete as possible, including biogenic emissions estimates. Modeling approaches that address a range of high O_3 conditions will be needed. Those modeling approaches need to include the effects of long-range transport and multi-day episodes where these phenomena occur. Control strategies need to reflect realistic goals in realistic time frames.

API PERSPECTIVE ON O_3

The role of the petroleum industry in the O_3 nonattainment problem is related to the control of precursor emissions from the extraction and refining of crude oil and the transportation, storage, and end-use of petroleum products. The petroleum industry is contributing and has contributed to the reduction of emissions of O_3 precursors.

The combustion of oil products (such as gasoline and diesel fuel) in vehicles is a major contributor to NO_x and VOC emissions. The USEPA reports that highway vehicles represented about 30% of total national NO_x emissions and 28% of total national VOC emissions in 1989. The combustion of oil products in electric utility, industrial, commercial, and residential equipment is responsible for a very small fraction (about 3% for NO_x and less than 0.1 percent for VOC) of total national NO_x and VOC emissions.[25]

Petroleum refining processes are also a very small contributor to total national NO_x emissions (about 1% in 1989). Crude oil production, petroleum refining, and product storage and transfer accounted for about 15% of total national VOC emissions in 1989, according to the USEPA.[25]

The petroleum industry has been spending large sums every year for air pollution control: $150–500 million annually for capital expenses and about $2 billion annually for operating costs.[26] These amounts will substantially increase with implementation of the 1990 CAA Amendments. Ashland Oil Company, for example, estimates that compliance with provisions for oxygenated fuels, reformu-

Table 1. Current NO_x and VOC Emissions Control Requirements for Petroleum Source Categories

Federal
New source performance standards
Industrial boilers (NO_x)
Bulk gasoline terminals (VOC)
Storage vessels for petroleum liquids (VOC)
Equipment leaks in petroleum refineries (VOC)
Petroleum refinery wastewater systems (VOC)
Gasoline volatility: Reid vapor pressure limits in summer months that vary by state and stricter (lower) limits on southern (warmer) states (VOC)
Phase I (effective in 1989)
Phase II (effective in 1992)
National emission standards for hazardous air pollutants for benzene emissions
Equipment leaks
Storage vessels
Transfer operations
Waste operations
State
Stages I and II vapor control (VOC)
Fuel combustion and industrial process emission limits (NO_x)

lated gasoline, low sulfur diesel fuel, and stationary source air toxics alone will exceed $42 billion.

There are several current NO_x and VOC emissions control programs that are applicable to the petroleum industry. Table 1 presents a list of some of these requirements.

The CAA Amendments of 1990 will greatly expand the number and magnitude of new ozone-related controls upon petroleum source categories. Table 2 lists some of the future NO_x and VOC emissions reduction programs that are applicable to the petroleum industry at the federal level. The Amendments require the further reduction of VOC emissions in O_3 nonattainment areas. These regulations will be prepared by the states as part of revised SIPs. New control requirements may also be required for existing and new major NO_x emissions sources in O_3 nonattainment areas. The types of emissions reductions applicable to the petroleum industry that may be required in these upcoming revised SIPs include:

- The use of low NO_x burners or selective catalytic reduction to reduce NO_x emissions from boilers, process heaters, and diesel engines
- Restricting the combustion of fuel oil and solid fuels in stationary combustion sources (as was approved in 1989 by the South Coast Air Quality Management District Governing Board in southern California)
- Vapor recovery or incineration systems at bulk terminal facilities to reduce VOC emissions
- Improved pressure relief valves, pump and compressor seals, valves, and flanges

The reduction of air toxic emissions is required by this recent legislation in all areas of the country, regardless of whether or not an area is in attainment with the O_3 NAAQS. Many of these hazardous air pollutants are VOCs.

Reformulated gasoline has been on sale for more than a year in some O_3 nonattainment areas. Reformulated gasoline has proven so effective in reducing

Table 2. Future NO_x and VOC Emissions Control Regulations

Reformulated gasoline sales beginning on January 1, 1995
Revised State Implementation Plans with new emission limits for stationary sources
Increased use of Stage II gasoline vapor recovery systems in serious and severe nonattainment areas
Maximum Achievable Control Technology requirements to reduce air toxic emissions for sources categories such as:
Exploration and production
Gas plants
Glycol reboilers
Refineries
Equipment leaks
Storage tanks
Process vents
Transfer operations
Waste water collection and treatment
Gasoline distribution
Storage tanks
Fugitives
Transfer operations
Fuel combustion (heaters, boilers, turbines, internal combustion engines)
Marine tank vessel loading and unloading VOC emissions standards
Control Techniques Guideline for petroleum/industrial wastewaters
California Low-Emissions Vehicle and Clean Fuels Program

tailpipe emissions that it has qualified under the new CAA to compete with other alternatives, such as methanol or natural gas. Reformulated gasoline offers the only immediate means of reducing automotive emissions. By using these emission-control gasolines, customers of one petroleum company have reduced emissions in the Los Angeles basin by more than 100 million pounds.[27]

The petroleum industry has been working with Chrysler, Ford, and General Motors to develop data to evaluate the potential reduction in total vehicle emissions and improvements in air quality, primarily O_3, from a variety of vehicle-fuel combinations. This Auto/Oil Air Quality Improvement Research Program was initiated in 1989.[28] More than 2200 vehicle emissions tests are being conducted with 53 cars and light-duty trucks and with 29 reformulated gasolines and methanol blends. Data from this joint study will play an integral role in aiding both business and government policymakers in reaching product development and control strategy decisions.

Initial analyses of mass exhaust emission results of hydrocarbons, CO, and NO_x from 16 reformulated gasolines have been released. There was a 2^4 factorial statistical matrix of fuels: four gasoline properties — aromatic content, olefins content, methyl *t*- butylether (MTBE) content, and 90% evaporation temperature (T_{90}) — set at two levels. The effects of these four gasoline composition parameters were tested in two fleets: one composed of 14 1983–1985 model year vehicles and one composed of 20 1989 model year vehicles. Responses to changes in olefin and MTBE content were similar for both fleets, while responses to changes in aromatic and T_{90} were different and, in certain cases, opposite in the two fleets. The largest percentage effect in the current fleet, the apparent impact of lower T_{90} on reducing exhaust hydrocarbon emissions, was not found in the older fleet.[29]

The CAA Amendments of 1990 require that, effective beginning January 1, 1995, reformulated gasoline be sold in the nine O_3 nonattainment areas having a 1980 population in excess of 250,000 and having the highest O_3 design value during the period 1987 through 1989. These nine areas are:

- Los Angeles, CA
- Baltimore, MD
- Chicago, IL
- Houston, TX
- Milwaukee, WI
- New York City, NY
- Philadelphia, PA
- San Diego, CA
- Hartford, CT

Other areas will also be required to use reformulated gasoline in the future if they are reclassified as a severe or extreme (O_3 design value is equal to or greater than 0.180 ppm) O_3 nonattainment area. In addition, other O_3 nonattainment areas may "opt in" to this program at the request of a state governor and with the approval of the USEPA.

Reformulated gasoline must be certified by the USEPA. One of the many requirements for reformulated gasoline is that it will reduce the aggregate emissions of ozone-forming VOCs by 15% (this may increase to 20–25% in the year 2000) from representative model year 1990 vehicles when using baseline gasoline.

Specific reformulated gasoline requirements to implement the CAA Amendments were the outgrowth of an historic regulatory negotiation with the petroleum industry, the USEPA, environmentalists, states, and other parties. A successful Agreement in Principle was signed by these parties on August 16, 1991.

In addition to the reformulated gasoline requirements, the CAA Amendments of 1990 also specified that oxygenated gasoline be sold in CO nonattainment areas during winter months beginning no later than November 1, 1992. In order to enforce these new regulations and identify reformulated gasoline, oxygenated gasoline, and conventional gasoline, the petroleum industry may be required to perform more frequent sampling and testing, and to add paperwork to shipping documents.

Owing to its O_3 nonattainment problem, California has been developing a low emissions vehicle (light-duty vehicles and light-duty trucks) program with stringent (more stringent than national standards) tailpipe emissions limits for nonmethane organic gases, NO_x, and CO beginning with the 1996 model year and tightened further with the 2001 model year. This program focuses on clean fuels, clean cars, or a combination of the two. In addition to reformulated gasoline, clean fuels include compressed natural gas, ethanol, methanol, liquefied petroleum gas, and electricity.

The 1990 CAA Amendments also provide that other states can voluntarily "opt in" to the California Pilot Program with incentives to encourage the sale or use of the same clean fuel vehicles and clean alternative fuels as are required in California. The incentives may include a registration fee of at least 1% of the cost of vehicles that are not clean-fuel vehicles, provisions to exempt clean-fuel vehicles from high-occupancy vehicle or trip reduction measures, and preference to clean-fuel vehicles in the use of existing parking places.

FUTURE STRATEGIES

Implementation of the 1990 CAA Amendments will be aided by a host of new regulatory authorities by which government officials can compel future O_3 precursor emissions reductions. These range from more restrictive tailpipe standards facing the auto industry, to technology-based air toxics controls for a myriad of source categories and changes in the formulation of consumer products such as fuels. Implementation of these measures will, at a greater cost, improve air quality in a number of cities. By themselves, however, they will likely fail to enable all metropolitan areas to attain or continue to attain the NAAQS for O_3.

Actual O_3 attainment will require going beyond the traditional command and control policies (including those identified in the preceding paragraph) to enable all participants in the O_3 debate (including the business community) to devise more innovative, effective, and less costly approaches. Three key criteria that would stimulate innovative ideas are the following: (1) recognition that, to the environment, it doesn't matter which source categories in a given area reduce O_3 precursor emissions, only that such reductions are verifiable and enforceable; (2) encouragement of the principle that any legitimate step that reduces emissions should receive preferential credit in the SIP planning process; and (3) support for market mechanisms to incorporate more cost-effective O_3 reduction alternatives. Adoption of these three criteria recognizes that attaining the O_3 standard in a cost-effective manner is the umbrella goal of the nation's O_3 policy, and that any idea dedicated to that goal should be encouraged.

Two examples can illustrate how the private sector, favoring market mechanisms in lieu of command and control decisions, can bring about faster ambient O_3 concentration reductions than can government regulation alone.

Older vehicles as an emissions source category contribute a disproportionate share of automobile pollution. Inasmuch as pollution control devices on older cars and trucks are less sophisticated than those on newer vehicles, older vehicles produce far more O_3 precursor emissions per mile traveled. Newer vehicles can also be heavy polluters (e.g., those with failed emission control systems) but average emission rates are much lower than for older cars and trucks.

Taken as a group, passenger cars manufactured before 1981 presently account for nearly three fourths of passenger car hydrocarbon emissions, yet they travel less than one third of total vehicle miles traveled. A similar relationship holds for light-duty trucks.

Emissions from older vehicles are a significant factor in the failure of many areas to meet the O_3 NAAQS. In the Los Angeles basin, for example, about 400,000 pre-1970 vehicles are still registered. These vehicles produce almost 15% of the total hydrocarbon emissions from all sources (mobile as well as stationary) in that area.

A variety of strategies could be used to reduce the number of older vehicles registered in nonattainment areas, to reduce use of older vehicles that continue to be registered, and to reduce emissions per mile of actual use. The best strategies appear to be those that target high-polluting vehicles regardless of vehicle age. A combination of policies that involve (1) charging higher registration fees, (2) monitoring of exhaust emissions with remote sensors coupled with repair or retirement of vehicles identified as high-polluting, (3) making available cleaner gasolines, and (4) instituting vehicle buy-back programs in the most severe nonattainment areas are likely to be cost effective.[30]

The second example builds upon the Administrator's 1991 initiative to encourage business to voluntarily reduce a host of toxic compounds from all media. This initiative, the so-called "33/50 program" (because it seeks company-wide commitments to reduce 17 compounds by 33% in 1992 and 50% by 1995) provides industry with clear goals and the flexibility to plan the reductions across facilities in ways that maximize opportunities for large reductions achieved in a cost-effective manner. Companies receive credit for reductions made from a baseline year, and future reductions can be banked and credited toward regulations yet to be developed. The response to the 33/50 program by the petroleum industry has been overwhelmingly positive. Some companies have decided to commit to a greater than 50% reduction.

States should allow businesses to achieve or exceed the new SIP emissions reduction targets in the most environmentally and cost-effective manner. Working from an agreed upon baseline year and accounting for future regulatory requirements, companies could receive credit for early reductions in the SIP planning process. The more reductions that are achieved, the more SIP credits can be obtained, and these credits could be marketed to others in the area.

Using this type of initiative, a whole host of new pollution prevention options becomes viable. These include the more widespread purchase and scrappage of higher-emitting cars, incentives to introduce and expand reformulated gasoline on an accelerated time scale, and speeding up plant modernization schedules. Instead of waiting for the more ponderous process of government regulation, companies can contribute to improved air quality for O_3 and other pollutants in a more rapid time frame.

SUMMARY

The O_3 nonattainment problem is the nation's most intractable air quality problem. Given the number of metropolitan areas currently in O_3 nonattainment status, the failure to develop more accurate emissions inventories (including those for biogenic emissions), the growing number of peer-reviewed scientific publications pointing to health concerns at levels below the current standard and the possible need for a new standard to protect against chronic health effects, the traditional affinity of policymakers to the ideology of command and control regulation, and the increasingly unfavorable relationship between O_3 controls and high costs/political pain, it is no wonder that the National Academy of Sciences recently concluded that O_3 control programs adopted over the past 20 years have been a failure.[1] *The first step to rectify this failure is to recognize that it has occurred.*

The passage of the 1990 CAA Amendments, coupled with some limited experiments using economic incentives and voluntary pollution prevention initiatives, provides a myriad of new policy tools and philosophical directions that constitute a break with the past.

The nation's policy goals should be the cost-effective attainment of the O_3 standard and improved environmental performance. The path to cleaner air for all our citizens is the adherence to these goals, supported by the twin approaches of more selective uses of command and control regulation and incentives for companies to go beyond a strategy of compliance.

REFERENCES

1. National Academy of Sciences, National Research Council. *Rethinking the Ozone Problem in Urban and Regional Air Pollution*, (Washington, D.C.: National Academy Press, 1991), pp. 7. See also Matthew L. Wald, "U.S. Lag in Controlling Smog Tied to Misplaced Emphasis," *New York Times*, December 14, 1991.
2. *National Air Quality and Emissions Trends Report, 1989*, U.S. Environmental Protection Agency, Office of Air Quality Planning and Standards, Research Triangle Park, NC, EPA-450/4-91-003, February 1991, pp. 3–25.
3. Lindsay, R. W., J. L. Richardson, and W. L. Chameides. "Ozone Trends in Atlanta, Georgia: Have Emission Controls been Effective?," *J. Air Waste Manage. Assoc.* 39:40–43 (1989).
4. "Tighter Ozone Standard Urged by Scientists," *Science* 240:1724 (June 24, 1988).
5. Review of the NAAQS for Ozone, Clean Air Scientific Advisory Committee, EPA-SAB-CASAC-89-1092 (May 1989).
6. Krupnick, A. J., et al, *The Cost-effectiveness and Energy Security Benefits of Methanol Vehicles*, Resources for the Future (Discussion Paper QE90-25), Washington, D.C., 1990, p. xv.
7. *Air Quality Criteria for Photochemical Oxidants*, U. S. Department of Health, Education and Welfare, March 1970.
8. *National Air Quality and Emissions Trends Report, 1989*, U.S. Environmental Protection Agency, Office of Air Quality Planning and Standards, Research Triangle Park, NC, EPA-450/4-91-003, February 1991, pp. 3–27.
9. *National Air Quality and Emissions Trends Report, 1989*, U.S. Environmental Protection Agency, Office of Air Quality Planning and Standards, Research Triangle Park, NC, EPA-450/4-91-003, February 1991, pp. 3–22.

10. Chock, D. P., and J. M. Heuss. "Urban Ozone and Its Precursors," *Environ. Sci. Technol.* 21:1146 (1987).
11. Holzworth, G. C. *Mixing Heights, Wind Speeds, and Potential for Urban Air Pollution Throughout the Contiguous U.S.* 1971.
12. Carter, W. P. L., and R. Atkinson, "Computer Modeling Study of Incremental Hydrocarbon Reactivity," *Environ. Sci. Technol.* 23:844 (1989).
13. Ingalls, M. N., L. R. Smith, and R. E. Kirksey. *Measurement of On-Road Vehicle Emission Factors in the California South Coast Air Basin. Vol. I: Regulated Emissions*, Report No. SwRI-1604 from Southwest Research Institute to the Coordinating Research Council, Atlanta, GA, June 1989.
14. Pierson, W. R., A. W. Gertler, and R. L. Bradow. "Comparison of the SCAQS Tunnel Study with Other On-Road Vehicle Emission Data," *J. Air Waste Manage. Assoc.* 40:1505–1509 (1990).
15. Trainer, M., E. J. Williams, D. D. Parrish, M. P. Buhr, E. J. Allwine, H. H. Westberg, F. C. Fehsenfeld, and S. C. Liu. "Models and Observations of the Impact of Natural Hydrocarbons on Rural Ozone," *Nature* 329:705–707 (1987).
16. Chameides, W. L., R. W. Lindsay, J. Richardson, and C. S. Kiang. "The Role of Biogenic Hydrocarbons in Urban Photochemical Smog: Atlanta As a Case Study," *Science* 241:1473 (1988).
16a. Sillman, S., J. A. Logan, and S. C. Wofsy. "The Sensitivity of Ozone to Nitrogen Oxides and Hydrocarbons in Regional Ozone Episodes," *J. Geophys. Res.* 95:1837–1852 (1990).
17. Novak, J. H., and J. A. Reagan. "A Comparison of Natural and Manmade Emission Inventories Necessary for Regional Acid Deposition and Oxidant Modeling," Paper 86-30 presented at the *79th Air Pollution Control Association Annual Meeting*, Minneapolis, MN (1986).
18. Penkett, S. A. "Changing Ozone," *Environ. Sci. Technol.* 25:631-635 (1991).
19. Altshuller, A. P. "Sources and Levels of Background O_3 and Its Precursors and Impact at Ground Level," in *Atmospheric Ozone Research and its Policy Implications*, T. Schneider, et al. Eds. (Amsterdam: Elsevier Science, 1989), pp.127–157.
20. *National Air Quality and Emissions Trends Report, 1989,* U.S. Environmental Protection Agency, Office of Air Quality Planning and Standards, Research Triangle Park, NC, EPA-450/4-91-003, February 1991, pp. 3–25.
21a. "Water Vapor Effect on Ozone Reference Methods," 29 December 1980 memorandum to T. R. Hauser from M. E. Bears and K. A. Rehme, U.S. Environmental Protection Agency, Research Triangle Park, NC.
21b. "Results of Tests for Water Vapor Interference in Dasibi Ambient Ozone Monitors," 6 May 1991 memorandum to E. J. Conley, L. Murphy, and W. G. Laxton from G. J. Foley, U.S. Environmental Protection Agency, Research Triangle Park, NC., citing *Comparison of Chemiluminescent and Ultraviolet Ozone Monitor Responses in the Presence of Humidity and Photochemical Pollutants*, by T. E. Kleindienst, E. E. Hudgens, D. F. Smith (all three authors from Mantech Environmental Sciences), F. F. McElroy, and J. J. Bufalini (these last two authors from U.S. Environmental Protection Agency, Research Triangle Park, NC).
21c. Meyer, C. P., C. M. Elsworth, and I. E. Gabally. "Water Vapor Interference in the Measurement of Ozone in Ambient Air by Ultraviolet Absorption," *Rev. Sci. Instrum.* 62:223 (1991).

21d. Grosjean, D. "Response of Chemiluminescent NO_x Analyzers and Ultraviolet O_3 Analyzers to Organic Air Pollutants," *Environ. Sci. Technol.* 19:862 (1985).
22. Wakim, P. G. "1981 to 1988 Ozone Trends Adjusted to Meteorological Conditions for 13 Metropolitan Areas," paper 90-97.9 presented at the *83rd Air and Waste Management Association Annual Meeting*, Pittsburgh, PA (1990).
23. Johnson, T. R., et al. "Estimation of Ozone Exposure in Houston Using a Probabilistic Version of NEM," paper 90-150.1 presented at the *83rd Air and Waste Management Association Annual Meeting*, Pittsburgh, PA (1990).
24. Pacific Environmental Services, Inc. and Systems Applications, Inc. *Detailed Analysis of Ozone State Implementation Plans in Seven Areas Selected for Retrospective Evaluation of Reasons for State Implementation Plan Failure*, American Petroleum Institute publication 4503, 1991.
25. *National Air Pollutant Emission Estimates 1940-1989,* U.S. Environmental Protection Agency, Office of Air Quality Planning and Standards, Research Triangle Park, NC., EPA-450/4-91-004, March 1991.
26. U.S. Department of Commerce, Bureau of the Census. "Current Industrial Reports, Pollution Abatement Costs and Expenditures," series MA-200, annual.
27. Dickerson, K., "Looking Ahead: At the Gasoline Pump," *EPA J.*, 17(1):49(1991). (Mr. Kenneth R. Dickerson is Senior Vice President of ARCO).
28. Pahl, R., and M. McNally. "Fuel Blending and Analysis for the Auto/Oil Air Quality Improvement Research Program," SAE Technical Paper 902098, October 1990.
29. Auto/Oil Air Quality Improvement Research Program Technical Bulletin No. 1, Initial Mass Exhaust Emissions Results from Reformulated Gasolines, December 1990.
30. Anderson, R., *Reducing Emissions From Older Vehicles,* Research Study #053, API order number 878-59595, American Petroleum Institute, August 1990, p. 33.

INDEX

B

C

D

E

F

R

S